Negotiating Disasters: Politics, Representation, Meanings

Ute Luig (ed.)

Negotiating Disasters: Politics, Representation, Meanings

PETER LANG

Frankfurt am Main · Berlin · Bern · Bruxelles · New York · Oxford · Wien

Bibliographic Information published by the Deutsche Nationalbibliothek
The Deutsche Nationalbibliothek lists this publication in the Deutsche Nationalbibliografie; detailed bibliographic data is available in the internet at http://dnb.d-nb.de.

Cover and Photo Design:
© Olaf Gloeckler, Atelier Platen, Friedberg

ISBN 978-3-631-61096-1
© Peter Lang GmbH
Internationaler Verlag der Wissenschaften
Frankfurt am Main 2012
All rights reserved.

www.peterlang.de

Contents

List of maps, diagrams and tables   vii

List of photos   viii

List of contributors   ix

I. INTRODUCTION   1

Negotiating Disasters: An Overview   3
Ute Luig

II. ENGAGING WITH THEORIES   27

II.1. THINKING ABOUT RISK AND RISK MANAGEMENT   29

Social-Ecological Change and the Changing Structure of Risk,
Risk Management and Resilience in a Pastoral Community in
Northwestern Namibia   31
Michael Bollig

The Perception of Natural Hazards in the Context of Human
(In-)security   59
Ingo Haltermann

II.2. VULNERABILITY AND RESILIENCE   79

"Flood Disasters". A Sociological Analysis of Local Perception
and Management of Extreme Events Based on Examples from
Mozambique, Germany, and the USA   81
Elísio Macamo and Dieter Neubert

Squatters on a Shrinking Coast: Environmental Hazards, Memory
and Social Resilience in the Ganges Delta   105
Arne Harms

*II.3 REFLECTING ON METHODS: HOW CAN WE MAKE
SENSE OF DISASTERS?*   129

Researching Coping Mechanisms in Response to Natural
Disasters: The Earthquake in Java, Indonesia (2006)   131
Manfred Zaumseil and Johana Prawitasari-Hadiyono

III. POLITICS OF SPACE: NEGOTIATING RECONSTRUCTION — 173

The Attabad Landslide and the Politics of Disaster in Gojal, Gilgit-Baltistan — 175
Martin Sökefeld

Representations and Practices of "Home" in the Context of the 2005 Earthquake and Reconstruction Process in Pakistan and Azad Kashmir — 205
Pascale Schild

The Anthropology of a "Disaster Boom" Economy in Western India — 235
Edward Simpson

IV. CONSTRUCTING LOCAL MEANINGS — 253

Lightning, Thunderstorms, Hail: Conception, Religious Interpretations and Social Practice among the Quechua People of the South Peruvian Andes — 255
Axel Schäfer

In the Shadow of an Unreconciled Nature: Muslim Practices of Mourning and/as Social Reproduction in Uganda — 279
Dorothea E. Schulz

Negotiating Culture: Indigenous Communities on the Nicobar and Andaman Islands as Focal Points in the Post-tsunami Media Coverage 2004/05 — 299
Brigitte Vettori

List of maps, diagrams and tables

Maps:

Resource Use and Mobility before 1950 (Omuramba area)   38

Borehole drilling in Kaokoland; Boreholes drilled in the 1950s;
Boreholes drilled until 1990   43

Resources and mobility in 1995 (Omuramba area)   45

Sketchmap of Gojal   178

Diagrams:

Scientific approaches to coping with disaster in local context   134

Framing the results: different ways / axes of coping & change   146

Playing with theory- using coping as an example   148

Hikmah: an example of mapping a specific concept   154

Semantic field of *hikmah*   155

Coping model according to Folkman   155

Local coping projected on appraisal-based theories   156

Qhaqia misa or *Rayusqa*: arrangement on the altar for lightning or the two
Santiagos   269

Tables:

The development of the regional herd between 1940 and 2005   44

Losses in Hunza and Gojal due to Attabad landslide and lake formation   182

VIII

List of photos

Pak P, Java   160

Bu L and children, Java   164

Travelling by boat across the lake (Photo: Martin Sökefeld)   183

Eighty-year-old religious specialist Alejo relaxing after the end of the ritual   270

Woven festive blanket from the Cuzco region depicting black and white Santiagos   273

List of contributors

Michael Bollig is a Professor of Social Anthropology at the University of Cologne. He is the speaker of the interdisciplinary research group "Resilience, Collapse and Reorganization in Social-Ecological Systems of Eastern and Southern African Savannahs". His main research interests are besides social-ecological dynamics, risk management and economic change, the anthropology of conflict and violence, common pool resource management and the anthropology of space and landscape. He published *Risk Management in a Hazardous Environment. A Comparative Study of Two Pastoral Societies (Pokot NW Kenya and Himba NW Namibia)* (New York, Springer) and two edited volumes *Interdisciplinary Perspectives on African Landscapes* with O. Bubenzer (New York, Springer) and *Production, Reproduction and Communication of Armed Violence* with A. Rao and M. Böck (Oxford, Berghahn).

Ingo Haltermann is since 2009 Associate Junior Fellow at the Kulturwissenschaftliches Institut in Essen, with a research focus on Climateculture (KlimaKultur). He completed his studies in Geography, Politics and Landscape Ecology at the University of Münster in 2007 and later on took a further training in Peace and Conflict Studies in 2007/08 at the University of Hagen. His main research interests are social-geographical climate studies, geographical risk research, human ecology, urban geography, peace and conflict studies as well as development studies. He is a member of the African Association of Germany (VAD) and of the Geographical Association of German Universities (VGDH).

Arne Harms (M.A.) has a background in Social Anthropology, Modern South Asian Studies and Religious Studies and is currently working on his PhD. His doctoral research is situated at the confluence of environmental anthropology, forced migration studies and social memory studies. Thus positioned he analyses genealogies of social vulnerabilities, interpretations of hazardous environments and the poetics and politics of displacement on marginalized coasts of the Gangetic Delta. Earlier he worked with coastal populations in South India and the Anglophone Caribbean, where he studied such diverse issues as political conflict, migration, gender and popular religion.

Ute Luig was a Professor of Social Anthropology at the Freie Universität Berlin till 2010. She is a member of the Cluster of Excellence "Languages of Emotions" as well as of Topoi "The Formation and Transformation of Space and Knowledge in Ancient Civilizations". Before her retirement she also taught at the University of Mainz (1980 – 1990) and at the University of Heidelberg (1972- 1978). Her main research interests are besides the anthropology of disaster, conceptualizations of nature, religious studies, memory, and gender. She edited a volume on *The Making of African Landscapes* (Steiner, Wiesbaden 1997) together with Achim von Oppen and in 2002 together with Hans-Dietrich Schultz *Natur in der Moderne. Interdisziplinäre Ansichten* (Berliner Geographische Arbeiten, Berlin).

Elísio Macamo is Professor of African Studies and Director of the Centre for African Studies at the University of Basel (Switzerland). He studied in Mozambique and England and carried out his doctoral and post-doctoral studies at the University of Bayreuth (Germany), where he taught in the development sociology department from 2000 to 2009. His "Habilitation" thesis with the title *The Taming of Fate* will be published by CODESRIA (Dakar) in the course of 2012. He edited an introductory volume to disaster sociology together with Lars Clausen and Elke Geenen with the title *Entsetzliche soziale Prozesse – Theorie und Empirie der Katastrophen* (LIT Verlag, Münster 2003). His current research interest is on technology in everyday life.

Dieter Neubert is since 2000 Professor for Development Sociology at the University of Bayreuth. He studied Sociology, Social Anthropology and Pedagogy in Mainz with degrees in Pedagogy (Diploma 1979) and Sociology (PhD 1986). He habilitated 1995 in Sociology at the Freie Universität Berlin. He taught at the University of Mainz, at the College of Education in Heidelberg, the University of Hohenheim and the Freie Universität Berlin. He was a fellow at the "Center for Interdisciplinary Studies" in Bielefeld (10/2010-3/2011). His main research interests are political sociology of Africa, sociology of violent conflicts, sociology of development, development policy, sociology of disasters and participatory methods. His regional focus is Africa (especially Kenya, Uganda, Rwanda) and additionally Southeast Asia (Vietnam, Thailand).

Johana E. Prawitasari-Hadiyono, PhD. is retired Professor of Psychology, Faculty of Psychology, University of Gadjah Mada, Yogyakarta, Indonesia. She now joins Christian University Krida Wacana in Jakarta. Her main task in her new institution is to develop interdisciplinary approach in medical psychology and health psychology. In her former institution and State University of Malang, Malang, East Java, Indonesia, she is still advising PhD. students. She has just published in Bahasa Indonesia a book entitled *Clinical Psychology: An Introduction to Micro and Macro Applications* (2011). With her former students she also published a book entitled *Applied Psychology* (2012) also in Bahasa Indonesia. Those books were published by Erlangga Publisher, Jakarta, a national company.

Axel Schäfer studied Comparative Religion and History of Ancient Americas in Berlin, Seville and Ayacucho. He specializes in popular religion in the southern Peruvian Andes. In 2006 he completed his M.A. at the Institute of Latin American Studies, Freie Universität Berlin. He worked with different NGO's supporting indigenous organizations. Since 2008 he is part of a research project on ritual landscapes of the Cluster of Excellence Topoi, "The Formation and Transformation of Space and Knowledge in Ancient Civilizations" in Berlin. In this context he conducted fieldwork in Cuzco and Apurímac/Peru and is currently writing his doctoral thesis on saint cults and agricultural rituals in the region.

Pascale Schild received her Master in Social Anthropology from the University of Berne, Switzerland, with a thesis on identity politics of Muslim migrant women in Switzerland. Since 2009 she is a doctoral student at the Department of Social and Cultural Anthropology of Munich University where she joined the research project on "Politics of Reconstruction after the 2005 Earthquake in Pakistan and Azad Kashmir" as a research fellow. She conducted fieldwork in Azad Kashmir from 2009-2011 on households' practices of disaster recovery and the reconstruction of homes. She is currently working on her PhD thesis which also includes ethnographies of "the state" examining how people deal with "disaster bureaucracy" and mitigate reconstruction policies and projects in everyday contexts of home after the earthquake.

Dorothea E. Schulz has previously taught at Cornell University, Ithaca, and Indiana University, Bloomington, and is currently Professor in the Department of Cultural and Social Anthropology at the University of Cologne, Germany. She has published widely on media practices and public culture in Sahelian West Africa, gender studies, and Islam in West Africa. She is currently researching Muslim practices of coming to terms with death and mourning in a situation of continued ecological and social disaster and irruption in Uganda. Her publications include *Perpetuating the Politics of Praise. Jeli Praise, Radios, and the Politics of Tradition in Mali* (Koeppe, 2001) and *Muslims and New Media in West Africa* (Indiana University Press, 2011).

Edward Simpson is Senior Lecturer in Social Anthropology at the School of Oriental and African Studies (SOAS), University of London. His research interests are in politics, natural disasters and social change in western India. He is the author of *Muslim Society and the Western Indian Ocean* (Routledge, 2006), editor with Aparna Kapadia of *The Idea of Gujarat: History, Ethnography and Text* (Orient Blackswan, 2010).

Martin Sökefeld is Professor of Social Anthropology at the Ludwig-Maximilian-University Munich. He wrote his PhD thesis on ethnicity in the town of Gilgit in Pakistan's high mountain area and worked subsequently on Anatolian Alevi migrants in Germany and on the (Azad) Kashmiri diaspora in the UK. His focus on disasters is a consequence of regional interest in northern Pakistan. In 2009, he started a research project on the politics of reconstruction after the earthquake that hit Azad Kashmir and Pakistan's Khyber Pakhtunkhwa province in October 2005. The study of the Attabad landslide in Gilgit-Baltistan is the most recent disaster research project. In accordance with his general interest in the anthropology of politics, Sökefeld's research focuses mostly on political implications of disasters.

Brigitte Vettori is Lecturer at the Institute of Social and Cultural Anthropology of the University of Vienna. She studied Social and Cultural Anthropology, International Development and African Studies in Vienna and completed her studies with the PhD Thesis *NGO Intervention and Interaction within the Framework of a Caritas Project for Iraqi Refugees in Syria*. Her research interests are urban anthropology, sociology of organizations, space and place as well as culture change. She has a long-term experience of work with NGOs in Syria as well as in emergency relief organizations. Her regional interests focus on Samoa, Fiji and the Nicobaresian islands.

Before his retirement in 2008 **Manfred Zaumseil** worked as Professor for Clinical and Community Psychology at Freie Universität Berlin. His present affilation is the "International Academy for Innovative Pedagogy, Psychology and Economics (INA) at the Freie Universität Berlin" (www.ina-fu.org). He worked about the field of mental health and cultural clinical psychology. He is now preparing a book with some colleagues about cultural psychology of coping with disaster which will be published by Springer, New York.

I. Introduction

Negotiating disaster: an overview

Ute Luig

1. Introduction[1]

1.1 Prolog

On 11 March 2011, at 14.46 p.m. local time, an earthquake which reached grade 9 on the Richter scale exerted its terrible impact on the east coast of Honshu, Japan. Minutes later it was followed by a tsunami which destroyed the nuclear power plants in Fukushima and devastated the surrounding area. When, after some days of anxiety the feared nuclear explosions happened, Fukushima became the very symbol of the maximum credible accident (*supergau*), to be paralleled only by the events in Tschernobyl in 1986. The term *supergau* implies that radioactive substances are set free and become a threat to humans and the environment. In the case of Fukushima, the exposure to radiation was very high and led to the evacuation of 100,000 to 150,000 people. The enormous destruction of lives, the pollution of the land and sea for many years to come and the following estrangement between civil society and the government turned Fukushima into an archetype of disaster.

Although Fukushima occurred locally, its impact was perceived and interpreted globally, sending shockwaves to the outside world. The devastated landscape appealed to the emotions of the world community, and sparked new discussions about humanity's relationship with the environment. The atmosphere in Europe after Fukushima in many ways resembles the period of the Lisbon earthquake in 1755. Susan Neiman (2006: 353ff.) has convincingly described how the destruction of this rich and important city deepened the ideological battles between representatives of the Enlightenment and their religious counterparts. The Lisbon earthquake not only destroyed one of the world's leading cities but also shook the philosophical convictions of the time. Similarly, Fukushima destroyed the belief of many people in the control of technology and aggravated the controversies between supporters and opponents of atomic power. That such

1   My deep thanks go to Carla Dietzel, who has diligently worked on the editing process and to Dr Robert Parkin for correcting the English in some chapters. I am also grateful for a critical reading of this introduction by Dorothea Schulz, Edward Simpson, Martin Sökefeld and Manfred Zaumseil.

a disaster of almost apocalyptic dimensions could happen in one of the most highly industrialised countries in the world undermined belief in unbounded progress.

Like the Lisbon earthquake, Fukushima also represents a mental crisis. This time the subject is not religion or morality but humanity's accountability for nature and its control of technology. As the very symbol of human vulnerability, Fukushima engrossed the disputes about ethics and technology; it intensified feelings of anxiety which Hoffman (2002: 136f) argued are more pronounced with regard to technological risks than to natural ones.

This profound shock had worldwide repercussions in countries that rely on nuclear power, especially in Europe, but also in Japan and Russia. Although Germany was the only country where the government revised its nuclear power politics by seeking to end its reliance on atomic energy, critical discourses in other countries opened up debates about the control of technology as well. These discussions are coinciding with the ongoing debates on climate change, perceived as another global threat. The insecurity and uncertainties resulting from climate change further support scenarios of a world in danger which is threatened by environmental migrations, hunger and deaths due to droughts and endless wars over resources (see Welzer, Soeffner and Giesecke 2010, Hastrup 2009). Many social scientists are aware of these challenges, in which the twin threats of technology and of disasters overlap. Ulrich Beck (1986) has already drawn attention to this problematic in his *risk society*, which was theorised later by Oliver-Smith and Hoffman (2002: 18) and others (see Paine 2002: 67). The (pre-modern) relationship between a dominant nature and the social world of humans has been turned upside down in the period of (postmodern) global capitalism. It is human activity which threatens the existence of nature: examples like the Exxon-Valdez disaster, the explosion of an oil platform in the Gulf of Mexico and the *supergau* of Fukushima, to name but a few, prove the immense devastation of natural resources through human activity in the era of global capital.

The actuality of climate change and disaster politics has already changed the scientific discourse on 'natural' disasters in fundamental ways. Since Fukushima the long-debated question whether 'nature' or 'humanity' is responsible for disasters (see Alexander 1997) has finally been settled. 'Natural' disasters cannot be perceived any longer as solely natural,[2] but as the complex intertwining of human interventions and environmental vulnerability. Due to the dynamics of globalisation social and natural forces are so closely entangled with one another that the old divide between culture and nature, originating in Western philosophy of the eighteenth century, has lost its foundation. The relentless technological progress as a powerful sign of modernity reveals its very ambivalence, which

---

2 Alexander (1997: 289) even speaks of this as a misnomer.

is characteristic of an increase in wealth, mobility and technological potentialities, but is at the same time also responsible for environmental destruction, famine, wars and an ever growing difference between access and entitlement to resources (see Sen 1981). Many social scientists are aware of these challenges and conceive the present situation as the necessary beginning of a new theoretical era for the social sciences (see Hastrup 2009). It may still be too early to propagate a turning point in history, but some of its foundations are laid by the present discussions. This volume aims to contribute to these challenges by offering some new perspectives for research. It takes up the idea of closer cooperation between the social and cultural sciences by offering a transdisciplinary approach. In order to understand the ever more complex changes, it has seemed necessary to leave the anthropological looking glass aside, though without abandoning it, for a widened perspective using new theoretical concepts. In contrast to interdisciplinary research, a transdisciplinary approach takes multiple perspectives on common problems which allow for more comprehensive answers. In an edited volume such an approach is only possible in a very rudimentary fashion, but it can nonetheless lead to interesting results, as the contribution by Zaumseil and Prawitasari-Hadiyono demonstrates.

1.2 An overview

The idea of this publication sprang from a colloquium at the Freie Universität Berlin in 2010. It aimed at acquiring an overview of the state of the art in disaster research, which had seen a tremendous intensification in recent years. Although this publication was conceived before Fukushima, some of the most salient questions and problems deriving from it are discussed by its authors. Some texts are a kind of stocktaking, reflecting earlier research results, comparing them with new empirical findings and questioning their empirical validity. Others open up new debates which reconcile social science approaches with the humanities. Topics like the politics of disaster, culture change, memory, rituals of mourning and good and bad deaths are covered in this book, while others, such as resistance and violence, spaces of death and the analysis of the emotions, will be just delineated in order to provide a trajectory for future studies.

A wide range of extreme events are nonetheless considered in this volume. The spectrum reaches from processes of environmental degradation, whether of pastures (Bollig) or coastlines (Harms), to sudden, incalculable events like lightning, hail (Schröder), earthquakes (Schild, Simpson, Zaumseil and Prawitasari-Hadiyono), landslides (Sökefeld), floods (Schulz, Macamo and Neubert) or the tsunami in 2004 (Vettori). With one exception, all the analyses are based on long-term fieldwork and accordingly on dense case studies. The events described outline the close entanglement between climate change, increases in vulnerability or resilience and a great variety of coping strategies and interpretations. The examples range from single and/or chronic crises (see Vigh 2008) to 'classical' disasters which, in contrast to crises, are experienced as shocking, overwhelming events which necessitate immediate help. Although some disasters are foreseeable, and in some cultures even prophesied in calendars and in notions of cyclical time, the possibility of their occurrence is quite often suppressed and then comes as a shock. In order to improve our understanding of their particular dynamics, which unfold over time and space and turn them into processes rather than punctuated events, it is necessary to analyse their specific (historical) conditions as part of a transnational context. This space-time-oriented approach also allows us to analyse a processual chain that divides the development of a crisis or disaster into a state before, during and after the event. This time frame is of particular importance when a crisis turns into a disaster (see Macamo and Neubert, Sökefeld this volume) or changes into a catastrophe, as in Fukushima.

The term 'disaster' is used in this introduction in a rather colloquial manner. The reason behind this 'loose terminology'[3] was the idea of taking up important discussions in recent years advocating the holistic analysis of disasters. This approach implied on the one hand showing the intertwining of risk perception, vulnerability and coping strategies which do not always focus on disasters but describe mere or chronic crises as well. On the other hand, since it was necessary to document how social practices and cultural meanings are inscribed on to a material environment over long periods of time, it seemed important to discuss the interconnectedness of risk perception and vulnerability and then to link them to interpretation and coping strategies from an emic point of view. Problematizing emic and etic perceptions of disasters is important in order to open up discussions that critically rethink our use of categories. Most of the case studies in this book bear witness to this problem since they are concerned with problems of classification and method. What is a disaster, how is it differentiated from a mere crisis and under what condition does one turn into the other are salient questions for some of the authors in this volume. Others (Zaumseil and Prawitasari-Hadiyono, Schulz) reflect on methods in order to understand how people can make meaning of their experiences in the face of disastrous events. They question Western scientific theories in comparison with local perceptions and interpretations. It is in these sections where the differences between different disciplines in accessing the same problems are most obvious and thus enrich the spectrum of theoretical trajectories in disaster research.

---

3   For an interesting discussion of possible definitions of disaster, see Oliver-Smith (1999a: 20). His conclusion "that disaster is a contested concept, with blurred edges, more a set of family resemblances ...rather than a set of bounded phenomenon to be strictly defined" (ibid.: 21) is good to think with.

2. Engaging with theories

2.1 Thinking about risk and risk management

Discussions in disaster research during the last twenty years have brought four terms to the fore to theorize about: risk, risk management, vulnerability and resilience. Most of these terms originated from other disciplines, like ecology or cultural psychology, and it was some time before they were introduced into anthropological discourses. **Michael Bollig** addresses the history of these terms in his introduction and shows how they can be fruitfully adapted by anthropological theory building. Bollig selects and differentiates three traditions in analysing risk and risk management: actor-oriented, ethnographic and interpretative approaches, and he discusses their merits and shortcomings. According to his reading, the actor-oriented approach mainly discussed this problematic with regard to risk management either in the form of rational choice models or as structuring social institutions and territorial behaviour. Institutions like food-sharing were and are the key strategies in foraging societies in minimising natural risks. Bollig criticizes the fact that, although these studies had clear hypotheses, they did not produce models of general applicability. The same criticism also applies to what he calls the ethnographic approach that focused on African drylands. Problems of desertification, food shortages and social marginalization have frequently arisen in the Sahel since the 1970s and require strategies of survival and community building. But despite fine-grained descriptions of "historically changing modes of risk management" (Bollig: 33) these studies were not embedded in any theory of risk perception, nor were they related to local belief systems. He is also critical of Mary Douglas's publications, which are considered the major anthropological contribution to risk perception by neighbouring disciplines, although their impact on anthropology has been limited. Douglas made the perception of risks the key aspect of her theory, which she discussed with reference to a vast range of case studies. Her main thesis that "risk perception is encoded in social institutions" (Bollig: 34) fits well with her analysis of its many dimensions, but as Bollig suggests it remains analytically diffuse. In his own approach he defines risks as "the culturally and socially embedded perceptions of future possible damage resulting from a variety of hazards" (Bollig:36), while risk management either reduces negative impacts by decreasing vulnerability or limits the impact of damage through a conscious decision. Bollig proposes an analysis that relates risks and risk management to a time- and space-specific structure of resources which is accordingly embedded in historical circumstances.

Ingo Haltermann, a social geographer by profession, shares this historically oriented, contextualised approach with Bollig, but differs from him in that he explicitly orientates his discussion of risk and risk management around the conceptualization and individual appropriation of space. His approach to space is based on a constructionist view of environment which is not limited to a geographical or spatial relationship but defined by human needs and desires that are historically and culturally constituted. He conceives of the environment as a system "representing a certain section of the external world to which the actions and perceptions of a subject give significance" (Haltermann: 64). These perceptions, together with former experiences, also shape subjective evaluations of risk. They have to be taken into account in order to make assumptions about the feasibility and extent of the dangers which may threaten in the future. The author stresses the concept of bounded rationality favoured by risk research in geography because of its findings that individual actions are not orientated towards cost-benefit factors, but are instead determined by individual interests, cultural values and trust in one's own ability. This psycho- cultural conceptualization of space recalls in certain ways Soja's redefinition of space as active and dialectical (see Keith and Pile 1993: 4). Haltermann goes on to question the ways of individual (and household) risk-taking by introducing the differentiation between acceptable and unacceptable risk-taking, which are discussed in terms of risk acceptance or damage acceptance. Decision-makers have to weigh several risks against several chances, and even risks against risks and chances against chances. Besides this balancing of reasons, the control of resources plays an important role in confirming safety, which is conceptualized as freedom from want and freedom from fear according to a UN convention. Despite this psychological argumentation, he stresses the direct connection between safety and control over wealth, knowledge and power. Social inequality thus becomes a key variable in explaining the unequal distribution of risks that manifests itself in limited possibilities to guarantee (human) security (see also Beck 1986: 55). Under precarious living conditions without access to security, "even extreme natural events lose importance and represent only one further aspect in a general state of continuous crisis" (Haltermann: 77). Later in his text, however, he revises this somewhat deterministic perception. Under situations of increasing danger, so his argument, people will change these limits of adaptation and restructure their culturally determined behaviour, provided that the structural disadvantages of their local households have undergone change. Along with their changing situations, their perceptions of risk will change as well and will lead to changes in risk management.

2.2 Vulnerability and resilience

Haltermann's argumentation in some respects comes close to the so-called risk and vulnerability discourse, which, due to the work of Wisner et al. among others, is one of the most frequently discussed in disaster sociology (see also Bollig, Zaumseil and Prawitasari-Hadiyono, Schulz this volume). Alexander (1997: 291) has pointed out that risk and vulnerability are different sides of the same coin. Risk is an active concept, whereas vulnerability is more passive. Risks can be taken, but vulnerability has to be endured, though it can also be changed (see Haltermann, above). According to Alexander it is therefore more appropriate to relate vulnerability to the susceptibility of damage or injury or, in Wisner's terminology, to "correlate it with past losses and the susceptibility to future losses" (quoted from Alexander 1997: 291). In order to avoid discussing vulnerability only in quantitative terms, as a form of material damage or loss of human life, Blaikie et al (1994) have added political factors as well. Alexander's own efforts to unpack this rather blurred category seem interesting enough to be cited in a much shortened version. He differentiates between 1) the total vulnerability of the poor and dispossessed, 2) the economic vulnerability of the marginally employed, 3) the technological or technocratic vulnerability of the rich, 4) newly generated vulnerability, that is, risks to property or other capital assets, 5) residual unameliorated vulnerability that includes risks to modern safety standards, and last but not least 6) delinquent vulnerability, which also refers to breaches of safety norms (see Alexander 1997: 292). Although these ascriptions consist mainly of references to social inequality and a lack of safety standards, they provide insights into the further differentiation and ramifications of the concept.

In their contribution to this volume, **Elísio Macamo and DieterNeubert** take issue with the widely accepted notion (see Wisner et al. 2004) that the greater the vulnerability of local people, the greater their exposure to risk and the "lesser are their chances to recover" (Macamo and Neubert: 83) by introducing two new categories into the debate: the notion of ordinary management expectation, and local relief management capabilities. These two characteristics allow them to understand disasters not from an individual or social-psychological point of view, but as a social phenomenon (ibid.: 85).

Their field of enquiry is the comparative study of coping with floods along the Limpopo (2000), Odra (1997) and Tennessee rivers (2003 and 2004), the latter being part of the so-called Bible belt in the USA. Their main aim is to understand the logics behind the different classifications of these events through an emic perspective which includes analysing local perceptions and religious beliefs, as well as the organization of support by social institutions.

Their interest in risk management is closely linked to the precision and refinement of existing categories to describe hazardous events, the differences of which have to be outlined for scientific reasons, but also for practical purposes. Only if planners and disaster experts understand the assessments of local populations are they able to offer improved measures of support in case of need. The centre of Macamo's and Neubert's endeavour is therefore a critical rethinking of the 'all-purpose metaphor' disaster, deploring its heterogeneous usage in very different contexts. They propose a kind of ranking between extreme events, threats or hazards, disaster and catastrophe, which they link to the functioning of the social order. Threats or hazards are characterized by a breakdown of normality within a certain time window, whereas in a disaster normality is permanently destroyed, but differs from the complete breakdown of the social order that characterizes a catastrophe.

Disaster management here is not linked to vulnerability per se but to the organizing capacities of the local people, which are shaped by their knowledge and technical means. Whereas the population of the Limpopo valley was used to the seasonal flooding of the river – which, if it arrived at the right times is even regarded as beneficial – their delayed run-off in retrospect turned the 'normal' crisis situation into a disaster. Their inability to plant their fields at the right time resulted in widespread hunger. It was therefore not the floods but their unexpected consequences that were responsible for the breakdown of their emergency management and the ensuing deaths. Despite their social vulnerability, the communities in the Limpopo valley have low expectations of outside help, and instead trust in their own management capabilities. This includes a greater tolerance of material and human loss and a lower threshold for interpreting these events as disaster. Resilience in their case was strong due to their understanding of normality as beset by material problems and the struggle for survival.

In contrast to these perceptions, the better off population in the Odra valley, as well as those in the Tennessee valley, have a much lower tolerance of these kinds of losses. Not surprisingly their expectation of outside management capabilities is much higher, despite their own well-coordinated emergency management. Normality for them is expressed in effective management expectations, which in case of need are supported by the smooth working of the management support structures. Compared to the Limpopo population their resilience is less developed, despite greater material security and elaborate techniques of prevention.

These comparisons lead Macamo and Neubert to the conclusion that the contextualization of key terms like crisis, disaster, vulnerability and resilience refutes standard explanations of disaster anthropology which were predominantly measured in quantitative terms, like the amount of damage and the number of casualties. Although the most vulnerable are those most impacted by hazardous events, their resilience is higher due to different expectations of normality. Vice versa, better off societies in the industrialised world are (also mentally) more vulnerable and less resilient because of their high expectations of safety in everyday life.

The authors opt for a de-construction or substitution (livelihood analysis instead of vulnerability) of these terms in order to be able to contextualize their meaning in divergent surroundings.

The ethnographic research of **Michael Bollig** in north-western Namibia is a good example of such a nuanced discussion. He undertook one of the very few long-term studies in environmental research in describing in detail the reciprocal relationships between physical vulnerability and human interaction. Using a framework of long-term research he was able to reassess his own judgements concerning the stability of the systems, which proved much more dynamic than expected. Comparing the impact of diverse socio-political situations on the environment enabled him to develop an understanding of resilience that points out the absorption of disturbances and reorganization through a constant process of learning and adaption to change.

The decisive factors in the functioning of the eco-system were local power relations, which regulated grazing patterns and access to water holes on a seasonal level. Colonial and postcolonial governments adapted them to varying political circumstances, but because of technological and political interventions the government itself became the greatest risk to a balanced ecosystem. It provoked a hunger crisis when it banned transnational trade, which was vital for the community in times of crises in the 1920s. The well-intentioned effort to improve pastures through a borehole drilling program in the mid-1950s led to a reversal of grazing patterns, overstocking and an increase in social conflicts. Although this alarming situation could be controlled for some time through the institutionalisation of chieftainships, some years later, under the postcolonial government, the dissolution of the system became obvious. The severe degradation of pastures as a result of the bureaucratization of grazing rules and ongoing modernization processes produced further conflicts which led to heavy outmigration. But the expected collapse of the system was prevented by its transfer to national conservancy units along the lines of traditional chieftaincies. Since they had the support of the traditional elites and the NGOs, these national conservancies became signs of hope for a sustainable future.

This long-term case study provides a detailed insight into the dynamic relationship between the eco-system, social and political configurations and the building up of resilience patterns. Bollig stresses that the alternate phases of stability and transformation[4] of the system required different forms of risk management and changing pattern of social resilience. He points to two different forms of resilience: social resilience as a form of communal investment in "collectively held social capital" (Bollig: 52), and the resilience of the eco-system, defined as a 'structural property' (ibid.). Both types of resilience are characterised by constant changes which impact on each other and lead to the overall dynamic of the system as a whole.

In contrast to Michael Bollig's presentation of social and ecological resilience, **Arne Harms** aims to liberate social resilience from its predominant employment in debates in ecology and political economy. He no longer understands the term as a personal or systemic capacity, but as a situated practice "related to culturally mediated interpretations of the present" (Harms: 108). The focus of his study is the squatters who populate the embankments of various islands in the Ganges Delta. They live under extremely vulnerable conditions due to a con-

---

4   See also Holling 1973 for a similar characterisation of eco-systems.

stant process of coastal erosion, which forces them to shift their houses from the shores of the embankment to the hinterland in search of habitable land. Harms' central argument is that in this shrinking life world social resilience is expressed in an 'ethic of endurance' articulated by 'narratives of loss' in which social memory is encapsulated. Arguing against conceptualizations which narrow memory to the application or intergenerational transmission of traditional knowledge (see Vettori, this volume), he opts for a culturally mitigated concept of memory in the sense of the remembered past defining the present (Harms: 109; also Antze and Lambek 1996).

In contrast to disasters which occur suddenly against all expectation, the creeping destruction of the environment in the Ganges Delta is not remembered as a series of special events but taken as an act of a "contingently present normality" (Harms: 110). Memory under these conditions of "chronic crisis" (Vigh 2008: 9) becomes an act of identity empowering the squatters to cope with their surroundings not only in a technical and functional manner, but in the form of cultural history. This history relates to memories of marginalisation, loss of land and familial conflicts, which are the results of colonial land and settlement policy in the Ganges Delta. In this constantly shrinking world without any security or stability, economic strategies strengthen resilience patterns beyond processes of memory. Thus, onshore fishing of tiger prawns seedlings (*meen*), labour migration into the Indian hinterland and work in seasonal deep-sea fishing are important methods of survival, as are catering for and transporting the pilgrims who visit a nearby annual festival. These strategies become strengthened through relations of solidarity among neighbours, who not only share their memories of loss and destitution, but generate from them a collective identity as a "community of loss" (see Butler 2003). Despite their difficult living conditions, most settlers preferred to stay on because this kind of communal support, trust in one's neighbour and the memory of a shared past prove to be more powerful than the hope of a better life, which may be very short-lived.

When we compare these different forms of resilience from a transdisciplinary perspective, we acquire a close up of differentiated contexts and practices. Bollig stresses the twofold notion of resilience as a structural property that is also a systemic capacity, a form of collective agency represented by hierarchical structures and political networks. In addition, Macamo and Neubert draw attention to managerial capacities which either strengthen or weaken the social resilience of groups. Whereas resilience in these chapters is seen as a function of adaptation in changing ecological environments under specific historical conditions, Harms describes it as a means whereby to live through these conditions. Sharing with Bollig as well as Macamo and Neubert the idea that resilience is constitutive of collective identities, he underlines its notion as a culturally mediated practice that "is certainly also a situated practice. In the latter sense, it therefore draws on and is enacted through a range of material relations, economic practices, social articulations and cultural interpretations" (Harms: 125). In Harms' understanding, resilience as a category no longer refers only to social or ecological processes but is embodied in a very dynamic understanding of culturally mitigated actions. The shared memory of the group, as a way to live through the past for a future present, plays an important role in his understanding of resilience. It therefore differs from notions of memory, which, as Harms labels it, have been incarcerated in the transmission of traditional knowledge. Although this critique is well taken, the concept of traditional knowledge is a complex one, especially with regard to what is understood as traditional. In his research on Indian fishermen, Hoeppe (2007) discovered that traditional knowledge draws its insights from many domains, science included, while for Stehr (1994) traditional knowledge means 'to do something'. Similarly, Lambek (1993) has drawn attention to this very close relationship between knowledge and practice. His differentiation of several layers of knowledge – sacred, objectified and embodied knowledge – is based on practice, not on knowing. The knowledge of an expert cannot be determined by asking him what he knows but only by analysing what he does (ibid.: 17). **Vettori's** interview (this volume) with a chief from the Nicobar Islands confirms this 'action side' of knowledge. Although when the tsunami set in he was able to save his group by remembering the advice of his ancestors, in the aftermath of the tsunami recourse to government relief operations was more important for his group. The situational selection of different knowledge systems is in his view decisive for the future of his people, traditional knowledge being an important means of upholding identity, which, however, is in a constant flux due to other influences. Since this context-orientated application of traditional knowledge comes near to what Harms described as situated practice, it is evident that knowledge in a 'liberated form' can be part of resilience in all its complexity, that is, not exclusively related to strategies of survival. This result confirms the polysemic aspects of social resilience, which, in addition to social memories and different types of knowledge, must incorporate religious beliefs as well (see the studies in 2.3 and 4).

2.3 Reflecting on methods: how can we make sense of disasters?
Disasters are, for the victims, a period of pain, loss and sorrow. After rescue, questions of why and how this has happened (see Macamo and Neubert: 99f.) are asked. In many societies the why questions are related to beliefs in transcendental powers, be they ancestors or God himself. It is the individual self as well as the community as a whole who have to make sense of the disaster. Why did God allow this, what does he want to tell us and what are the proper reactions? Whereas these questions were part of elaborate philosophies in Western languages and have been well researched by anthropologists, much less is known about how people make sense of the disaster for themselves. The cultural psychologists **Zaumseil and Prawitasari-Hadiyono** posed this question as a meth-

odological problem during their research on the earthquake in Java in 2006. What kind of theory, they ask, is suited to penetrating into individual cosmologies of suffering – or, put differently, what insights do we gain when we refer to specific theories? Their interest concentrates on the analysis of "the retrospective psycho-spiritual processing of the direct consequences of the earthquake two to five years afterwards" (Zaumseil and Prawitasari-Hadiyono: 135) and on the ways in which cultural meanings and practices mediate subsequent resilience and disaster preparedness. Their problematic is not far from Harms' intellectual pursuits, but instead of memory as cultural practice, they focus on spiritual coping as a means of social resilience. The very intricate process of coping and aid, which has not been well studied from an individual, psychological point of view, deems it necessary to unpack them in order to understand how strategies of external help can be integrated into local cultural logics. In order to understand this process, they discuss "how the risk and vulnerability approach in disaster research, research into psychological coping and knowledge of local and cultural specificity in cultural anthropology interrelate with each other" (ibid.: 134). The aim of this very ambitious program is the production of knowledge in the sense of the grounded theory of Glaser and Strauss (1967) and the hope of improving coping strategies in disaster management.

After a short discussion of the risk and vulnerability approach by Wisner et al. (2004), which focuses on the socio-economic, political and ecological aspects of coping but is mute on how different groups of people experienced the disaster and how it altered their well-being, the authors introduce the different conceptualizations of psychological coping research. The various appraisal-based approaches are presented and evaluated with reference to their explanatory value in reducing stress. They dismiss the acclaimed differentiation between problem-focused and emotion-focused coping because they often overlap. Instead they favour a meaning-centred coping within appraisal-based approaches, which "is seen as positive cognitive restructuring, examining beliefs and values, reordering life priorities, infusing new meanings and finding benefits in adversity" (Zaumseil and Prawitasari-Hadiyono: 138). In this way, the contents of spiritual and religious beliefs are not discarded and can be integrated into universal models of coping. From the various appraisal-based approaches in the field of religious psychology, Zaumseil and Prawitasari-Hadiyono mention among others Hobfol's theory of resource conservation (COR) and the neuro-culture-interaction model advanced by Kitayama and Uskul, who stress that "cultural influence [on the brain] is mainly exerted by doing and practising what is relevant in the cultural context" (quoted in Zaumseil and Prawitasari Hadiyono: 140) In contrast to trauma theory, which they criticize as being dominated by Western assumptions, they conclude that the cultural psychology of religion offers a rich perspective for understanding coping strategies in specific cultural contexts. The decisive question is thus to specify what is culturally specific about Java? Is there a Javanese culture, and what are its characteristics? Turning to anthropological theory, they opt for the deconstruction of the notion of culture, as well as for de-essentializing the image of Java. It is not only religious plurality in Java which makes it difficult to make statements about Javanism, but also the dynamics and inner variety which are constitutive for this society.

Their very complex theory-generating method, based on insights from Clarke's *situational maps*, which are an enhancement of the methods propagated by Glaser and Strauss, enables them to confront these different theoretical approaches, which all originate from Western scholarship with the interpretation of the local people. Through intensive participatory methods, including village theatrical performances, they obtained a wide range of answers regarding the local interpretation, which they then projected on the theoretical explanations just mentioned. These newly generated theory blocks are then superimposed again on the findings of the various appraisal-based approaches they formerly discussed. As a result, they gain an analytically saturated insight into their own findings by looking at them from different theoretical perspectives and local semantic fields.

3. The politics of space: negotiating reconstruction

Disasters symbolise social disorder, representing a time of chaos, sometimes of upheaval, but also of hope for a better future. Many local people want to reconstruct the status quo ante. Others hope for a new beginning, for the opportunity to have a second chance in life (see Hoffman 1999: 150). Bureaucrats and entrepreneurs dream of creative destruction, that is, of building new cities and enforcing radical concepts of modernization. Ashkabad, Algiers and Tashkent are examples of such a rigid top-down approach which in many ways offended people's wishes to protect their homes and neighbourhoods. Phases of reconstruction thus implicate the politics of space and identity. They are arenas of negotiations, of cooperation in forging new alliances, but also of conflicts and bitter struggles.

Disasters reveal structures of hierarchy and inequality in societies which are based on class, gender or age, on social ascription or achievement. It is in these situations that social vulnerability can best be studied with regard to access to resources or its frustration. To achieve normality again can become a long drawn out process, taking months or even years. The phases of reconstruction are stretched out and fragmented into times of rescue, relief and reconstruction. On the occurrence of the event follow phases of consolation, of mutual support and solidarity, which come close to the idea of Turner's *communitas*. Social differences are ignored, feelings of unity and comradeship celebrated (see Hoffman 1999, Schlehe 2006). Oliver-Smith (1999b) has named this transient period the 'brotherhood of pain', which, however, gives way to rivalry and envy, often when outside support comes in. These feelings dominate the phase of reconstruction in many, albeit not in all societies, leading to the social isolation of the victims, their psychological estrangement and to their possible fragmentation. Negotiations between the local population and different strata of government personnel may become as conflict-ridden as the relationship between the various NGOs and the government, as is the case in many societies.

Martin Sökefeld gives a detailed account of the phases of relief and reconstruction that focuses on the negotiation between local communities, their respective local and national government representatives and outside aid workers. Although the landslide in Attabad had made itself known some time previously, people did not pay attention to the first signs and cracks in the mountains. They ignored this warning and were taken by surprise by the landslide, the debris of which "created a huge barrier of more than hundred metres height and one kilometre width which completely blocked the flow of the Hunza-River and also buried the Karakorum Highway (KKH)" (Sökefeld: 178). The demolition of this part of the highway destroyed the north's lifeline to the rest of Pakistan and neighbouring countries like China. Through the slow, but continuous expansion of the lake which resulted from the blocked Hunza River, the population of Gojal was cut off from food, trade and communication. The lake flooded several villages, destroying houses, fields and plots. It shattered the basis of the local economy, which, in addition to agriculture, rested on tourism and led to a dramatic setback of the well-developed system of education. This 'man-made disaster' whose end was incalculable brought unrest to a region that was already characterised by 'political marginalization because it is affected by the Kashmir dispute'.

Since the area belongs to neither Kashmir nor Pakistan because it lacks constitutional status, its politically marginalized positions have given rise to a long history of political strife. Questions of ethnicity, which were directly linked to the former feudal system of the Mir, as well as to religious tensions between the Ismailiyya, Sunni and Shia Muslim communities, came as much to the fore as different loyalties to rival political parties which either favoured independence or constitutional integration into Pakistan. Government became the common enemy. The deep-seated mistrust of the ruling establishment expressed itself in rumours, public criticism and accusations of corruption. Despite the politician's visits and (small) donations, most of the locals were not satisfied with these politics of symbols but wanted practical action. Because their own resources were insufficient to deal with the challenges of the lake, they insisted on outside help from China, which they considered technologically more advanced. But before this could be organised, past tensions among ethnic and religious groups, as well as between different generations, broke out and complicated the situation further, which culminated in protest demonstrations and mass rallies. The newly created spaces of resistance brought into being new symbols which celebrated the memory of the disaster's anniversary, like the so-called Black Day. Accompanied by intense media coverage, this symbolic commemoration 'culturalized' the 'natural disaster', which seems to be a typical reaction of survivors (see Hoffman 1999: 143). At the same time, it was a political demonstration of the seriousness with which a fraction of the local population fought for the re-appropriation of their living space and right to belong. This politics of space turned more violent in the course of events. Not only have protesting youngsters been arrested under §144 and turned into 'terrorists', but the rallies, agitation and protest demonstrations culminated in the shooting of two protesters, both internally displaced persons who were fighting to return to their homes.

The occurrence of violence during the phases of reconstruction is not altogether new. The aftermath of hurricane Katrina, when racial conflicts were sparked off, exposing long-established relations of inequality and superiority, is a case in point. The increase in vulnerability and suffering apparently activates old conflicts, and memories of loss and deprivation under certain historical and political conditions. Coping strategies in Gilgit and Baltistan are thus determined – as in the Ganges Delta – by social memories which, albeit under different conditions, refer to former histories of inequality and political conflict.

Pascale Schild's description of the earthquake in Azad Kashmir, which caused at least 80,000 deaths and left over three million people homeless, is another example of the intricate relationship between the state and local survivors. In her research in Muzzaffarabad, the capital of Azad Kashmir, one object of enquiry was the negotiations over the compensation for destroyed houses. Since the earthquake was considered one of the most severe in Asia during the last hundred years by the US Geological Survey, the enormous impact on the social fabric had to be mitigated by government aid. Supported by international donors, the Pakistani government created a new bureaucratic organisation (ERRA), which actively provided a form of compensation according to the principle of one house, one family. Despite the small payments involved, competition for these financial resources aggravated existing tensions. But in contrast to the public conflicts in Gojal, these conflicts were not openly discussed, but rather resembled hidden transcripts. What is of special interest is that they were framed in idioms of ideal kinship relations. Due to the patrilineal ideal of joint families in this area, a house can include several households or nuclear families. However, as rumours had it, many families no longer fulfilled this ideal but people had moved away when their fathers had died and established households of their own. This practice had already begun before the disaster but became more common after it due to the scarcity of houses and rooms. Schild argues that government policies not only neglected the complexity of social realities and their ever-changing forms but that they increased social hierarchies by creating winners and losers. Furthermore, this interference of the state "in the minute texture of everyday life" (see Gupta 1995: 375) raised people's awareness of the importance of place as an integral part of their identity. This politics of space is well documented for disasters in other societies as well (see Bode 1989; Hoffman 2002), since most victims want to stay near their old homes.

Schild's contribution is an intriguing analysis of the re-negotiations of the social fabric which articulated itself in the dialectical relationship between houses and homes. The materiality of the house and its close links to a location coalesces with the sociality and emotional belonging of the home. Spaces turn into places which convey feelings of trust and security under the conditions of severe suffering.

To analyse the symbolic and emotional meanings of material objects, which embodies a somewhat new trend in anthropology, opens up new paths in disaster research to understand better the complexities of identity which are salient for the successful process of reconstruction. How these changes are mastered depends on the relationships between the state, the different strata of civil society and to a great degree on national and international capital. An outstanding example of the hubris involved in such efforts of modernization is described by **Edward Simpson** in his discussion of the earthquake in Kutch in 2001. He vividly describes a process of "creative destruction" (see Schumpeter 1994: 81ff.) as a combined form of political and economic interventions by insiders and outsiders. The reconstruction of the greatly destroyed city of Bhuj represents a tour de force on the part of several government agencies and economic entrepreneurs who, by distributing of money and promises to the poor, fought for their idea of a modernized city. In order to change the old city into a modern capital, politicians and planners alike did not hesitate to ignore social and religious rules. They disregarded caste regulations, bulldozing Muslim graveyards and other religious sites. The aim was to create a capital like Singapore dominated by vast alleys, traffic lights and empty spaces, which according to Simpson became the very symbol of a fake city, de-humanised in its construction of space and estranged from its inhabitants. Resistance to these policies of 'forced modernity' took manifold forms, including intense negotiations and struggles between bureaucrats and local residents and ended in some cases in such desperate acts as suicide. But not all the city's inhabitants shared a nostalgic view of their home town. Others welcomed the possibility of seemingly unlimited consumption. Instead of using their money to rebuild their houses or buy new plots, they invested it in short-lived consumer goods.

The reconstruction of Bhuj and the neighbouring villages came close to what was locally referred to as a 'second earthquake'. It erased the familiarity of home and the values attached to it. The modern Bhuj was much more than the development of a postcolonial town which had shunned its colonial past, since "its effect was to demand a new type of citizen: a mobile suburban consumer" (Simpson: 253). This case study reminds one of Frederic Jameson's view of "the spatial logic of multinational (postmodern) capitalism" that "is simultaneously homogeneous and fragmented" and creates "a kind of 'schizo-space'" (quoted in Keith and Pile 1993: 2). Simpson's own conclusion is rather more sociological, reminding "readers that post-disaster booms are old and regular features of catastrophe, and that these cannot simply be attributed to the universal expansion of capital but must be understood as parts of the moment of sublime destruction and the collective emotional and sociological responses of the beleaguered" (Simpson: 240).

4. Constructing local meanings

Cultural interpretations structure the understandings of natural disasters and impinge upon strategies of coping. Since disasters create spaces of death, they naturally demand questions and answers which go beyond technological or scientific reasoning. Macamo and Neubert call them the 'how' and 'why' questions. In their comparative analysis of the floods in three structurally different societies, they came to the conclusion that their importance depended on the technological know-how in these societies. 'Why questions', which are questions of causation, were of great importance in the Limpopo valley, with its less developed technological expertise, whereas in those societies with highly developed management capabilities, the 'how questions' dominated. Even in the Bible belt, both explanations were used according to circumstance. The simultaneity of moral/religious and scientific/technological explanations has been confirmed by many anthropological studies of the perception and interpretation of disasters. During the floods in Hamburg in 1962, religious interpretations were evoked side by side with technological explanations, although less frequently (see Engels 2003: 125). The complexity and multi-dimensionality of disasters is certainly one of the reasons for the varieties of causes mentioned. The differing constructions of nature are as much implicated as differences in social organisation, especially with regard to political configurations. Frömming (2006: 50ff), Schlehe (1996, 2010) and Bode (1977) have drawn attention to the widespread imagination that a punitive or revengeful God (or spirits) sends a disaster because of people's immorality or the bad political behaviour of the elite, which also triggers off discussions about tradition and modernity (see Schlehe 2010). In Bode's example, it was the abasing treatment of the local Indian population by the *mestizos* which was made responsible for the combination of earthquake and avalanche that caused between 50,000 and 100,000 deaths. These examples, which could be multiplied, highlight the moral ascription and highly symbolic meanings that are attributed to disasters under certain political constellations. They are read as a cultural critique of social and political relations and as moral agencies which are in a position to punish or to do justice.

The chapter by **Axel Schäfer** draws attention to a hazardous event which is typical for the Andean mountains: the deaths of people and animals by lightning.

He describes in detail such an incident in which the Andean mountain gods of lightning and hail killed the only two cows of a poor family. In the Andes, as in many other societies, nature is still imbued with religious meaning as an expression of invisible forces, like the ancestors, spirits or gods. In these "local topographies of meanings" (Hastrup 2009: 28), nature is in no way a natural given separate from culture, as Lévi-Strauss postulated, but rather a part of it (see Luig 2002: 2). Andean constructions of nature intersect with cultural values and moral evaluations. The perception of the mountains as the seats of powerful gods reflect their societies' vulnerability, which is expressed in disasters like earthquakes, avalanches and landslides (see Bode 1989), which always cause great numbers of deaths and great damage. In addition, lightning and hail, which most often kill individuals, are widespread threats in the mountains, surrounded by a rich folklore and embedded in old belief systems. Lightning can take different forms, ranging from bolts of lightning to 'left or right, male or female lightning strokes' which, according to context, are interpreted as punitive or as revelatory. Revelatory knowledge is empowering knowledge: it transforms the victim into a healer (see Rösing 1990). It is this very relationship between man and God which forbids the rescuing of victims of lightning or giving them medical attention. If the person survives he can either become a healer or pursue his normal life. In both cases, however, the question why this particular person was affected has to be resolved.

The example Schäfer describes is by no means a disaster since it caused the deaths of only two cows of a single family. However, under the conditions of extreme poverty in which the family lives, the loss of their cows threatens their economic survival. It is therefore necessary to find out the reasons why this individual family was hurt in order to placate the gods. In a ritual rich with symbols passed down from different Andean cultures (pre-Inca, Inca, Christian), the victims are purified and their relationship to the punishing God of lightning is transformed through different ritual acts. The aim of the ritual is to restore the balance between the gods and the humans, which include the safety of their animals, as well as the fertility of the land. These ritual practices are important contributions to the resilience of the family and the agricultural system as a whole. Being passed down even from before the times of the Inca, these rituals were only slightly transformed through Christian beliefs, and their practice continues until the present. They thus exhibit a surprising parallel with beliefs about lightning and hail in Europe, which were also thought of as embodiments of God in Germanic times. Only (slightly) changed during the Enlightenment, these beliefs are still reproduced in narratives and fables (see Bächtold-Stäubli 1987). They reveal a stability of structures despite consecutive waves of modernization. A case in point is the famous wayside shrines (*Marterl*) in the Bavarian and Austrian mountains consisting of crucifixes or Madonna statues. Reminders of past disasters, in which lightning plays a prominent part, they are symbols of gratitude and of the fortunate rescue of the victim. The many prayers of thanks for escape from dangerous situations scribbled on leaflets and fixed to the walls in pilgrimage chapels are further proofs of religious vitality. Schäfer's contribution in this volume does not reflect the 'exoticism' of a pre-modern world view, but demonstrates its *coevalness* – to cite a term of Fabian's (1983) – with beliefs in a highly complex society.

A similar world view regarding the dependence between humans and the invisible forces of nature is described by **Dorothea Schulz**. Taking as her point of departure the triple disasters that Ugandans have had to endure in recent decades – civil war, natural disasters like landslides, floods and droughts, as well as the AIDS epidemic – she asks how people deal with these premature, often violent deaths. The main issue at stake is the question whether the experience of civil war and natural disasters are perceived as the same form of violence and to what extent these experiences are differently constructed.[5] Rituals of mourning are a helpful tool in understanding the conceptualization of death and the dead, and their transformation into spiritual beings does have great importance in Africa and elsewhere. They can even become political issues, as discussion concerning the death of the Congo dictator Mobutu made apparent (see Jewsiewicki and White 2005). As Butler (2004) has shown for other parts of the world too, the differentiation between good and bad deaths is of prime importance for the living. A good death involves proper mourning and mortuary rituals in order to allow for the transformation of the dead person into a 'living' ancestor who will decide the fate of his relatives. In contrast, improper rituals or no rituals at all will haunt the living, causing illness, infertility, misfortunes etc. As Schulz points out, we do not know much about how disaster-related deaths are perceived in the Muslim community in and around Mbale in eastern Uganda, nor elsewhere in Uganda and beyond. How can threats from the living be averted and the harmony between the world of the spirits and humans restored? It is a merit of this chapter that it addresses the importance of the physical presence or absence of corpses, or even of particular intimate parts of it, like skulls, to the proper performance of mourning rites. In most rituals of mourning, the presence of the corpse is necessary for a proper disposal,[6] but due to the multitude of violent deaths in Africa (see de Boeck 2005) people have devised substitutes for them (see Luig 2009). Yet we still do not know to what extent such pragmatic measures are applied in the case of disasters as well.

Schulz's chapter dicusses new lines of inquiry into new research options which put rituals of mourning and their accompanying emotions high on a future research agenda. The theoretization of emotions which have been conspicuously

---

5 To regard natural disasters as a war of nature against humans is a familiar metaphor in disaster research (see Engels 2003: 124).

6 See Cohen and Atieno-Odhiambo 2001 on the war over the corpse of the Kenya foreign minister.

absent from discussions in disaster research,[7] allows us to analyse the painful experience of loss and sorrow, of beloved victims who are commemorated in ritual celebrations. The analysis of emotions, embodied in religious rituals, mediates individual and social spheres of action. Emotions translated into actions are important sources in building up resilience, possibly shaping strategies of prevention, as well as specific measures of coping.

The yet underdeveloped discussion of disaster and violence opens up a window not only into discourses on war but equally into the spaces of death that are closely linked to social memory, body politics and suffering. In the process of remaking their world (see Das, Kleinman, Locke et al. 2001), religious beliefs – or spiritualism, as Zaumseil and Prawitasari-Hadiyono ultimately call it – can be of great importance in relieving the sufferings of the victims. In her personal account of the Oakland firestorm, Susanna Hoffman (2002: 118ff) recounts that prayers and ritual practices helped the victims find a new relationship with themselves and to overcome their depression. The experience of total loss was so fundamental for their self-image that nobody felt the same person as before. This essential change of identity is also documented in some of Zaumseil and Prawitasari-Hadiyono's case studies. While narrating their subjective maturation process, villagers referred among other topics to questions of guilt and to the various ways of overcoming sorrow and depression. Phases like acceptance, surrender and gratitude, which are embedded in rich semantic fields of meaning, were helpful steps after inner protests in order to come closer to God. Their interviews show how victims were able to constitute new identities for themselves to master their lives after the earthquake. But the narration of their disaster experience also laid open the dynamics of cultural change in Indonesia and the various ways in which they constructed a new moral order.

The construction and perception of change after disasters is also the concern of **Brigitte Vettori**. She questions the frequent assumption that disasters eradicate whole cultures and allow for new beginnings like a tabula rasa. Her argument is not only well illustrated by the articles on coping strategies in this volume (Sökefeld, Schild, Simpson), but by the very interview she was able to have with a leading village chief in the Nicobar Islands. In an unpretentious manner, this village chief provides insights into the everyday fabric of his society, the emotions and doubts he had during the time of rescue, the mistakes of his relatives and neighbours during the tsunami and the kind of support he was looking for. Through this very individual and intimate picture, basic assumptions about cultural changes during and after disasters are refuted, as is the naïve trust in local knowledge which for long belonged to one of the cherished items in disaster research. The interview discloses a world in which new and old traditions compete for acceptance. The very practical behaviour of the chief, who selected the

---

7 For an exception, however, see Oliver-Smith 1999.

item he needs and thinks valuable according to his individual convictions, belies the images in the Indian as well as the European press of the pure but innocent 'natives' being deeply embedded in traditional knowledge and old customs. Vettori's article, which analyses in detail the worldwide media reception of the survival of the Nicobar and Andaman islanders during the tsunami, contains a biting critique of clichéd images of these 'natives' among Indian and European journalists. It affirms yet again the well-known fact that this kind of romanticisation and idealisation of the 'noble savage' says more about the author's way of thinking than about the subjects of his or her studies. Vettori's article comes close to Oliver-Smith and Hoffmann's (2002: 5ff) suggestion that disaster anthropology reveals the fundamental questions of anthropology and of its diverse challenges in a technologically complex world. Although globalization theory has made big inroads into anthropology and changed conventional ideas about culture and cultural change, the inequality of these changes, their different dynamics and their unequal positioning in global relations of power come to the fore again with every new disaster. Disaster anthropology therefore still has a long way to go to come to terms with the theorization of its increasing complex conditions.

5. Bibliography

Alexander, David (1997): The Study of Natural Disasters, 1977-1997: Some Reflections on a Changing Field of Knowledge. In: Disasters 21, 4, pp. 284-304.

Antze, Paul and Michael Lambek, (eds) (1996): Tense Past: Cultural Essays on Trauma and Memory. London: Routledge.

Bächtold-Stäubli, Hanns (1987): Handwörterbuch des deutschen Aberglaubens. Berlin, New York: de Gruyter.

Beck, Ulrich (1986): Risikogesellschaft. Auf dem Weg in eine andere Moderne. Frankfurt: Suhrkamp.

Blaikie, Piers, Cannon, Terry, Davis, Ian and Ben Wisner (1994): At Risk: Natural Hazards, People's Vulnerability, and Disasters. London: Routledge.

Bode, Barbara (1989): No Bells to Toll. Destruction and Creation in the Andes. New York: Macmillan.

——— (1977): Disaster, Social Structure and Myth in the Peruvian Andes: The Genesis of an Explanation. In: Annals of the New York Academy of Sciences 293, pp. 246–274.

Bohle, Hans-G., Downing, T.E. and Michael J. Watts (1994): Climate Change and Social Vulnerability: Towards a Sociology and Geography of Food Insecurity. In: Global Environmental Change 4, pp. 37 - 48.

Butler, Judith (2004): Precarious Life: The Power of Mourning and Violence. London, New York: Verso.
———— (2003): Afterword: After Loss, What Then? In: Eng, David L. and David Kazanjian (eds.): Loss: The Politics of Mourning, edited by. Berkeley: University of California Press, pp. 467.474.
Cohen, David William and E.S. Atieno-Odhiambo (2001): Wer begräbt SM. Münster: Lit Verlag.
Das, Veena, Kleinman, Arthur, Lock, Margaret, Ramphele, Mamphela and Pamela Renolds (eds.) (2001): Remaking a World: Violence, Social Suffering, and Recovery. Berkeley: University of California Press.
De Boeck, Filip (2005): The Apocalyptic Interlude: Revealing Death in Kinshasa. In: African Studies Review 48, 2, pp. 11-32.
Douglas Mary (1992): Risk and Blame: Essays in Cultural Theory. London: Routledge.
Engels, Jens Ivo (2003): Naturbild und Naturkatastrophen in der Geschichte der Bundesrepublik Deutschland. In: Groh, Dieter, Kempe, Michael and Franz Mauelshagen (eds.): Naturkatastrophen: Beiträge zu ihrer Deutung, Wahrnehmung und Darstellung von Text und Bild von der Antike bis ins 20. Jahrhundert. Tübingen: Gunter Narr Verlag, pp. 119-142.
Fabian, Johannes (1983): Time and the Other: How Anthropology makes its Object. New York: Columbia University Press.
Frömming, Urte Undine (2006): Naturkatastrophen. Kulturelle Deutung und Verarbeitung. Frankfurt a. Main: Campus.
Glaser, Barney G. and Anselm L. Strauss (1967): The Discovery of Grounded Theory: Strategies for Qualitative Research. Chicago: Aldine.
Gupta, Akhil (1995): Blurred Boundaries: The Discourse of Corruption, the Culture of Politics, and the Imagined State. In: American Ethnologist 22, 2, pp. 375- 402.
Hastrup, Kirsten (ed.) (2009): The Question of Resilience: Social Responses to Climate Change. Copenhagen: The Royal Danish Academy of Sciences and Letters.
Hoeppe, Götz (2007): Conversions on the Beach: Fishermen's Knowledge, Metaphor and Environmental Change in South India. New York: Berghahn.
Hoffman, Susanna M. (2002): The Monster and the Mother: The Symbolism of Disaster. In: Hoffman, Susanna M. and Anthony Oliver-Smith (eds.): Catastrophe and Culture: The Anthropology of Disaster. Santa Fe: School of American Research Press, pp. 113-141.
———— (1999): The Worst of Times, the Best of Times: Toward a Model of Cultural Response to Disaster. In: Oliver-Smith, Anthony and Susanna M. Hoffman (eds.): The Angry Earth: Disaster in Anthropological Perspective. New York: Routledge, pp. 134-155.

Holling C.S. (1973): Resilience and Stability of Ecological Systems. In: Annual Review of Ecology and Systematics 4, pp. 1-23.

Jewsiewicki, Bogumil and Bob W. White (2005): Introduction. In: African Studies Review, Focus: Mourning and the Imagination of Political Time in Contemporary Central Africa, 48, 2, pp.1-9.

Keith, Michael and Steve Pile (eds.) (1993): Introduction Part I: The Politics of Place. In: Place and the Politics of Identity. London and New York: Routledge, pp. 1-21.

Lambek, Michael (1993): Knowledge and Practice in Mayotte: Local Discourses of Islam, Sorcery and Spirit Possession. Toronto: University of Toronto Press.

Luig, Ute (2009): Reinheit und Unreinheit in afrikanischen Beerdigungsritualen. In: Malinar, Angelika and Martin Vöhler (eds.): Un/Reinheit, Konzepte und Praktiken im Kulturvergleich. Paderborn: Wilhelm-Fink-Verlag, pp. 103-122.

——— (2002): Einleitung. In: Luig, Ute and Hans-Dietrich Schultz (eds.): Natur in der Moderne: Interdisziplinäre Ansichten. Berlin: Berliner Geographische Arbeiten, pp. 1-22.

Neiman, Susan (2006): Das Böse denken: eine andere Geschichte der Philosophie. Frankfurt: Suhrkamp.

Oliver-Smith, Anthony (1999a): What is a Disaster? Anthropological Perspectives on a Persistent Question. In: Oliver-Smith, Anthony and Susana M. Hoffman (eds.): The Angry Earth: Disaster in Anthropological Perspective. New York: Routledge, pp. 18 -34.

——— (1999b): The Brotherhood of Pain: Theoretical and Applied Perspectives on Post-Disaster Solidarity. In: Oliver-Smith, Anthony and Susanna M. Hoffman (eds.): The Angry Earth: Disaster in Anthropological Perspective. New York: Routledge, pp. 156-172.

Oliver-Smith, Anthony and Susanna M. Hoffman (2002): Introduction. Why Anthropologists Should Study Disasters. In: Hoffman, Susanna M. and Anthony Oliver-Smith (eds.): Catastrophe and Culture: The Anthropology of Disaster. Santa Fe: School of American Research Press, pp. 3-22

Paine, Robert (2002): Danger and the No-Risk Thesis. In: Hoffman, Susanna M. and Anthony Oliver-Smith (eds.): Catastrophe and Culture: The Anthropology of Disaster. Santa Fe: School of American Research Press, pp. 67-89.

Rösing, Ina (1990): Der Blitz. Drohung und Berufung. Glaube und Ritual in den Anden Boliviens. München: Trickster Verlag.

Schlehe, Judith (2010): Anthropology of Religion: Disasters and the Representations of Tradition and Modernity. In: Religion. Special Issue on 'Religions, Natural Hazards, and Disasters', 40, 2, pp. 112-120.

————— (2006): Nach dem Erdbeben auf Java: Kulturelle Polarisierungen, soziale Solidarität und Abgrenzung. In: Internationales Asienforum 37, pp. 213-238.

————— (1996): Reinterpretations of Mystical Traditions: Explanations of a Volcanic Eruption in Java. In: Anthropos 91, pp. 391-409.

Schumpeter, Joseph A. (1994[1943]): Capitalism, Socialism, Democracy. London: Routledge.

Sen, Amartya (1981): Poverty and Famines: An Essay on Entitlement and Deprivation. New Delhi: Oxford University Press.

Stehr, Nico (1994): Knowledge Societies. London: Sage Publications.

Vigh, Henrik (2008): Crisis and Chronicity: Anthropological Perspectives on Continuous Conflict and Decline. In: Ethnos 73, 1, pp. 5-24.

Welzer, Harald, Soeffner, Hans-Georg and Dana Giesecke (eds.) (2010): KlimaKulturen: Soziale Wirklichkeiten im Klimawandel. Frankfurt: Campus-Verlag.

Wisner, Ben, Blaikie, Piers, Cannon, Terry and Ian Davis (2004): At Risk: Natural Hazards, People's Vulnerability and Disasters. New York: Routledge.

II. Engaging with theories

II.1 Thinking about risk and risk management

Social-ecological change and the changing structure of risk, risk management and resilience in a pastoral community in northwestern Namibia

Michael Bollig

1. Introduction

The idea of this paper is to relate risk, risk management and resilience to social-ecological change. In many anthropological accounts risk, risk management and resilience are analysed in historically non-sensitive ways. Risk management strategies are often depicted as "traditional" and unchanging and "traditional" ways are often juxtaposed to current vulnerabilities in a simplistic manner. In this paper I would like to show how pastoralists adapt to challenges brought about by profound technological and institutional changes and how resilience is communally produced as a kind of common good for a period of time. In order to make my point clear I will focus on the manifold implications for risk, risk management and resilience brought about by state-led changes of the hydrological regime in a pastoral context between the 1950s and 1970s and the governmentally promulgated reorganisation of land tenure and water management since the 1990s.

I will start with a short conceptual exploration of risk, risk management and resilience in social and cultural anthropology, then sketch my case study and finally try to summarize what this case may contribute to the study of risk, risk management and resilience.

2. Explorations on risk and risk management in anthropology

Risk, risk management and resilience are relatively new fields in the social and environmental sciences. More recent advances in cultural ecology (Moran 1979), evolutionary biology and ecology (Borgerhoff-Mulder 1991; Boyd and Richerson 1985), economic theory (Ensminger 1992) and political economy (Watts 1983, 1991) have given risk a prominent status in theory building. These new orientations point out that risk minimisation is central to our understanding of individual economic and social strategies, social institutions and social change and not just a peripheral and transient moment in a group's history. As yet there is only little anthropological literature on resilience: whereas archaeologists and geographers have embraced the concept with some sympathy, anthropologists are still somewhat reluctant to do so. I will first touch upon literature pertaining to risk and risk management and then shortly report on the uses of the resilience concept within the social sciences. I distinguish at least three distinct anthropological approaches to the topic of risk: an actor-oriented, an ethnographic and an interpretative approach.

2.1 Actor-oriented approaches to the study of risk

Anthropologists interested in forager societies have emphasized risk management strategies as a major force shaping hunting and gathering routines and structuring institutions of food sharing and territorial behaviour (Acheson 1989; Wiessner 1977, 1982; Winterhalder 1990; Cashdan 1985, 1990a, 1990b; Kaplan, Hill and Hurtado 1990). Pertinent features of social structure are explained as a sequel of risk minimizing behaviour. The Kung !hxaro exchange is portrayed as a key institution lowering the risk of shortfalls of food production. These contributions have in common that they either apply rational choice-based models to empirical data and/or ascribe social institutions specific functions in a highly variable environment. They often set off with some clearly defined hypotheses with relevance for a larger theoretical framework; quantify the variation of key-resources (which they equate with risk) and then single out specific strategies of risk minimization. Finally they test why specific risk management strategies are superior to other, alternative ways of handling unreliable food supplies. Kaplan, Hill and Hurtado (1990) have explained the specifics of Aché (Paraguay) hunting strategies and institutionalized sharing by referring to subsistence risks. Instead of relying on widely available food sources such as palms, the Aché go for the high variance game resource, thereby accepting a much higher subsistence risk. Institutionalized food-sharing levels out the unpredictable supply of meat. Hames (1990) takes up the subject of sharing as a strategy of lowering subsistence risks of food amongst the Amazonian Yanomami. He finds that the higher the variability of a resource (i.e. the higher the risk) the more intensive the sharing, the larger the spatial scope of sharing and the lower the importance of a kin bias in sharing is. The social institution "sharing" directly answers to a set of risks.

These studies did not result in models of risk and risk management with a more general applicability. While their concentration on individual coping strategies rhymes well with game-theoretical approaches, we learn little about the emergence of institutionalised efforts of risk management. Risk analysis is often offered in a historically de-contextualised way (but see Wiessner 1994 for an attempt to overcome this shortfall), emic perspectives on the variability of resources are neglected and actors are presented as constrained only by natural conditions but not by existing norms and values and/or political dynamics (cf. for a comprehensive critique of these highly formalised studies see Baksh and

Johnson 1990). Nevertheless, it has to be conceded that this branch of risk management analysis has developed a set of tested hypothesis and articulates well with some social science theories; unlike other anthropological explorations of risk and risk management it has a clear theoretical agenda.

2.2 Ethnographic approaches to risk management

The literature on hazards and risk management in Africa's dry belts has increased rapidly since the middle of the 1970s. Starting with the Sahel crises of the early 1970s anthropological reports on pastoral societies have dealt with economic marginalization and impoverishment (e.g. Hogg 1986, 1989; White 1997; Broch-Due and Anderson 2000), environmental degradation (Spooner 1989), the dissolution of communal lands (Galaty 1994; Ensminger 1992; Hitchcock 1990; Stahl 2000), and land loss (Arhem 1984; McCabe 1997), internal stratification (Hogg 1986; Little 1987, 1992) social disintegration (Hogg 1989) and manifold reactions of local people to react to these challenges. Risk here is often equated with hazards (e.g. drought) and complex social (e.g. marginalization) phenomena. A number of excellent descriptive studies on hazards and risk management in Africa's dry-land areas have been written since the 1970s (Colson 1979; White 1984; Johnson and Anderson 1988; Downs, Reyna and Kerner 1991; Hogg 1986, 1989; Hjort and Salih 1989; Spittler 1989a, 1989b; Fratkin 1991; see Shipton 1990 for an overview of recent literature). These ethnographically condensed accounts clearly indicated that risk management is a major key to understanding African rural communities. These contributions are strong on ethnography and history and provide substantial descriptions of historically changing modes of risk management. However, we usually learn little about how people perceive hazards and how they relate them to their belief system. Furthermore, no attempt is made to establish more generalizing hypotheses: the contributions are first of all excellent ethnographic descriptions of social and ecological systems and risk management within these environments.

2.3 Interpretative approaches to risk

While Mary Douglas's approach to risk analysis (Douglas 1985, 1994) has not had much influence on the two paradigms discussed above, it had considerable impact on the study of risk analysis beyond anthropology and is widely accepted as *the* anthropological contribution to the study of risk. Her major contributions 'Risk Acceptability according to the Social Sciences' (1985) and 'Risk and Blame' (1994) embrace the concept of risk perception as a general concept of social analysis. Douglas (1994: 15) pinpoints the relevance of the concept in western thought aptly: "The idea of risk could have been custom made. Its universalizing terminology, its abstractness, its power of condensation, its scientificity, its connection with objective analysis, makes it perfect. Above all, its forensic uses fit the tool to the task of building a culture that supports a modern industrial society." Her approach to the topic is demanding: she develops a theory of risk perception which includes all sorts of risks ranging from nuclear waste pollution to witchcraft in various types of society (Douglas, 1994: 22). Her main message is: risk perception is encoded in social institutions. Although Douglas's approach brings risk perception, as a new sub-field, into the study of risk, its analytical focus is vague. The typology Douglas proposes is simplistic and apparently has not influenced empirical studies very much (see Boholm 1996 for a general critique).

2.4 Resilience
As yet the resilience concept has not been widely used in anthropological accounts. Its main applications are within ecology. However, social scientists have started to make use of the concept. Folke's definition (2006:259) of the term points to new perspectives for the social sciences. He defines resilience as "the opportunities that disturbance opens up in the terms of recombination of evolved structures and processes, renewal of the system and emergence of new trajectories. In this sense, resilience provides [is] adaptive capacity that allows for continuous development, like a dynamic adaptive interplay between sustaining and developing with change." In order to circumvent naturalistic assumptions underlying the concept resilience other authors have offered resistance or social resilience as alternative terms. Social scientists such as Anderies, Janssen and Ostrom (2004) have argued that the resilience concept may be difficult to apply to social-ecological systems, which are in part designed by humans. Instead, they propose the concept of robustness, defined as "the maintenance of some desired system characteristics despite fluctuations in the behavior of its component parts or its environment" (ibid.: 18). Unlike resilience the robustness framework incorporates institutions and technology (both addressed as infrastructure), and a distinction between resource users and infrastructure providers into the analysis (see also Janssen 2006 for a longer discussion of the robustness concept). While the resilience/robustness framework has been fruitfully applied to relatively simple and bounded SES we still lack a convincing methodology to describe resilience/robustness in complex systems driven largely by anthropomorphic influence. In this paper I will stay with the concept resilience but will regard resilience as socially constituted.

3. Theoretical approach of this study

A general conceptualization and cross-disciplinarily applicable definition of the term risk is lacking (Bechmann 1993: 240). The concept *risk* is used in colloquial as well as in various scientific discourses and we are confronted with various meanings and definitions. Jungermann and Slovic (1993) present six scientific definitions of risk of varying complexity.[1] It was especially the definition of risk as the product of the probability and the extent of specific damage that has dominated insurance studies and sociology for some time. While such clear-cut definitions of risk may be necessary in insurance studies in order to allocate premiums differentially, they do not capture the core of the problem in the social sciences. Jungermann and Slovic (1993: 171) point out that risk is *not* a directly perceivable phenomenon. They conclude (1993: 201):

> In short, there is not an 'objective risk'. Risk is a **multidimensional construct**. 'Risk' exists as an intuitive concept, which for most people **means more than the 'expected number of future damage'**. Its mental presentations are shaped by **knowledge** on the subject matter, by **characteristics of the cognitive and motivational system** and finally by **social reality** with its inherent interests and values. (translation and emphasis by author)

In contrast to the sociological approximations of the term, a great deal of anthropological ideas on the concept were tied to the observable. Wiessner (1977: 5) defined risk as the "probability of loss or the possibility (or probability) of an unfortunate occurrence. ... An unfortunate occurrence can be considered to be anything which alone, or in combination with other occurrences, can be detrimental to the survival and reproduction of an individual and his family." Cashdan, the editor of one of the most frequently cited anthropological volumes on risk management (Cashdan 1990b:2/3), defines risk in a similar way: According to her terminology risk is the "unpredictable variation in some ecological or economic variable (for example, variation in rainfall, hunting returns, prices etc.) and an outcome is viewed as riskier if it has a greater variance".

A further, but strongly related problem arises with the terms *unpredictability* and *uncertainty*. Generally *uncertainty* is defined as a lack of information about the world whereas *unpredictability* is a feature of a hazard itself. Only if future damage is unpredictable, i.e. actors are uncertain about their occurrence, do we speak of risks. The lack of information may relate to the temporal framework (we do not know when we have to cope with specific damage), the spatial framework (we do not know what area will be affected) and the extent of damage (we do not know the effect of damage). Hence, *unpredictability* is a salient feature of hazards and *uncertainty* is a defining criterion for risk, something with

---

1 They list the following definitions: (a) risk as the probability of damage, (b) risk as the extent of damage, (c) risk as a function (usually the product) of probability and extent of damage, (d) risk as the variance of probability distribution of all possible outcomes of a decision, (e) risk as the semi-variance of the distribution of all negative outcomes with a definite point of reference, (f) risk as a weighed linear combination of the variance and the expected value of a distribution of all possible consequences (Jungermann and Slovic 1993). (translation by the author).

which risk is inextricably linked. It is a matter of perception and not of objective, quantitative measurements. If risks are socially and culturally embedded perceptions of future damage, uncertainty is the perception of unpredictability. Below I list the working definitions for these key concepts as used in this text.

(1) Hazards are defined as "naturally occurring or human-induced process(es) or event(s) with the potential to create loss, i.e. a general resource of danger". (Smith 1996: 5)

(2) Risk relates to an unpredictable or hardly predictable event impacting a temporally, spatially and socio-culturally specific social-ecological system. The consequences of this event for the social-ecological system (or some part of it) are perceived negatively. Risks are the culturally and socially embedded perceptions of future possible damage resulting from a variety of hazards. Risks are neither directly observable nor are they directly measurable.

(3) Risk management is based on the culturally and socially embedded assessments and perceptions of past and future damage. The analysis of prior personal experiences or consensus based models is a necessary first step for developing risk management strategies. Risk management may be based on conscious decisions or may be embedded in custom and refers to (a) attempts at eliminating the occurrence of negatively evaluated events and (b) to strategies to decrease vulnerability and (c) to limiting the impact of damage once it has occurred.

(4) I will start of to work here with the definition of resilience given by the resilience alliance quoted above. However, I will extend the concept and understand it less as a property of a social-ecological system but more as common good produced by a group of actors within a social-ecological system and under specific political conditions.

4. Changing land tenure in northwestern Namibia

Human-environment relations in north-western Namibia have been depicted as static in popular media and occasionally in academic accounts as well. Pastoral nomadism has been depicted as well adapted to the hazards of an arid environment and socio-cultural features were often interpreted as being determined by the vicissitudes of resource scarcity (for a critical account of this rather static discourse on human-environment relations in north-western Namibia see Bollig and Heinemann 2002). Drought is conceptualized as the major risk within this semi-arid environment. These accounts glance over the fact that human-environment relations in north-west Namibia changed rapidly throughout the past one hundred years. While climatic perturbations are clearly a key driving force resulting in profound damages in herds and the environment, a number of other challenges has arisen over the past 100 years and has necessitated a continuous readjustment of risk management strategies.

A short note on fieldwork: Between 1994 and 1996 I gathered information on pasture management among the Himba of north-western Namibia. The management of common pool resources and risks linked to them seemed to be fairly settled and successful in this time period. This was not only my impression but also the impression of other specialists working in the wider area (Behnke 1998a, 1998b: 23; Schulte 2002: 101). I described the coping mechanisms as traditional and long lasting (Bollig 1997). In 2004/2005 I restudied the communal management of natural resources and to my astonishment not much of the stability I had observed a decade earlier remained. Uncontrolled access to pasture, rapid environmental decline of pastures and the assumed propensity of the government to intervene into grazing matters were described as major risks. In-depth research on the historicity of resource management showed that the "resilient" system of resource tenure and risk management I had observed in the mid 1990s and which informants had dubbed as "traditional" had only existed for about 30 years. What I had perceived as a traditional and successful state of CPR management, now seemed to be a transient phase of stability and resilience within a larger sequence of social-ecological change characterized rather by ruptures and recurring phases of reorganization than by continuity (for a general introduction to the ecology, history and contemporary economy of the region see Bollig, Brunotte and Becker 2002).

4.1 Tenure and mobility before the 1950s

The pre-1960 system of grazing rotation and land tenure was strikingly dissimilar to the system observed in the 1990s, as interviews conducted on the change of settlement patterns all over the northern Kunene Region showed: Livestock camps moved out to graze distant pastures during the rainy season (and not during the dry season as today) when pans and other seasonal water sources offered enough water. Cattle grazed in orbits around these seasonal pools and once these rain-dependent water-resources fell dry, they had to retreat to settlements near permanent water reservoirs along rivers and permanent wells. In the Omuramba area households from Ombuku river migrated to Omuramba with their large cattle herds during the rainy season and on to Ondova with their oxen-herds (see Map 1).

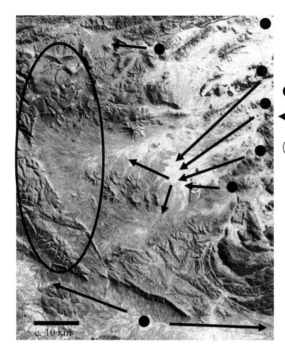

Map: 1 Ressource Use
and Mobility before
1950.

● Dry season settlement

← Migration to rainy
season pasture

○ Rarely used pasture

Once the water-sources in these areas had been used (or fell dry due to the on-set of the dry season) they returned to the Ombuku river, where water could easily be found in the river's sandy deposits, to stay for the remainder of the dry season. Roughly a quarter of the year was spent on outlying rainy season pastures (February to May), whereas the major part of the year people and herds congregated around the few permanent water sources. The group of users of specific resources was clearly defined. Households held tenure rights in specific places that had permanent water.

Places with affluent permanent water sources were often dominated by powerful leaders. Vedder (1914) meets the local leader Tjongoha at Kaoko Otavi's permanent well and reports that Tjongoha was an acknowledged leader in the environs of the place. Remarks like "At Khairos waterhole live two Herero chiefs, Langman and Herman, both wealthy ….for they possess hundreds of cattle and large flocks of sheep" made in Deneys Reitz' travel account (Reitz 1943: 103, travelling in the mid 1920s) are revealing in that they link water holes with political leaders. Leaders migrating into the Kaokoveld either from Angola or from the settler frontier in the Kamanjab/Outjo region in the early 20th century, targeted permanent water-points in the Kaokoveld to settle there. In many cases an acknowledged leader was surrounded by relatives and clients.

Typically these men were wealthy in cattle, owned horses and guns. Leaders competed for the more affluent wells in the area (on the political economy and organization of these returnees see Bollig 1998a, b, 2006), sometimes violently and often in battles of words and through intrigues in which they tried to involve the incipient colonial administration (Rizzo 2009).

I will shortly address salient structural features of the pre 1950s common pool resource (CPR) management: Households "owned" specific places and a clearly defined and numerically small number of households managed pastures together. The heads of these place-owning households are addressed as *oveni vehi*, owners of the earth/land. The absence of water in large tracts of land narrowed down choices for mobility during the major part of the year significantly. Water points were usually associated with a big man (*omuhona*). Patrilineal relatives were often attached to his homestead and lived in the close vicinity of the acknowledged leader. Additionally clients joined the homesteads of big men and squatted in the environs of the waterhole. Typically clients were poorer herders and/or foragers who often acted as shepherds to more endowed households. It is especially the larger water-points which allowed for some agriculture and a sedentary lifestyle. For the larger wells political domination through big-men (e.g. "many guns", "good relations to the colonial administration") is emphasized in oral accounts, for other water sources (especially in the Himba context) ritual domination (e.g. "presence of ancestral graves at which rituals addressing paternal or maternal ancestors are conducted") is privileged in explanations why some people have access rights to a well and others not.

A senior male member of one of the families with a long settlement history in the area was regarded as the *omuni wehi*, the owner (or guardian) of the land. This relation between a household and a place was given a historical dimension of longevity by numerous graves at which ancestors were worshipped. Access to water and pastures was inherited through different channels: the typical pattern was that a son inherited rights in a place from his father. Matrilineal channels were used to obtain access to grazing too: young men frequently herded their animals in the realm of their mother's brother, the person they would possibly inherit their cattle from later on in their lives. To accept client status with a wealthier patron was another way to attain access to a place. Until the 1930s the number of stockless or near stockless pastoralist foragers was still quite high and it was especially individuals and households with such a background who sought access to wealthier households.

We know little about rules of resource protection for the pre 1950 period. Several informants claimed that such rules were not necessary at all as there were very few people and much less livestock than today. Additionally we may add that the water sources in the rainy season grazing areas were usually finished some few months after the rains. This brought a rapid and non-negotiable end to the exploitation of pastures. Without water these pastures could not be used any longer and, of course, there was no need to develop rules of protection or equality of access for such pastures. Informants also stressed that there were few conflicts over grazing, while conflicts over places between leaders are prominent in oral lore. Informants did not report any specific rules governing the mode of use of water-points. Access to water-points was channelled through access to the community and possibly the affiliation to a local big man.

What role did the colonial government play in this context? Botha reports that before 1950 there was hardly any effort to intervene into the farming of reserve dwellers in Namibia/South-West Africa (Botha 2005: 183). During the first thirty years of South African rule the impact on resource management in north-western Namibia indeed was moderate or to be more precise solely targeted pastoral mobility. Three chiefs were given reserves with more or less fixed boundaries in the northern parts of former Kaokoveld. Migrations across chiefdom boundaries were prohibited and controlled for some time. For about 20 years (1925-1945) settling at the Kunene river was prohibited as well. While the delineation of chiefdom boundaries did not have a great impact on mobility and was circumvented frequently, the prohibition of settlement along the Kunene severely upset transhumant cycles. Generally the incipient South African administration acknowledged and sustained local authority structures. Where no such structures existed they were ambitious to push for them. However, the administration did not impact greatly on resource management as such. Such attempts at modernization would have contradicted the ideology of the administration. The long-serving Native Commissioner for the region, Cocky Hahn, was adamant that what he thought to be the traditional political-economic system should be retained to maintain 'a healthy tribal situation' (Hayes 2000). Such a situation was thought to be the basis for a disciplined African labour force which then could be used in the mines and the commercial ranching area to the south.

What were the risks in this system and how did risk management answer to such risks? A major risk which is of crucial importance today – the non-availability of fodder on dry-season pastures due to prior grazing by competitors – was simply not there. Homesteads returned to permanent wells during the dry season and left outlying pastures once the last pools there had dried up. They did not have to care whether these pastures were still productive and offered enough grazing for the dry season: the system was grafted upon the premise that such pastures were only used during the rainy season. Of course, also during the rainy season, they could only be, if there had been sufficient rains sparking off grass growth. Risks were more specifically connected with the dry season grazing around the few permanent wells. There was a number of risks directly linked to them: (1) insufficient grazing; (2) the denial of access to key resources due to a lack of stable links to people "owning" the resource, (3) the eviction of one's own group by another, militarily stronger and/or politically better connected group of resource users. Another major risk was that the colonial administration inhibited the exchange of livestock for food and other commodities with producers and traders living across the river in southern Angola or in neighbouring Ovamboland. Bollig (2006) documents a number of cases in which the governmental inhibition of trade across boundaries contributed significantly to turn a drought into a famine.

There seems to be a number of risk management strategies directly answering to these challenges. Generally the number of livestock has been restricted. A major percentage of the Kaokoveld's population continued to exist well into the 20th century as foragers or as pastoralists with few livestock. It is unlikely that such restrictions have been intentional but the fact that only a small number of homesteads in each community owned substantial numbers of cattle made control easier. Access to a permanent waterhole and its surrounding resources were guaranteed by social strategies effectively tying individuals into patron-client networks. Oral traditions speak a lot about such strategies. Kinship ties to patrons who after all hailed often as mercenary leaders from southern Angola or from Herero communities in central Namibia were eagerly forged. Patri-lineal and especially matrilineal links to such patrons were emphasized and obligations of a moral economy based on kinship evocated. While since the 1950s local authority structures were based on administrative chiefs who controlled a specific territory this political system was grafted upon patrons who controlled numbers of clients. The thread to be out-competed through other groups could be countered by a number of strategies: (1) enlarging the number of clients: the larger the number of clients of one big man the lesser the chance that they were driven off the land, (2) improving the military strength of the community: cattle and especially ivory were invested in guns and horses, (3) creating reliable linkages to the administration, or "capturing" the resident native commissioner in a patron-client network by making him the super-patron, (4) creating symbolic ties to the land through ancestral graves which in conflicts could serve as reference points for claims of legitimacy. The governmental inhibition to trade livestock across boundaries was circumvented by the establishment by smuggling routes across the Kunene and into Ovamboland. However, as documented in other publications (Bollig 1998a, 1998b, 2006) for about three decades (the 1920s to the 1940s) governmentally dictated isolation of the region was dominant as the administration successfully cut down trade through highly repressive measures.

Was the social-ecological system of the 1920s to 1950s resilient? In many ways yes: its basic structure could be maintained despite massive droughts in 1915, 1929-32 and 1941. The system also withstood the imposition of colonial rule and the forced removal of about 1,000 herders from the southern Kaokoveld to the central areas of the region. A researcher working with the resilience concept in the 1930s would have regarded local pastoralism as highly resilient. Major subsistence crises in the wake of droughts were countered by a number of buffering mechanisms: e.g. drought food (roots, tubers, berries), sharing of food and far ranging migrations. Stock losses were buffered within extensive networks of livestock donations of wealthy patrons to clients. Resource tenure was also embedded within patron-client relations. Resilience was apparently produced within such hierarchical structures, which guaranteed some (if unequal) supply of livestock and prevented the emergence of open access situations through a clear definition of access rights. The major challenges for resilience were not climatic perturbations but rather governmental measures. There are a number of complaints of local actors documented in the archives which claim that the prohibition of trade was damaging and during droughts contributed significantly to famine.

The structure of risk and consequently risk management changed massively when the availability of grazing resources was enlarged in a comprehensive borehole drilling programme since the late 1950s.

4.2 Post 1950 attempts at modernization: the borehole-drilling program

After the official introduction of Apartheid in 1948 the policy towards the African reserves changed profoundly. The Long-Term Agricultural Commission (LAPC) in 1948 found that "the limit of carrying capacity ... has been reached" (Botha 2005: 177) and recommended soil conservation, appropriate stocking rates and improved livestock breeds. Reports of catastrophic degradation of southern African savannahs due to overexploitation haunted the public media (Botha 2005: 176). Reserve economies were to be modernized to produce enough food for a growing population. Traditional authorities were regarded as important partners in the policy of modernization (see also Alexander 2001: 216 for similar developments in Zimbabwe). However, the decision on how and where to start modernization programs was given to experts, which emerged as a new group of administrative actors: where a native commissioner had ruled until then, he had now to share his powers with experts send from the centre. Whereas the area had been cordoned off beforehand, after 1950 a number of scientific studies were undertaken in the area: geological and hydrological (e.g. Abel 1954), anthropological (van Warmelo 1951, Malan 1974) zoological (Viljoen 1988) to name just a few. In 1957 the Kaokoveld became the "independent" reserve *Kaokoland* under the direct authority of the Chief Native Affairs Commissioner in Windhoek (van Wolputte 2004: 163).

Scientists, especially geologists, hydrologists and agriculturalists became influential and grounded much of the political decision making process with data. While in the 1930s some few waterholes had been blasted, a regular borehole program was launched in the mid 1950s. In a letter to the Native Commissioners the Chief Bantu Commissioner in Windhoek argued in 1955 that "*ontwikkeling*" (development) in reserve areas was necessary and that borehole drilling was an important step into that direction. He asked Commissioners to identify areas which could be "opened up" for human exploitation by boreholes. Water was regarded as basic resource which just had to be exploited with new technological means (Odendaal 1963/64: 411). Throughout the 1960s, the 1970s and 1980s boreholes were drilled in large numbers in the Kaokoveld. In the 1950s some 43 boreholes were drilled in total, these were all around Opuwo or to the southeast of Opuwo; during the 1960s another 136 boreholes, and in the 1970s some 128 boreholes were drilled. In the 1980s another 57 boreholes were added and in the 1990s some 40 (see Map 2a and Map 2b). Boreholes were carefully registered and fully maintained by the Department of Water Affairs. This rhizomatic extension of the state penetrated the local economy and the pastoral landscape effectively and irrevocably.

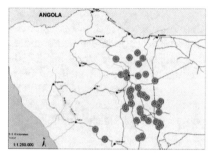 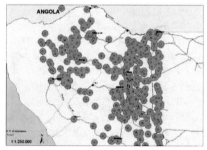

Map 2: Borehole drilling in Kaokoland, 2a, boreholes drilled in the 1950s, 2b boreholes drilled until 1990, radius around point indicates a 6 km grazing radius.

The development of a network of boreholes in former dry season grazing areas between the 1960s and the 1970s led to a reversal of the mobility pattern. Herds could now stay in the former rainy season areas during the dry season as they were no longer dependent on rain-water. Much wider areas than before were now accessible for permanent grazing. This resulted in a profound increase of cattle herds (Figure 1).

A new set of rules had to be developed to address questions pertaining to the regulation of access and to sustainability. The net effect was that herds stayed at the permanent water-points along the rivers during the rainy season and then moved out to outlying pastures during the dry season where water was now provided by a borehole – a complete change of the mobility pattern. This change also made more labour investment into agriculture possible. Whereas before, the major number of people had been far away from arable lands along rivers in the rainy season, now they converged on these lands during the rainy season.

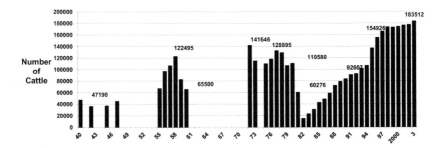

Figure 1: The development of the regional herd between 1940 and 2005.
Source: Directorate of Veterinary Services; Page, 1976; van Warmelo, 1951.

Also in the wider Epupa area, in the northernmost area of the Kaokoveld, the situation changed with the development of water resources in the area: boreholes were drilled at Omuramba, Otjikango and at Omuhandja in the 1960s and 1970s and a dam was built at Okombanga near Omuramba. Pastures which had been rainy season pastures in the past now became dry season pastures (Map 3). The major focus of settlement became Omuramba which replaced Ombuku. Only during the height of the dry season would people fall back on the Ombuku valley where they had dwelt before the entire dry season.

Rapidly increasing cattle herds and the reliance on extensive dry-season pasturing brought about the problem that even mild droughts diminished the availability of grazing profoundly. Whereas before, with cattle numbers not exceeding 30,000 to 50,000 animals, there had been vast areas of unused pasture every year, now a regional cattle herd exceeding 150,000 needed all available grazing. The droughts of the mid 1970s and especially the centennial drought of 1980-82 showed how vulnerable the system had become through the intense increase in herds made possible by the governmentally endorsed borehole programme. We may conclude that in the long-run the hydrological revolution of the 1960s and 1970s on the hand led to a massive increase in livestock wealth but on the other hand it also made local pastoralists more vulnerable to drought. Other risks were brought about by the borehole programme and the opening up of vast stretches of land for dry season grazing. Competition for access to dry season pastures became a potential source of conflict. While in the past, when these outlying pastures had only been grazed when water was available there, now everybody who could make primary use of such dry season pastures had distinct advantages. Whereas grazing such stretches during the rainy season did not bring about the need to limit access as stretches grazed had still considerable re-growth, now stretches once grazed could only be reused in the next year. The coordination of grazing efforts and a detailed set of rules was the only way to circumvent such conflicts. A third risk widely commented upon was the failure of boreholes. If a distant borehole had technical problems this could potentially result in the loss of herds. At least such borehole failures necessitated sudden and uneasy moves with large herds of animals. Only if there were close and binding ties to the administration such problems could be ameliorated quickly. How did herders and their communities react to this new structure of risk?

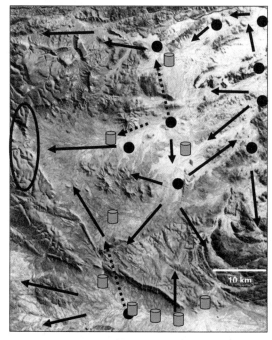

Map 3: Resources and mobility in 1995.

● Rainy season settlement

▢ Boreholes and Wells

← Migration to dry season pasture

◀·· Early dry season migration

4.3 Changing patterns of mobility and tenure as a way to lower new risks
The community managing pastures had been fairly clear cut before the 1950s. A community around a permanent water source controlled the key dry season pastures whereas rainy season pastures needed less control – at least for the major part of the year. Within the new set-up the borehole-fitted dry season pastures needed control of access. The South African administration had continuously developed chieftaincy structures. The practice that chiefs were to apply for boreholes in the name of their communities gave them even more power. The emergent resource tenure system became closely linked to the institution of chieftaincy. All households that were under one chief were allowed to use the

pastures that fell within this chiefdom. Behnke (1998b: 4) observed that most Himba chieftaincies were organized along drainage systems and combined river frontage and upland grazing areas. A chieftaincy in this sense formed a large and loosely integrated herding and resource management unit (ibid.: 12). In contrast to the prior state, a clear link between individual households and specific grazing areas was replaced by a clearer specification of the political constituency dealing with a resource – a water-point was not longer under the control of a local big-man but rather part and parcel of a chiefdom with rather clear-cut administrative boundaries. Increasingly chiefs and their councillors as paid semi-officials and intermediaries between local community and the state became prominent. Membership in a resource managing community was still established via kinship links. Graves and rituals conducted at these graves gave proof of the legitimacy of tenure rights in a specific place.

The rules of good grazing (*ozondunino yomaryo*) – nowadays listed by many seniors as evidence of sustainable indigenous pastoralism – apparently evolved as a reaction to the implementation of the borehole programme and the altered structure of subsistence risks. The change which was brought about by this hydrological revolution necessitated the development of a set of rules to guarantee equal access and some degree of sustainability:

- grazing in the dry season grazing areas is prohibited during the rainy season,
- cattle camps must move a considerable distance away from the main settlement areas,
- livestock camps must move together in a group,
- herders should look for dry season grazing near to their major settlement,
- once settling in a specific place cattle camp owners decide in which "direction" each herd is herded; i.e. in the morning each herder takes out his herd from the camp in a specific direction;
- too much movement to and from (*okukandakanda*) and frequent changes between camp associations are not appreciated and
- special areas should only be used during droughts and not during normal dry seasons.

These rules first of all tried to keep conflicts over grazing at bay: The prohibition to settle ahead of others for example ensured that all herders had the same chance to exploit a pasture area (see Behnke 1998a: 15). Rules narrowed down the choices for cattle camps and made moves more predictable. A second goal was to maintain grazing near the main settlements to ensure the supply of milk to elderly folks and children during the dry season when the majority of cattle was taken to outlying pastures; a third intention may have been to ensure a sustainable management of pastures. Grazing guards (*ovatjevere vomaryo*) were named. These were instituted into their office by a meeting of the community.

The fact that these men were also addressed as *ovapolise vomaryo* (grazing police) shows that they were regarded as an extension of the homeland bureaucracy. Formally these men were entitled to screen the area for homesteads and camps which did not adhere to the rules. However, everybody could bring up complaints against a neighbour or herder he thought was behaving contrary to the rules. At first a meeting took place at the neighbourhood level. The accused was summoned and he was given time to explain his case in an ad hoc meeting. Usually there was no punishment at these meetings and, if possible, men tried to reach a consensus on how to change the situation for the better. If these minor meetings at a neighbourhood or cattle-camp level did not lead to a decision the case was then referred to a meeting at the chief's home.

The South African government encouraged chiefs to punish wrongdoers 'according to tradition'. Chiefs occasionally resorted to flogging which was rather in the tradition of the colonial administration than legitimized within the local authority system. Furthermore trespassers had to pay a fine, usually an ox. When several oxen had come together these were driven to the Commissioner in Opuwo. In the 1990s physical punishment was no longer practiced and fines were invariably paid in livestock. These were either paid to the neighbourhood council, or to the meeting at the chief's home and not any longer transferred to the administration although such fines were still often addressed as 'government animals'. Usually animals paid as fines to a meeting were immediately slaughtered to provide those attending the meeting with meat.

Again I would like to put forward the question of resilience? In 1995 the social-ecological system made the impression of being fairly resilient! The social-ecological system had been confronted with some major perturbations in the 1980s: civil war and army rule had negatively affected pastoral mobility, a major drought in 1980-82 had killed 90 per cent of the region's cattle – and still the system preserved without collapsing into a qualitatively different state controlled by a different set of processes. It was mainly a set of institutions developed in the 1950s and 1960s, grafted upon earlier institutional solutions to resource access, which ensured equitable and to a certain extent also sustainable access to grazing areas. Extensive networks of livestock loans provided security. These networks still hinged around patron-client relations but due to the overall increase in livestock wealth many more exchange relations were engaged in between equals. A densely knit network of exchanges with some centrality of key actors superseded a patron-client structure.

However, the change of the original mobility pattern brought about a shift of the vegetation structure. The pre-1950s system with intense grazing on outlying pastures only during the rainy season (i.e. for only three to four months during the growth period) favoured perennial grasses. In contrast, the new mobility pattern which implied intense stress through grazing during the dry season disadvantaged perennial grasses and advantaged annual grasses. The new rules stipulated that outlying pastures should not be grazed during the rainy season, thereby ensuring that annual grasses were not disturbed during their main vegetation period. Annual grasses were factually protected until they had developed seeds and their reproductive success was granted. In contrast perennial grasses were massively disturbed during a period of time when due to the lack of soil moisture they could not sufficiently recuperate. This led to a rather rapid change from a pasture dominated by perennials to a pasture dominated by annuals. The major drought induced livestock mortalities of the early 1980s – the number of cattle dropped from about 120,000 to c. 15,000 animals in a period of only two years – may have brought about some reversals of this trend, but at least since the early 1990s grazing pressure increased and by the end of the 1990s the cattle number had reached c. 180,000 animals. A loss in biodiversity in the grass and herb cover as well as in the tree and bush cover (Bollig and Schulte 1999) was observable in many areas. In the late 1990s remaining perennials such as Stipagrostis uniplumis and highly valued annuals such as Willkommia newtonii, Entoplocamia aristulata, Anthephora schinzii were decreasing, whereas lower-valued annuals like Enneapogon desvauxii and Aristida adscenionis were increasing. The almost complete change from perennial grasses to annual grasses made the local grazing system more vulnerable to drought: whereas perennial pastures may still yield some fodder in bad rain years, annual pastures rarely produce any fodder during drought. Very high livestock numbers altered the state of the vegetation further. Since the late 1990s a distinct trend towards the replacement of high yielding annuals towards more robust but low yielding annuals was observable.

4.4 Deinstitutionalisation and the search for the reorganisation of common pool resource management

In 2004/2005 I focused research on grazing management and much to my surprise a lot had changed within a decade. During interviews in 2004 respondents emphasized the high degree of detrimental vegetation change they had experienced during the past decade. It is interesting to note that ten years earlier in 1994/95 herders had related degradation to the imponderability of rainfall and had rejected the idea that the phenomenon was related to high livestock numbers. In 2004 many respondents were sure that high grazing pressure caused severe forms of degradation in the grass/herb layer.

In several interviews informants claimed that the local system of pasture management had virtually crumbled. Several indicators for the demise of grazing control were given:

- In several grazing areas disputes were pertinent as to who was allowed to make use of specific pastures. While on the one hand politically dominant figures succeeded in reserving pastures for their herds (without fencing them), in other areas young herders moved into pastures which were far away from their regular settlement areas.

• All informants stated that the sanctioning of misbehaviour had virtually been abandoned. Several reasons were given: one informant said that it was simply hard to put enough pressure on a culprit to pay his fine; others claimed that the death of some elders who had dominated the local community as wealthy cattle patrons and who had been central for the handling of cases had weakened grazing control; yet another informant stated that many people reasoned that it was not the cattle camp herders who were to be fined but the owners of these herds who profited from such moves.

• Some significant shifts of pasture-use had taken place. Some previous dry-season pastures were now permanently settled by households.

Several factors were apparently contributing to these institutional dilemmas. The number of cattle had nearly reached the 200,000 level, an all time high, and to manage such large numbers of livestock within a restricted area was simply much more difficult (in the sense of increased transaction costs). Large livestock numbers and the demise of grazing control had apparently contributed to a demise of pasture quality. Informants claimed that nomadic livestock husbandry in the region was in disarray. The demise of a grazing system which was perceived as successful in guaranteeing equal access to resources (and diminishing competition) and in safe-guarding the quality of pastures was deemed to be the major risk of communal grazing. The issue was much debated locally and many meetings were held how to curb the situation best.

How did local herders react to this crisis? It is noteworthy that interviewed herders were unanimous in their judgement that the system was actually going through a crisis. The degradation of pastures they framed as "the country has become weak" and then listed more detailed observations. Despite the fact that the decade between 2000 and 2010 had generally seen good rain years, they judged the risk of livestock losses as higher than before. They also reported on a great number of meetings which had taken place to find new institutional solutions to the grazing problem, i.e. to provide more resilience for the social-ecological system. Some, however, did not want to wait for new institutions to emerge. During the past years a remarkable out-migration of households from the research area had taken place: some households had migrated south of Ouzo (some 200km away), one reaching the Sesfontein area (some 350km away), three other households went to the wider Ekoto area (c. 300km), some six households had left for Angola and three had more or less permanently settled in a neighbouring mountainous area. Some of the household heads of out-migrating households could be interviewed on the reasons for their decision. The answers were unanimous: the pastures had been bad in the research area, losses of livestock due to emaciation had been frequent and other areas seemed to offer better grazing and they had seen no way how to improve the situation.

The co-management of resources had become another answer to the crisis; NGOs and governmental bodies were seen as potential allies for the solution of problems which seemed unsolvable at the local level. It was no longer the relation between state and local community but the triangle of state institutions, non-governmental organisations and local communities, which mattered. The region's herders resorted to the conservancy movement in astonishingly high numbers and in an amazingly rapid manner: the first conservancy in the wider region was inaugurated in 1998, only 12 years later almost the entire region is parcelled in already endorsed or in emergent conservancies (see also Bollig 2004 for this process until 2003).

Many conservancies in the Kunene region are (more or less) cut along the boundaries of chieftaincies: On the one hand it was much easier to resort to chieftaincy boundaries than to define completely new ones when negotiating boundaries with neighbouring communities. On the other hand a number of traditional elites were keen to see the territorial structures they were relating to being transformed into entities of natural resource management. However, while chieftaincy boundaries are often ill defined and leave space for negotiation and overlap, conservancy boundaries need to be well defined. In order to prepare the documents for registration the boundaries of conservancies were to be clearly marked making use of modern mapping equipment. While traditional authorities had comparatively little say in a number of newly instituted governmental administrative procedures, they now became partners of non-governmental organizations. While the participation of modern elites – traders, teachers and other salaried workers – was taken for granted, traditional authorities were actively wooed. Nevertheless, the boundary demarcation process was often also contested and drawn out. It often took years to find consensus on boundaries as each community got the idea that it would cede access rights to natural resources once a specific area was going to be gazetted under a neighbouring conservancy.

Another crucial issue a conservancy had to solve was to define the relation of its elected committee to the traditional authorities in the area. *Conservancies* in north-west Namibia have solved this issue with great enthusiasm for constitutional experiments. *Conservancy* constitutions have found very different ways to define the linkage to traditional authorities: in some committees the traditional authorities are ex-officio members, in some they are only allowed to send representatives and in yet others they establish a sort of second chamber together with their councilors. A land management plan drawn up by elected committee members who specified pastures for seasonal grazing, settlement areas and core conservation areas was part and parcel of the application for conservancy status with the Ministry of Environment and Tourism. The land management plan then becomes an officially endorsed document which is deemed to be the blueprint for future land use. Many people interviewed voiced their hope that the establishment of conservancies, the combination of traditional and modern elites in committees and a governmentally endorsed land use plan would make common pool resource management again effective and provide more resilience in the face of droughts and unwanted immigration.

Water-management became a second instance of co-management of natural resources. In contrast to conservancies, non-governmental organisations were of little importance in this field. A new legislation on water (Water and Sanitation Policy [WASP] of 1997, Water Resources Management Act of 2004) stipulated that the control over boreholes in rural areas was to be handed over to communities. At the local level the Act "encourages and provides guidelines for local water point users to establish water point user associations and local water user associations. Members of the water point user associations will elect water point committees to oversee the daily management of water resources and financial issues." (Government of the Republic of Namibia, 2005: 16, see especially Part 5 Sect. 16). Of crucial importance for the overall application of the Act, was that water-point associations were meant to collect user-fees and use them to manage the water-point.

In many discussions I learnt that local communities were searching for new ways of social-ecological regulation incorporating new governmental rules and the approaches of NGOs in a meaningful way.[2] Both conservancies and water point committees are coveted because they seemingly provide a way out of institutional dilemmas: (1) user groups are redefined and membership in them is determined in a clearer way; (2) boundaries of user-groups and resources are redefined and (3) organisational structures are – at least in theory – instituted that monitor the adherence to certain rules.

5. Instead of a summary

The preceding passages have shown that land use in north-western Namibia developed rapidly over the past century and with it the structure of risk and risk management changed. While the pre 1950 set up relied on home ranges and regulation within neighbourhoods, the second half of the 20th century was characterized by the regulation of pasture-access within the framework of chieftaincies. During the last two decades this mode of regulation was replaced by a more fragmented pattern of access and land use, in which different stakeholder groups apply various forms of management which in turn relate to different government policies. Each period lasted about 30 to 40 years. Stability dominated these phases but transitions between phases have been swift. During stable peri-

---

2   In fact since 2010 the anthropology departments of the Universities of Hamburg and Cologne are conducting a joined research project on the emergence of institutions related to water management.

ods of each mode of regulation a certain set of institutions was developed and supported by a specific configuration of social networks and power structures. Each mode of regulation can also be linked to a specific vegetation structure of pastures. In other words: communal investments into collectively held social capital – institutions controlling access to resources – provided resilience. Indeed, each phase of stability also saw major crises; but due to these very institutions structures of land use were maintained.

Currently herders are experiencing a transitional phase. A number of alternative regulative approaches compete for influence and the local community is searching for a predictable and fair approach to the management of a set of common-pool resources, which they regard as rapidly degrading due to uncoordinated use. Whether local communities will again succeed in adopting an efficient and internally undisputed system remains to be seen. For anthropological theory building it is highly interesting to consider that order is created rapidly and not in an incremental process. Periods of rapid transformation are followed by periods of stability and cultural elaboration. It is these periods of stability which are characterized by specific modes of risk management. Due to the fact that most anthropological studies concentrate on a rather thin time-slice of one to two years or – in the better case – rely on two such time slices some twenty, thirty or more years apart, anthropologists usually have little to say about the speed of changes and about the relation between phases of stability and those of transformation. Anthropology will need some kind of laboratories where specific cases are followed up through longitudinal studies and where data on e.g. institutional change is gathered systematically and consistently over long periods of time. Only such long term studies will help us to develop theories of change within complex social-ecological systems.

What do we learn about risk, risk management and resilience from this case study? I will summarize the gist of my argument in some few bullet points conceding that much more could be said.

- The structure of risk and risk management is very much tied to specific historical circumstances. In the case study the temporal and spatial availability of resources mattered, the social system as much as historical experiences of actors acting within this system and the wider governance structures mattered as well; this implicates that when talking about risk and risk management we need to be very specific about the concrete historical context; risk and risk management always relate to a time and space specific structure of resources. This may seem trivial; the assumption however helps to conceptualize risk and risk management as dynamic concepts.
- What is risk management? Is risk management the amalgamation of individual risk management strategies? Or something like the average risk management strategy of a community? Do we really understand how individual risk management strategies become the consensual approach of a

community towards risks? It is clear from the cases that it is individuals who adapt to risk, but they usually do so in a coordinated manner. The case study shows that the emergence of risk management is not necessarily a slow and incremental process but a rather rapid move. We have seen that risk management strategies like the rules of good grazing or the rapid engagement with conservancy structures were adopted within a period of ten years – what happened in this time, and why are communities able to reach consensus so quickly.

* How does risk management relate to resilience? I start from the assumption that risk management is linked to sets of actors whereas resilience describes a structural property within a social-ecological system. We learned that pastoralists reacted swiftly to new challenges and that they actively created social-ecological systems that were resilient. Resilience is a common good which people actively have to invest social and economic capital in. However, the social-ecological system described here was only resilient within certain threshold values. Apparently the risk management which came about as a consequence of borehole drilling depended on a limitation of livestock numbers, a functioning common pool resource management and a political set up sustaining local traditional elites. Once one of these framing threshold values was out-passed a new framework for creating resilience emerged. While we are well equipped with good descriptions of such strategies we still have little understanding of how resilience/robustness emerge or decline as structural properties of SES and how individual risk minimizing strategies translate (or do not translate) into resilience of the overall system.
* Moments of crises or the collapse of a social-ecological system or some of its parts are apparently the key to understand the emergence of risk and risk management. Power, land use but also epistemologies are being renegotiated. Under what circumstances such a process takes place (and when it is inhibited) is an open question.

6. Bibliography

Abel, Herbert (1954): Beiträge zur Landeskunde des Kaokoveldes (Südwestafrika). In: Deutsche Geographische Blätter 47, pp.7-123.

Acheson, James M. (1989): Management of Common-Property Resources. In: Plattner, Stuart (ed.): Economic Anthropology. Stanford: Stanford University Press, pp. 351-378.

Alexander, Jocelyn (2001): Technical Development and the Human Factor: Sciences of Development in Rhodesia's Native Affairs Department. In: Dubw, Saul (ed): Science and Society in Southern Africa. Manchester: Manchester University Press, pp. 212-237.

Anderies, John M., Janssen, Marco A. and Elinor Ostrom (2004): A framework to analyze the robustness of social-ecological systems from an institutional perspective.   In: Ecology   and   Society   9,   1. http://www.ecologyandsociety.org/vol9/iss1/art18/.

Arhem, Kaj (1984): Two Sides of Development: Maasai Pastoralism and Wildlife Conservation in Ngorongoro, Tanzania. In: Ethnos 49, iii-iv, pp.186-210.

Baksh, Michael and Allen Johnson (1990): Insurance Policies among the Machiguenga: An Ethnographic Analysis of Risk Management in a Non-Western Society. In: Cashdan, Elizabeth A. (ed.): Risk and Uncertainty in Tribal and Peasant Economies. Boulder: West View Press, pp.193-228.

Bechmann, Gotthard (1993): Risiko als Schlüsselkategorie in der Gesellschaftstheorie. In: Bechmann, Gotthard (ed.): Risiko und Gesellschaft. Grundlagen und Ergebnisse interdisziplinärer Risikoforschung. Opladen: Westdeutscher Verlag, pp. 237-276.

Behnke, R. (1998a): Grazing Systems in the Northern Communal Areas of Namibia. A summary of NOLIDEP Socio-economic Research on Range Management. Windhoek. Ms (unpublished).

————— (1998b): Range and Livestock Management in the Etanga Development Area, Kunene Region. Northern Regions Livestock Development Programme. Ministry of Agriculture, Water and Rural Development. Windhoek. Ms (unpublished).

Boholm, Åsa (1996): Risk Perception and Social Anthropology: Critique of Cultural Theory. In: Ethnos 61, 1-2, pp. 64-84.

Bollig, Michael (2006): Risk Management in a Hazardous Environment. A Comparative Study of Two Pastoral Societies. (Pokot NW Kenya and Himba NW Namibia). New York: Springer.

————— (2004) Die Landreform in Namibia. In: Förster, Larissa, Henrichsen, Dag and Michael Bollig (eds.): Namibia - Deutschland: eine geteilte Geschichte. München: Minerva, pp. 304-323.

————— (1998a): Power and Trade in Precolonial and Early Colonial Northern Kaokoland. In: Hayes, Patricia, Silvester, Jeremy, Wallace, Marion and Wolfram Hartmann (eds): Namibia under South African Rule. Mobility and Containment, 1915-1946. London: James Currey Ltd, pp. 175-193.

————— (1998b): The Colonial Encapsulation of the North-Western Namibian Pastoral Economy. In: Africa 68, 4, pp. 506-536.

————— (1997): 'When war came the cattle slept…'. Himba Oral Traditions. Köln: R. Köppe.

Bollig, Michael, Brunotte, Ernst and Thorsten Becker (eds.) (2002): Interdisziplinäre Perspektiven zu Kultur- und Landschaftswandel im ariden und semiariden Nordwest Namibia. Kölner Geographische Arbeiten 77. Köln: Geographisches Institut der Universität zu Köln.

Bollig, Michael and Heike Heinemann (2002): Nomadic Savages, Ochre People and Heroic Herders: Visual Presentations of the Himba of Namibia's Kaokoland. In: Visual Anthropology 15, pp. 267-312.

Bollig, Michael and Anja Schulte (1999): Environmental Change and Pastoral Perceptions: Degradation and Indigenous Knowledge in two African Pastoral Communities. In: Human Ecology 27, pp. 493-514.

Borgerhoff-Mulder, Monique (1991): Behavioral Ecology of Humans: Studies of Foraging and Reproduction. In: Krebs, J.R. and N.B. Davies (eds.): Behavioral Ecology. 3rd edition. Oxford: Blackwell, pp. 69-98.

Botha, Christo (2005): People and the Environemnt in Colonial Namibia. In: South African Historical Journal 52, 1, pp. 170-190.

Boyd, Robert and Peter J. Richerson (1985): Culture and the Evolutionary Process. Chicago: University of Chicago Press.

Broch-Due, Vigdis and David M. Anderson (2000): The Poor are not Us: Poverty and Pastoralism in Eastern Africa. Oxford: James Curry.

Cashdan, Elizabeth (ed.) (1990a): Risk and Uncertainty in Tribal and Peasant Economies. Boulder: West View Press.

——— (1990b): Introduction. In: Cashdan, Elizabeth (ed.): Risk and Uncertainty in Tribal and Peasant Economies. Boulder: West View Press, pp. 1-16.

——— (1985): Coping with Risk: Reciprocity among the Basarwa of Northern Botswana. In: Man 20, 3, pp. 454-476.

Colson, Elizabeth (1979): In Good Years and in Bad: Food Strategies of Self-Reliant Societies. In: Journal of Anthropological Research 35, 1, pp.18-29.

Douglas, Mary (1994): Risk and Blame. Essays in Cultural Theory. London: Routledge.

——— (1985): Risk Acceptability According to the Social Sciences. New York: Russell Sage Foundation.

Downs, R.E., Reyna, Stephen P. and Donna O. Kerner (eds.) (1991): The Political Economy of African Famine. Food and Nutrition in History and Anthropology, Volume 9. Philadelphia: Gordon and Breach.

Ensminger, Jean (1992): Making a Market: The Institutional Transformation of An African Society. New York: Cambridge University Press.

Folke, Carl (2006): Resilience: The emergence of a perspective for social–ecological systems analyses. In: Global Environmental Change 16, pp. 253-267.

Fratkin, Elliot (1991): Surviving Drought and Development: Arial Pastoralists of Northern Kenya. Boulder: Westview Press.

Galaty, John (1994): Rangeland Tenure and Pastoralism in Africa. In: Fratkin, Elliot, Galvin, Kathleen A. and Eric Abella Roth (eds.): African Pastoralist Systems. Boulder: Lynne Rienner Publishers, pp. 185-204.

Government of the Republic of Namibia (2005): Strategic Options and Action Plan for Land Reform in Namibia, Ministry of Lands and Resettlement, The Permanent Technical Team (PTT) on Land Reform. Windhoek.

Government of the Republik of Namibia, Ministry of Agriculture (2004): Water and Rural development. 2004. Technical Summary. Natural Resource Accounting Programme. Windhoek.

Hames, Raymond (1990): Sharing among the Yanomamö. Part I. The Effects of Risk. In: Cashdan, Elizabeth (ed.): Risk and Uncertainty in Tribal and Peasant Economies. Boulder: West View Press, pp. 89-106.

Hayes, Patricia (2000): Camera Africa: Indirect rule and landscape photographs of Kaoko, 1943. In: Miescher, Giorgio and Dag Henrichsen (eds): New Notes on Kaoko. Basel: Basler Afrika-Bibliographien, pp. 48-76.

Hitchcock, Robert K. (1990): Water, Land, and Livestock. The Evolution of Tenure and Administration in the Grazing Areas of Botswana. In: Galaty, John G. and Douglas L. Johnson (eds.): The World of Pastoralism: Herding Systems in Comparative Perspective. London: Belhaven, pp. 195-215.

Hjort, Anders and Mohamed Salih (eds.) (1989): Ecology and Politics: Environmental Stress and Security in Africa. Uppsala: Scandinavian Institute of African Studies.

Hogg, Richard (1989): Settlement, Pastoralism and the Commons: The Ideology and Practice of Irrigation Development in Northern Kenya. In: Anderson, David and Richard Grove (eds.): Conservation in Africa: People, Policies and Practice. London: Cambridge University Press, pp. 293-306.

——— (1986): The New Pastoralism: Poverty and Dependency in Northern Kenya. In: Africa 56, 3, pp. 519-555.

Janssen, Marco A. (2006): Historical institutional analysis of social-ecological systems. In: Journal of Institutional Economics 2, 2, pp. 127-131.

Johnson, Douglas and David Anderson (1988): The Ecology of Survival: Case Studies from Northeast African History. London: Westview Press.

Jungermann, Helmut and Paul Slovic (1993): Die Psychologie der Kognition und Evaluation von Risiko. In: Bechmann, Gotthard (ed.): Risiko und Gesellschaft. Grundlagen und Ergebnisse interdisziplinärer Risikoforschung. Opladen: Westdeutscher Verlag, pp. 177-208.

Kaplan, Hillard, Hill, Kim, and Ana Magdalena Hurtado (1990): Risk, Foraging and Food Sharing among the Aché. In: Cashdan, Elizabeth (ed.): Risk and Uncertainty in Tribal and Peasant Economies. Boulder: West View Press, pp. 107-143.

Little, Peter D. (1992): The Elusive Granary: Herder, Farmer, and State in Northern Kenya. Cambridge: Cambridge University Press.

———— (1987): Land Use Conflicts in the Agricultural/ Pastoral Boderlands. The Case of Kenya. In: Horowitz, Michael M., Little Peter D. and Endre Nyerges (eds.): Lands at Risk in the Third World. Local Level Perspectives. New York: Westview, pp. 195-211.

Malan, St. (1974): The Herero-Speaking Peoples of Kaokokland. Cimbebasia B 2, pp.113-129.

McCabe, Terrence B. (1997): Risk and Uncertainty among the Maasai of the Ngorongoro Conservation Area in Tanzania: A Case Study in Economic Change. In: Nomadic Peoples N.S. 1, 1, pp. 54 -66.

Moran, Emilio F. (1979): Human Adaptability: An Introduction to Ecological Anthropology. North Scituate, MA: Duxbury Press.

Rizzo, Lorena (2009): Gender and Colonialism. A History of Kaoko (north-western Namibia) between the 1870s and 1950s. PhD Dissertation. University of Basel.

Reitz, Deneys (1943): No Outspan. London: Faber and Faber.

Odendaal, Frans Hendrik, Commission of Enquiry into South West Africa Affairs (Republic of South Africa) (1963/1964): Report of the commission of enquiry into South West Africa affairs 1962-1963 (Odendaal Commission). Pretoria.

Schulte, Anja (2002): Stabilität oder Zerstörung? Veränderungen der Vegetation des Kaokolandes unter pastoralnomadischer Nutzung. In: Bollig, Michael, Brunotte Ernst and Thorsten Becker (eds.): Interdisziplinäre Perspektiven zu Kultur- und Landschaftswandel im ariden und semiariden Nordwest Namibia. Kölner Geographische Arbeiten 77. Köln: Geographisches Institut der Universität zu Köln, pp. 101-118.

Shipton, Parker (1990): African Famines and Food Security. In: Annual Reviews in Anthropology 19, pp. 353-394.

Smith, Keith (1996): Environmental Hazards. Assessing Risk and Reducing Disaster. London: Routledge.

Spittler, Gerd (1989a): Dürren, Kriege und Hungerkrisen bei den Kel Ewey (1900-1985). Stuttgart: Lang.

———— (1989b): Handeln in einer Hungerkrise. Tuaregnomaden und die große Dürre von 1984. Opladen: Westdeutscher Verlag.

Spooner, Brian (1989): Desertification: the historical significance. In: Huss-Ashmore, Rebecca and Solomon H. Katz (eds.): African Food Systems in Crisis. Part One: Microperspectives. New York: Gordon and Breach, pp. 111-162.

Stahl, Ute (2000), 'In the end of the Day we will Fight' - Communal Landrights and Illegal Fencing amongst Herero Pastoralist in Otjozondjupa Region. In: Bollig, Michael and Jan-Bart Gewald (eds.): People, Cattle and Land: Transformations of a pastoral society in Southwestern Africa. Köln: R. Köppe, pp.319-346.

van Warmelo, Nicolaas Jacobus (1951): Notes on the Kaokoland (South West Africa) and its People. Ethnological Publications 26. Government Printers, Pretoria.

van Wolputte, Steven (2004): Subject Disobedience: The Colonial Narrative and Native Counterworks in Northwestern Namibia, c. 1920-1975. In: History and Anthropology 15, 2, pp. 151-173.

Vedder, Heinrich (1914): Reisebericht des Missionars Vedder an den Bezirksamtmann von Zastrow. Namibia National Archives J XIIIb5. Geographische und Ethnographische Forschungen im Kaokoveld 1900 – 1914.

Viljoen, P.J. (1988): The Ecology of the Desert-dwelling Elephants Loxodonta africana (Blumenbach 1797) of the western Damaraland and Kaokoland. PhD Thesis, University of Pretoria, South Africa.

Watts, Michael (1991): Heart of Darkness: Reflections on Famine and Starvation in Africa. In: Downs, R.E., Reyna, Stephen P. and Donna O. Kerner (eds.): The Political Economy of African Famine. Food and Nutrition in History and Anthropology, Volume 9. Philadelphia: Gordon and Breach, pp. 23-68.

———— (1983): Silent Violence: Food, Famine and Peasantry in Northern Nigeria. Berkeley: University of California Press

White, Cynthia (1997): The Effect of Poverty on Risk Reduction Strategies of Fulani Nomads in Niger. In: Nomadic Peoples N.S. 1, 1, pp. 90-107.

———— (1984): Herd Reconstitution: the Role of Credit among the WoDaaBe herders in central Niger. Pastoral Development Network Paper 18d. London Overseas Development Institute.

Wiessner, Polly (1994): The pathways of the past: !Kung San Hxaro exchange and history. In: Bollig, M. & F. Klees (eds). Überlebensstrategien in Afrika. Köln. Heinrich-Barth.Institut. pp. 101-124.

———— (1982): Risk, Reciprocity, and Social Influences on !Kung San Economics. In: Leacock, Eleanor and Richard Lee (eds.): Politics and History in Band Societies. Cambridge: Cambridge University Press, pp. 61-84.

———— (1977): Hxaro. A Regional System of Reciprocity for Reducing Risk among the !Kung San. University Microfilms. UMI, Ann Arbor.

Winterhalder, Bruce (1990): Open Field, Common Pot: Harvest Variability and Risk Avoidance in Agricultural and Foraging Societies. . In: Cashdan, Elizabeth (ed.): Risk and Uncertainty in Tribal and Peasant Economies. Boulder: West View Press, pp. 67-88.

The perception of natural hazards in the context of human (in-)security

Ingo Haltermann (translated by Jessica Holste)

1. Introduction

Any geographer would have to admit that his research in a certain way relates to matters of space. Accordingly, in this article the consideration of space – especially the three-dimensional section of the surface of the earth that each individual constructs for himself as "his environment" – will serve as a point of departure even if it is not its main topic. At the outset, we will shortly discuss how humans acquire space, perceive and mentally construct it, put it to their individual use and, where necessary, adjust it. Subsequently, we will consider which phenomena the individual perceives as endangering exactly this environment. In this article, the individual's approach to hazards is our starting point because the way they are handled primarily depends on how the respective individuals perceive them. At the individual level not the question how likely it is that a threatening event will take place or how big the assumed damage will actually be determines an individual's actions. Rather, the central question is how the individual assesses both probability and damage. In addition, costs and benefits of measures of adaptation to be executed or neglected are evaluated, the hazard has to be weighed up against other chances and risks and further learned and/or culturally mediated factors are taken into account. The individual's perception of hazards also is of major importance for actions occurring at a collective level. On the one hand, in this way statements on how decisions might be accepted by the population can be made. Stating the obvious, on the other hand any discourse on hazards can only develop if someone perceives something as a hazard. Without the perception of hazards no discourse on hazards, without discourse on hazards no management of hazards.

In this article, the question how hazards are evaluated will be given particular attention. Of special importance is the question which significance natural hazards have in the eyes of those concerned in comparison to other hazards of daily life. This issue seems to be particularly interesting when considering the context of precarious housing situations because it appears that here environmental concerns frequently are of minor importance and natural hazards, so to speak, only play second fiddle in the concert of dangers (Pelling and Wisner 2009). Why this is so has as yet not been sufficiently investigated, especially as regards research investigating the lowest level of life support systems. On the one side, however, urbanisation and the growth of the population force an increasing number of people to live in precarious housing situations while, on the other, in many regions of the world climate change promotes the development of natural hazards: The intensity and frequency of extreme weather events are increasing and sea levels are rising so that more and more people live in the danger of being damaged by their immediate environment. For this reason, the research deficit must be covered urgently.

Naturally, individuals or households, which together constitute the lowest level of livelihood systems, only have a limited amount of resources at their disposal so that the answer to the question who has how many and which resources at hand is decisive for the ability to adapt to natural hazards (Müller-Mahn and Rettberg 2007). In this respect, the *Human-Security* approach offers valuable options how exactly natural hazards might be handled on the lowest level of livelihood systems.

2. Environment and risk

No matter how much I would like to avoid the obligatory definitions of terms at this point of the discussion, I cannot justifiably do so since the terms in question are too differently used in both academic and daily contexts. Hence, to assume a shared understanding of the terms without defining them at the outset seems an inadequate way to proceed. Although it is not necessary to start from scratch, in the following I would like to give an overview on the interpretations of the central terms *environment* and *risk* as used in this article.

2.1 Environment

At first sight, environment might simply be defined as everything surrounding us – the world around us, so to speak. Everything that is, but is not I or we. Yet, *us, not I* and *not we* already indicates that the environment as such does not exist, but only because it is defined by a subject. It is the section of space that is meaningful to the subject that puts itself in relation to it (Ittelson et al. 1977: 162). People use it, for instance, to provide for and accommodate themselves, to control their exchange relationships and to dispose of their waste. Humans need their environment to secure, firstly, their survival and, secondly, their well-being. Which section of space with which boundaries and contents functions as the environment of a given subject, therefore is primarily determined by its needs. Hence, the external does not exist in contrast to the internal world, but is defined by the latter. Jacob von Uexküll, the founder of the particular notion of *Umwelt*, understands it to be a system "that is established by the relationship of subject and external world" (Uexküll qtd. in Hellbrück and Fischer 1999: 23; translated from the German original). Environment, therefore, means a lot more than is implied by the usual concept of environment which primarily focuses on the natural environment of the commonality and which is expressed in terms such as environment protection and environmental policies. Environment in the sense used here applies to the natural as much as to the social and cultural environment (Hellpach qtd. in Hellbrück and Fischer 1999: 26). However, it should be kept in mind that this dissolution of the environment into its elements is an exclusively theoretical process that does not correlate with the human perception of the environment as the reference system of human action. *Gestalt* psychology argues that people do not perceive their environment as the sum of its elements but as a whole and see themselves as part of this whole (Hellbrück and Fischer 1999: 120; Ittelson et al. 1977: 139). As a consequence, the perception of the environment is not constituted by the perception of objects and surfaces – as in a jigsaw puzzle. It is rather the other way round: The perception of objects is part of the perception of the environment (Ittelson et al. 1977: 137). Only the specific type of a relationship between an object and its environment does imbue it with meaning. Hamm and Neumann (1996) point out that this is a process of interpretation depending on the situational context. They give the example of a razor lying in a bathroom, in the street or featuring as a picture in a news report on a murder. While the semantic label "razor" remains the same, "I relate the sign 'razor' to other signs present in the context and my interpretation then may be: 'Razor' in a bathroom setting, 'waste' in a street setting or 'weapon' a news report setting" (ibid.: 254; translated from the German original). However, this only works when these interpretations have been previously learnt, i.e. when they have been culturally mediated. A good example further illustrating the point is the Botswanan film "The Gods must be crazy" in which a bottle of Coke falls from a plane into a San-village. Since its genuine environment is missing, the San population is not able to relate it to its own environment. Eventually, the San interpret it to be a riddle sent by the gods – an interpretation which results in utter chaos. The same type of relationship exists for social or cultural elements of people's environments. Their meaning, "read" by us, develops only in relation to their specific environment and according to our learnt ability to interpret. This fact also plays a decisive role for the definition of natural hazards – as will be my aim to show in more detail in 2.3.

Furthermore, the limits between, for example, what is natural and what is artificial cannot be reasonably defined anymore and would be more or less arbitrary. Nature, understood as something existing completely independently of humans and/or their actions, as a matter of fact almost does not exist anymore. Even if forests and meadows to us may seem natural and untouched, these landscapes have long since and comprehensively been reshaped by cultures. This also holds true for the sea, the perpetual ice regions (*das ewige Eis*) and the so called virgin forests which, in the face of global consequences, can no longer be regarded as unaffected by human action.

Our environment has a purpose for us and we see it primarily from the vantage point of utility – regardless of the human influence that actually has a bearing on its actual status (cf. also Hamm and Neumann 1996: 51). To us, the environment is a resource. Depending on the situation, it may be a space for recreation, for traffic, for accommodation, for social contacts. Our respective needs determine, to put it metaphorically, which colour of glasses we wear to perceive specific information provided by our environment while blanking out other aspects. The process of perception is far more than an immediate absorption of the external into the internal world. Rather, it is a highly selective and constructive event. Ittelson et al. (1974: 124) comment on this:

> Environment perception provides the means whereby we can establish environments within which to carry out our purposes, environments that will accommodate the range of behaviour we have chosen to enact and which are most likely to provide the consequences to action which we anticipate.

Which preferences we use as our means of selection to some degree depends on our socialisation but also on specific features of our natural, cultural and social environment (Stengel 1999: 176; Löw 2001: 197). Also, what we require of our environment partly results from dispositions and values which our natural, cultural and social environment endowed us with.

Furthermore, the relationship between human being and environment is no one-way street. People shape and use their environments according to their ideas. They modify them, for example, when they plough a field, build a house or simply install a sign saying "Keep driveway clear". To sum up, 1. environmental influences affect our behaviour, 2. the individual perception of the environment influences our actions and 3. in turn, our behaviour influences our perception of the environment.

Considered from an internal point of view, i.e. from the point of view of the subject, two interacting systems relate to each other with the human body as their medium. Considered from the external point of view, the assumption that a system and its environment existing on one level, on the next higher level also form an independent system and another environment (Green 2004: 324), results in a system of two corresponding elements that determine each other's existence. To sum up, environment can be defined as a system representing a certain section of the external world to which the actions and perceptions of a subject give significance.

In the context of this article, two points are of major importance: Firstly, we perceive the environment as a whole. Resources as much as hazards, objects as much as other people are aspects of it and gain their ultimate significance only because they are framed by the environment constructed by us. Secondly, the environment only insofar is relevant to our actions as we react to it and shape it to secure our survival and well-being.

2.2 Risk

According to a simple definition a risk is a possibly occurring damage caused by a certain event or action, with the term possibly meaning "possibly damaging" as well as "possibly occurring". In mathematical terms such a risk can be expressed as the product of the probability of the event times the extent of the damage. According to this understanding, risks are realities objectively existing in the world and as such can be accurately observed and calculated. This *realistic* understanding of risk is typical for the insurance industry and in technical or approaches of the natural sciences, for instance, when the dam height necessary to prevent a flood disaster has to be calculated that statistically is likely to occur every 200 years. Considered from this vantage point, people are only relevant because they are the recipients of the events.

Social-scientific risk analyses open this human black box and extend the range of risk research. They scrutinise the results of natural science risk analyses critically, for instance, when investigating the fact that experts and laypeople fundamentally differ in the assessment of the probabilities and the extent of damage caused by specific risks. This question is relevant here because perception and estimation is a basic part of how to deal with risks.

Niklas Luhman (1993) points out that risk is a necessary part of human action. At bottom line every decision made by a person constitutes a risk. A simple game of dice may be used as an example illustrating the point: The dice may be cast any number of times and all pips are added. If a six turns up, the result is erased and it is the next player's turn. With every cast there is the risk of a six turning up. If we do not cast the dice, there is no risk. If we do decide to cast the dice once more, we risk something. However, if we do not cast the dice, we run the risk that another player has a better result. The future always is in the dark while our decisions have to be made in the present. This is only possible when we apply what we have learnt experientially in the past to the future. For example, we take for granted that the seasons will occur again in the following year. We do not know this for certain, but we can safely assume it, and this suffices to give us the feeling of security necessary to make decisions in the present. Since we cannot know for certain what the future holds, however, we cannot make right or wrong decisions, but only decisions which prove successful or which do not (Hellbrück and Fischer 1999: 513). We assume that our decisions will have the results we anticipate. For example, I have to store enough fire wood if I do not want to feel cold in the winter. Correct decision, it might be argued, but this is not so much due to the decision than to what the future holds. The bark beetle, a broken wood-burning stove or an inherited house somewhere in Spain might change this assessment. Yet, our method of using experiences as a point of orientation restricts our imagination as to what might happen in the future and consequently prevents us to feel faced with an overwhelmingly complex future (Luhmann 1993: 152). Therefore, we also use individual and social experiences to make predictions on the probability and the extent of risks that, being the result of our actions or of our interactions with our environment, will threaten us in the future. If we act in accordance with these assumptions (when casting dice, the six statistically only turns up once) the threat to which we are exposed becomes a risk that we take – as the example of the game of dice illustrated.

Which individual and social experiential values we use for the transformation of threats into risks depends on considerations which are of a thoroughly social and cultural nature. In this context, the possibility to use statistical calculations of probabilities as a reference system is only one option amongst many (Renn et al. 2007: 35).

Yet, to use experiential values as a point of orientation also has its disadvantage. We find it difficult to accept things about the future that have no corresponding representation in the past. Robert Kates (1976: 151f.) calls this limitation of our imagination *prison of experience*. It implies that experienced catastrophes set the upper limit of what decision makers presume may happen to them. Their actions are accordingly limited. Especially as regards the dangers related to climate change, this phenomenon might well gain importance.

Before we will turn to risk perception in the next chapter, I would like to briefly quote Luhmann's tongue-in-cheek statement on the relationship of danger and risk:

> Before the umbrella was invented, there was the danger to become wet when leaving the house. It was dangerous to go out. Usually, people had a sense of the danger but not a sense of risk in such situations because it was practically impossible to always stay at home because there was a chance it might start to rain [...] When the umbrella was invented, all this changed fundamentally. These days it is not possible at all to live without risk. The danger that we might get wet becomes a risk we run when we do not take our umbrella with us. However, if we take it with us, we run the risk to forget it somewhere. (Niklas Luhmann qtd. in Hubing 1994: 311; translated from the German original)

2.2.1 Risk perception

The natural scientific or technical understanding of risk has little bearing on human actions. Humans do not act according to what is, but according to what they assume (Hellbrück and Fischer 1999: 644). This becomes obvious when statistical data on occurrences of death or the frequency of catastrophes is compared to the fears of the public (cf., for example, Burton, Kates and White 1993; Geipel, Härta and Pohl 1997). Considered from a statistical point of view, the most frequent or dangerous threats are, for instance, the consequences of hypertension or of the consumption of alcohol or tobacco. Yet, seen from people's point of view, these are not necessarily the most dangerous threats (Renn et al. 2007: 86). "The risks that kill you are not necessarily the risks that anger and frighten you" (Sandmann, qtd. in Jungermann and Slovic 1993: 80). If the fears and worries of most people are considered, rather "spectacular" risks such as radioactive contamination or the possibility to fall victim to a plane crash are more prevalent. Hence, the perception of risks and their consequences is not so much based on statistical data on probabilities, amplitudes and damage potentials but on how humans perceive these. We understand perception to be the process of the absorption and processing of environmental stimuli. Theoretically, the process of risk perception might be further differentiated into the individual processes perception/identification (in the sense of pure absorption of information) and evaluation (intuitive, heuristic assessment of probability and extent of damage) (Rowe 1993: 46). However, this differentiation is of a purely theoretical nature and if, then only relevant to neuroscientists. We perceive information provided by the environment not unprocessed, i.e. as if it was nothing but a stimulus. Our experiences, our ideas and our memory are always involved as the example of the razor illustrated. Perception as we know it is always combined with interpretation (Hellbrück and Fischer 1999: 118).

Up to now, the results of psychological risk research have enhanced our knowledge of risk perception considerably (cf., for instance, Jungermann and Slovic 1993; Graf Hoyos 1987). Relevant to the perception of risk are, first and foremost, qualitative risk characteristics, the risk context and individual factors:

> The evaluation and assessment of a risk to a large degree depends on the specific types of its sources as well as on the knowledge and values of the group of people in question. (Jungermann and Slovic 1997, qtd. in Weichselgartner 2001; translated from the German original)

The specific types of risk sources can be further subdivided into characteristic features of risk results and characteristic features of risk causes (Jungermann and Slovic 1993: 85ff.). The former include aspects such as the controllability of the risk, the sustainability and reversibility of potential damages, the question in how far a hazard can be perceived and accounted for as well as the radius and the lethality of the expected damages. These concepts are often subsumed under the terms awfulness (*Schrecklichkeit*), limited awareness (*Unbekanntheit*) and exposedness (*Ausgesetztsein*) (Weichselgartner 2001: 35; Dikau and Weichselgartner 2005: 100). The characteristic features of causes describe, for instance, if a risk is taken deliberately. Is this the case, for example when someone decides to parachute, the corresponding risk is assessed considerably lower than it would be if someone was involuntarily exposed to that same risk source. In addition, the just distribution of profit and risk of an event/an action and the evaluation of a risk as new/well-known or as increasing/decreasing contributes to the assessment of the probability and the extent of the damage. However, in the quote above Jungermann and Slovic also point out that the risk source constitutes only one component contributing to the overall perception of risk. Additionally, personal interests, values, interpretations of reality, trust in regulative institutions and, last but not least, personal experiences with risk-taking shape risk perception as much as do the situational context and qualitative risk characteristics. According to empirical research, these basic mechanisms at work when risks are mentally processed are effective cross-culturally. They may be partially shaped by a given culture, yet the basic intuitive processes of perception "remain consistently present and measurable in spite of all cultural influences" (Renn et al. 2007: 86; translated from the German original).

2.2.2 Risk assessment

Risk perception must not be confused with the process of risk assessment (the acceptance/refusal of a risk). Risk assessment, the more detailed discussion of which shall be the focus of this chapter, can clearly be distinguished from information absorption and processing.

In principle, the aim of the process of risk assessment is to evaluate a given perceived risk as acceptable or not. In order to do so, risks related to certain actions or events are compared to the corresponding personal values, criteria or value judgments, related to associated chances and risks as regards their costs and benefits, and the "profitability" of different strategies of risk reduction or risk-prevention is estimated. In real life, it is likely that different assessment strategies are combined depending on the type of risk. Acceptable risks are those that lie below the so-called *action limit* (*Handlungsgrenze*) and, therefore, do not necessarily require that measures must be taken to reduce them – although these may well be considered. Risks that do not justify administration lie below the so-called *no action limit* (*Nichthandlungsgrenze*). Inacceptable are all risks that necessitate measures to alleviate risk, i.e. which lie above the action limit (Rowe 1993: 47; cf. also Renn et al. 2007: 88f.).

2.2.3 Risk management

So far, we have clarified why we consider what to be a hazard or risk and how we ascribe a certain meaning to it. However, merely knowing that a certain hazard or risk is judged to be inacceptable does not automatically result in an action that aims at reducing this risk. From the point of view of the decision maker, the first step to take is to evaluate the possibilities that may minimize the perceived risk. This does not necessarily imply all the options which, as a matter of fact, exist since the decision maker only chooses between those options which are known to him. In a given situation, the resources (of an individual, a household, a small group etc.) at hand are applied in an action – in accordance with Max Weber's statement "regardless whether external or internal action, forbearance, sufferance" (Weber 1985: 542) to minimise an identified risk. These may be roughly divided into four categories: *Acceptance* of damage, *reduction* of damage and *change* of usage or of location (cf. Beyer 1974: 269ff.; Burton, Kates and White 1993: 57ff.; Hewitt 1997: 172ff.).

a) Acceptance of damage

The decision maker settling in an environment that is considered to be hazardous possibly has already decided to allow for potential damage. Property in hazardous areas often is considerably cheaper than property in safe environments so that a decision maker might reasonably choose the insecure option. Potential damages might be overcompensated by savings made when purchasing the property or provisions might be made which can be used in the event of damage. In addition, an event of damage is only a possibility. If it does not occur, the gain is even higher than it would have been had a safer but more expensive property been bought. Also, arrangements which aim at a socialization of damages are a form of an acceptance of damage. Insurance companies institutionalise this principle and governmental and international disaster relief aims to distribute the damage of some onto the shoulder of many (tax payers or donors).

Last but not least, the feeling to have no alternative or to be powerless might result in the plain acceptance of damage. In such cases, empirical evidence must be used to differentiate the acceptance of *damage* from the acceptance of *risk*. Hence, it must be distinguished between missing necessity and missing possibility as a reason for the plain acceptance of damage. This will be further discussed in the fourth and fifth part of this article.

b) Reduction of damage

When discussing the reduction of damage, measures that aim at reducing the degree of damage caused by an event must be distinguished from measures that focus on the prevention of an event. Whereas, for example, the damaging effect of an earthquake may be reduced by adequate construction work or regular emergency trainings, this measure cannot prevent the earthquake from occurring. Avalanches, on the other side, at least to some degree might be prevented easily when the accumulation zone is stabilised or the snow removed from areas which have the potential to cause an avalanche. Likewise, the construction of dams or, even more importantly the designation of flood-prone areas as a means of flood control serve to prevent a potentially damaging event, in this case a flood. Still, such preventive measures are rather collective efforts that have to be handled politically than individually executable instruments of risk reduction. On the level of households, damages can, first and foremost, be reduced by reducing the vulnerability of a household, for example, by implementing the already mentioned measures of building security, emergency trainings or evacuation plans. In these cases, early warning systems bear a high potential of risk reduction on yet another, superordinate level. Measures to reduce potential damages therefore intend to prevent the damage of the chosen environment, its utilisation and the subject constituting it. As will be illustrated elsewhere in this article, such measures fundamentally depend on the availability of economic, social and political resources.

c) Change of location or utilisation

This is the category of risk reducing measures that demands the most extensive changes from the decision maker. Accordingly, the options to move or to substantially modify the basic modes of economic activities on site as a strategy of risk reductions rather seem to be perceived to constitute the absolutely last resort. People usually defend decisions once made. We tend to confirm rather than refute the hypotheses on which an action is based ("this place is safe") (Hellbrück and Fischer 1999: 514).

This *heuristics of confirmation* also applies to the perception of risks because these are adapted to individual attitudes and preferences. Accordingly, influences of the environment which threaten the choice of location are rather pushed into the background. Catastrophes can act as catalysts of change when they sever these mechanisms of self-affirmation, reveal that the former decision how to treat the threat was unquestionably wrong and thereby make room for a realignment how a threat ought to be handled.

Which decisions are eventually made and how the acting people reach their decisions once again depends on situational, individual, social and cultural aspects. Yet, at least geographic risk research increasingly seems to favour the concept of Bounded Rationality. It holds that decision makers do not strictly follow the Rational Choice Theory, according to which the weighting of all possibilities results in the choice of the option which is likely to yield the best benefit-cost factor. Instead, decisions are limited by individual and situational contexts. An example for this is the precondition to absolutely avoid running up debts or the premise not to put people's lives at risk. Discussing these concepts, Burton, Kates and White (1993: 65) write:

> It is rare indeed that individuals have access to full information appraising either natural events or alternative courses of action. Even if they were to have such information, they would have trouble in processing it, and in many instances they would have goals quite different than maximizing the expected utility. The bounds on rational choice in dealing with natural hazards [...] are numerous. The process is stressed here because there is a tendency that people left to themselves with enough information will select the optimal adjustments. Any doubt about this is dispelled by examining the kinds of experience with natural hazards provided by both developing and high-income nations.

2.3 Environment as hazard

The premises outlined in 2.1 maintained that we are only able to perceive individual aspects of our environment in relation to their broader setting and that the constitution and perception of our environment is shaped by motivations, attitudes and values. They also apply to the concept of environmental risk. It is not that hazards are realities that lie in wait for us in our environment. Rather, they are constructs representing an aspect of a given environment and consisting of information presented in the external world that we absorb and process according to our individual predisposition.

Leaving aside the concept of environment for a moment, let us consider an initially uninhabited, theoretical space representing a simple three-dimensional section of the earth's surface. It is shaped by a particular constellation of geographic, geological, atmospheric and biological conditions. Like any other place on earth it is subject to certain natural processes. Processes on the earth's crust (volcanoes, earthquakes), on the surface of the earth (e.g. avalanches and landslides) or in the atmosphere (storms, heavy rain) can set free enormous amounts of energy that may cause damage – i.e. if they occur in regions where there are societies and their (cultural) assets (cf. Haltermann 2011: 351). Geographical risk research calls this combination of natural phenomenon and potential damage a *hazard*. In other words: At a place uninhabited by people neither dangers nor risks or hazards exist. Parallel to the potentially damaging aspects of a place certain aspects may also be considered from the vantage point of their utility. In this case, one would talk about *resources* (Burton, Kates and White 1993: 31f.). Here, it should be noted that both damage and utility are genuinely human categories and that their identification as the one or the other once again depends on the already mentioned situational, individual, social and cultural factors. To illustrate this, imagine that a cherry tree in your garden was hit by a stroke of lightning. If it was the cherry tree in which you liked to climb about when you were a child, which yielded buckets full of cherries every year and which, in the summer, so generously gave shade, it might well be that you consider this event to be a damage – on the one side a loss (the tree as representative of memories), on the other side as an advantage lost in the future (cherries, shade). If the same event occurred five years later, with the tree by now rotten, hardly any cherries left and the shade so abundant that not a single sunbeam reached the terrace any more, would it then still be damage? And if the tree had to be cut and the expenses for the woodcutter could be saved? Maybe there is even a use since the store of firewood for the domestic fireplace is diminishing? The example shows that it is the individual who transforms his environment into resources and hazards in accordance with its intended usage. If an individual settles in the imagined place, he has to weigh up the chance to use certain resources against the risk to jeopardize certain cultural assets as well as life and limb. He risks something because he has the chance to acquire a desired resource and, therefore, risk becomes an aspect of his actions.

In real life, this process of the decision maker is, of course, much more complex. In addition to the chances and risks that result from the kind of relationship between an individual and his natural environment, there are also those that result from the individual's relationships with his social environment. As was mentioned at the beginning, every decision comprises chances and risks and, consequently, it is never merely *one* risk that is weighed up against *one* chance but *many* risks against *many* chances. Furthermore, it has to be taken into account that chances are weighed up against other chances and risks compared to other risks. Is the risk to become the victim of a traffic accident worse than the risk to be too late for a job interview? Or does the answer to this question not even depend on the chances and risks that the potentially new job entails?

3. Human security and vulnerability approach – lacking property rights as limits to adaptation

In order to minimise an identified risk, the available resources (of an individual, a household, a small group etc.) are employed to realise a given action in a given environment – as was described in 2.2.3 when discussing the way to proceed when reducing risks. We take possession of a territory and use the resources we were able to identify in it to warrant our security and thus reduce risks. If applied thus, the concept of security is understood in its broader sense, namely as the United Nations has defined *human security* as *freedom from want* and *freedom from fear*. This definition may easily be related to the concept of risk provided in this article. Deprivation means nothing else but the danger not to be able to satisfy our needs, and danger becomes a risk because of our future oriented action. The concept of fear cannot be imagined without the concept of a threat. Even if a threat only exists in our imagination – as, for instance, the fear of spiders – we act accordingly. The walk into the basement for the person in question bears the risk to encounter a spider. If, therefore, planned actions serve to secure human security, this is a kind of risk-related action and hence we require resources.

Yet, the times in which we could simply take possession of the territory which we expect to provide us with the necessary resources are (if they existed at all) long past. On our planet, hardly a spot of land exists of which nobody would claim the right to use or own it. The foundation of these claims is, at bottom line, power, represented by the four dimensions wealth, knowledge, organisation (hierarchy) and association (selection). These four institutionalised dimensions may be assigned the four institutionalised means of barter money, qualification, social status and affiliation (Kreckel 1992, qtd. in Löw 2001:

210f.). Their availability determines the options a subject has to acquire space and, thereby, exploit its resources.

This process also applies to the constitution of space. Space may be defined as the relational order of social goods and creatures at a given place (Löw 2001: 159f.). Social inequality, in turn, means the continuously restricted possibility to attain economic (wealth), cultural (knowledge) and social capital (organisation and association) or the corresponding commodities. The fewer social commodities someone has at his disposal, the more limited are his options to act in a spatially relevant manner. In this way, lacking rights of disposal when constituting space represents a form of powerlessness that manifests itself in our environments as may be illustrated by the slums, especially by those in the cities of the global south. By analogy, access to territories can be facilitated by money, qualification, affiliation and social status. Social inequality thereby manifests itself in space and is continuously reproduced (Löw 2001: 210).

Applied to the discussion of risks this means that social inequality implies an uneven distribution of risks. The more – and the more valuable – money, qualification, social status and affiliation we can dispose of, the better are the chances to a) have access to risk-poor places and b) act in a spatially relevant way which also enables us to further minimise risks. This is why some own an earthquake-proof, two hundred square-metre large house in a gated community while those who do not have access to economic, cultural and social capital must make do with what is left. These are places full of various and/or high risks so that no one would like to live there who could afford to make his living somewhere else. As a consequence, it is the poorest of the poor who gather in flood-prone areas, live at mountainsides exposed to landslides or on the approach paths of the major airports without having much of a chance to change anything about these risks (cf. Haltermann 2011). These are the people who are most vulnerable to the hazards of nature or the threats of their social environment. Moreover, they often have little means to react to external shocks such as catastrophes. The possibilities to compensate damages caused by a catastrophe decrease when no resources which could be used for reconstruction are available. The relationship between the availability of resources and catastrophes is characterized by two mutually dependent aspects: Lacking resources lower the resilience of a household. For instance, the lacking access to clean water may cause diseases which, in turn, reduce the work capacity of the household members. As these no longer or only partly can assist in the prevention of or protection against catastrophes, the proneness to fall victim to catastrophic events increases. Vice versa, catastrophes destroy humans' resources (if we stick to the example of human capital, for instance, when a member of the family dies) which further decreases resilience (Pelling 2003: 55).

To sum up: In order to secure our livelihood, we need resources. Their spatial distribution as well as the socially or hierarchically organised rights of disposal of space generate and reproduce social inequality. From the point of view of marginalised households, this inequality manifests itself as the limit of the production of (human) security. Therefore, social inequality limits their possibilities to constructively cope with hazards. As a consequence and because there are no alternatives, risk management which initially meant the reduction of damage increasingly becomes the acceptance of damage. Vice versa, an increasing availability of resources implies that the acceptance of damages as one option of risk management loses importance. In other words: The more resources are available to us to reduce damage, the more likely it is that we can declare something to be damage. This may be illustrated when we take a look at how similar events are perceived in different parts of the world with differently equipped households. Elísio Macamo and Dieter Neubert investigated the interpretation of two flood disasters in Germany and Mozambique. Whereas the flood event in Oderbruch 1997 was declared a flood disaster by the local population, officials and the media – although it caused primarily (insured) material damage (360 million US $) and although of the 5,200 people concerned no one was severely injured and no one died –, the flood event in the Limpopo valley in 2000 was hardly considered a catastrophe by the local population although several hundred people died, it affected 4.5 million people and caused 420 million US $ mostly uninsured material damage (cf. Macamo and Neubert 2008, 2009, see also Macamo and Neubert this volume). Macamo and Neubert in this context speak of the "usual experiences in everyday life". Against the backdrop of a "permanently precarious situation" and since no governmental prevention or other institutional, security warranting services exist, the responsibility to avoid and cope with events of damage lies with the individual households or, at best, with the local social community. However, the causes of floods such as the opening of the South African dams at the upper reaches of the Limpopo or the deficient or missing infrastructure of dams along the river are beyond the reach of the affected population. In other words, the fundamental causes are beyond the local population's influence and their resources. Therefore, individual cases of death in the local environment, the destruction of dwellings and the loss of livestock are perceived as consequences of floods that are to be expected in the normal course of events. In accordance with our definition of the environment as a system which – by way of the actions and perceptions of an individual – ascribes a subjective meaning to a certain section of the external world, the causes of the flood cannot be traced back to the environment of the local population because they are beyond its limits of action. The threat is, so to speak, of an external nature. Without the option to react adequately to a hazard, the threat cannot be transformed into a risk, but remains a threat which might become a reality any time.

4. Risk perception under precarious living conditions – suppression as a means of adaptation?

At this point, the question poses itself how people can react to such a situation. Life in constant fear of a catastrophe cannot be lived without an impairment of one's physical and psychological health. Likewise, making decisions in daily life is seriously affected by the constant fear of danger, for example, when a decision means that you must leave your family for a couple of hours. This phenomenon follows a psychological dynamic that – under positive circumstances – warrants the optimal adaptation to our environment, namely when we absorb and process information provided by the environment according to our individual experiences and our social and cultural background so that they guide our actions. If it is not possible to realise the idea of optimal behaviour thus generated (for example because resources are lacking or because conflicting goals complicate the assessment of risks), a *cognitive dissonance* is experienced. In order to restore our inner harmony, we try to reduce the information on the environment that requires us to act. We either try to influence our environment or, if this is not an option or does not seem to make sense, try to block it out, reinterpret it or ignore it. As we thus maintain our ability to act, the latter is by no way pointless and a means of our adaption to our environment (cf. Hellbrück and Fischer 1999: 525, 564; Weichselgartner 2001: 37). If we include the statements made in the previous part, we can now sum up that the deficit of economic, social and political capital – or put differently, the deficits of Human Security as defined by the UNDP – restricts the perception of natural hazards. In the face of missing rights of disposal as regards the constitution of resilient space, the suppression of those risks that appear to be the least manageable constitutes a means of adaptation.

I have chosen the term *appear* because the barriers to adaptation described here are by no means immutable factors. What may be regarded to constitute a barrier to adaptation or what should be considered a resource enabling us to cope with risks, is subject to processes which are basically of a social and cultural nature. I quote Neil Adger et al. (2009: 345):

> For individuals, and the societies they are members of, actions are shaped in part by deeply-embedded (but not static) cultural and societal norms and values. Some characteristics operate at the individual level and include beliefs, preferences, perceptions of self-efficacy and controllability. These, together with perceptions of risk, knowledge, experience, and habitual behaviour, norms and values, determine what is perceived to be a limit to adaptation – at both individual and social levels in any particular society – and what is not. These limits are therefore not absolute and insurmountable but rather socially constructed, subjective and mutable. It depends on individuals' underlying values and their enacting by societies whether a limit is perceived as such.

If, in the context of permanently precarious situations, one inquires about potential courses of actions, frequently the following answer is given: There is nothing we can do about it. We have no alternatives (cf. Haltermann 2011; Rohland et al. 2011). However, what is considered to be an alternative and what is not, is a dynamic social and cultural variable that is constantly negotiated. Which type of capital is relevant because it can reduce vulnerability; in which combination, which quality and quantity a risk may be reduced to what extent is part of a continuous process of negotiation. The social geographer Hans-Georg Bohle emphasises that the value of capital (economic, cultural as well as social) as a factor reducing the vulnerability of people is determined by the prevailing social, institutional and organisational environment:

> Within the context [of vulnerability] they [the poor people] have access to certain assets ("capitals") or poverty reducing factors. These gain their meaning and value through the prevailing social, institutional, and organisational environment. This environment also influence the livelihood strategies – ways of combining and using assets – open to people in pursuit of beneficial livelihood outcomes that meet their own livelihood objectives. (Bohle 2007: 12)

Apart from this, the development of a dangerous situation may shift the limit between what is considered an alternative and what is not. A situation that is increasingly precarious might ultimately entail the consideration of alternatives that previously had not been taken into account because of social rules, individual values and attitudes or because of a generally different risk calculation. An example for this are the boat people at the borders of the EU whose living conditions in their countries of origin have become worse to such a drastic extent that they accept leaving their home country and their families and are ready to risk their lives. What has to happen to perceive something as an alternative if this alternative leads us to expect that we will never again see our families and that there is a ten-percent chance of our death?

Our central statement is therefore: The boundaries of adaptation to an environment perceived as dangerous are socially and culturally constructed and as such open to change. Whenever these boundaries are shifted, simultaneously the boundaries of what we perceive as a risk changes so that we can incorporate it into our actions.

5. Conclusion

Even if the fact that natural hazards only play second fiddle in the orchestra of threats seems a banal truth, for the management of these risks the basic mechanisms must be understood if the loss of social and material commodities as well as the loss of lives is to be countered effectively. Especially if seen against the backdrop of climate change and the danger of increasingly worse natural hazards, the discussion of how natural hazards are perceived is a matter of growing urgency. Unfortunately, it is especially those who are affected by the consequences of climate change that have the least chances to adapt to it. On the one side, they do not have access to the necessary resources. On the other side, the deficits in the realm of human security contribute to the suppression of these, apparently least manageable, risks. Therefore, it might be concluded that in the context of constantly precarious living conditions even extreme natural events lose importance and represent only one further aspect in a general state of continuous crisis.

It should have become apparent that this is a structural problem. A purely technical approach or the provision of emergency aids can result in a symptomatic and punctual alleviation of the acute problem, yet, if the fundamental chronic problems are to be resolved, a more comprehensive approach is required. Emergency relief usually is only available at places where extreme events lead to a concentrated manifestation of physical and material damages and the corresponding attention by the media. The major and, as regards the extent of damages, chronically underestimated part of everyday hazards and small disasters for the most part must come to terms without external aid (Pelling and Wisner 2009: 33f.). Likewise, technical measures are primarily taken at places where hazards are discussed broadly by the public so that political pressure can build up. This is primarily the case with cities especially in those regions in which sufficient amounts of social capital exist so that people can make themselves heard and influence political decisions.

However, in the case of poorer regions, an approach must be developed that incorporates the basic requirements of households as central institutions of livelihood. Households must be equipped to acquire resources sufficient to sustain their livelihood – including the appropriate means of defence against hazards or else the production of human security. This can only work out if the relationship with the actual environment is maintained, since the environment of a subject a) determines the need of resources and b) the value and the meaning of a resource. Therefore, a punctual and temporary emergency relief that exists independently of local coping strategies is insufficient. What is required is a renewed negotiation of attitudes and values which form the basis for the evaluation of what we consider to be capital and which value we ascribe to it. I, once again, quote Adger et al. (2009: 347):

> We suggest that individual and social characteristics interact with underlying values to form subjective and mutable limits to adaptation that currently hinder society's ability to act [...] these limits are indeed relative and could be overcome, although on a large scale they may necessitate deep cultural and social change. Such changes, however, may imply modifying existing structures and value systems, thus threatening deeply held cultural, historical or ethnic identities.

To me, the *Human Security* approach seems to be suitable to reach this aim. The orientation toward the equipment of a household with resources (economic, cultural and social) is fundamental to any society's ability to adapt. It offers good chances to once again highlight the consideration of people in the analyses and to make them the addressees of the effort to create a secure world.

> The capabilities of the poor to participate in decision-making processes and the rights available to claim options for coping and adaptation are the prerequisites for attaining livelihood security. From this perspective social vulnerability and human security are not just physical, material, and financial support, but rather a life in self-determination, freedom, and dignity. (Bohle 2007: 350)

6. Bibliography

Adger, Neil W., Dessai, Suraje, Goulden, Marisa, Hulme, Mike, Lorenzoni, Irene, Nelson, Donald R., Naess, Lars Otto, Wolf, Johanna and Anita Wreford (2009): Are there Social Limits to Adaptation to Climate Change? In: Climate Change 93, pp. 335-354.

Beyer, Jacquelin L. (1974): Global Summary of Human Response to Natural Hazards: Floods. In: White, Gilbert F. (ed.): Natural Hazards: Local, National, Global. New York, London, Toronto: Oxford University Press, pp. 265-274.

Bohle, Hans-Georg (2007): Living with Vulnerability: Livelihoods and Human Security in Risky Environments. Interdisciplinary Security Connections (InterSecTions), Publication Series of UNU-EHS No. 6/2007. Bonn: United Nations University Institute for Environment and Human Security.

Burton, Ian, Kates, Robert W. and Gilbert F. White (1993): The Environment as Hazard. 2nd ed. New York, London: The Gilford Press.

Dikau, Richard and Jürgen Weichselgartner (2005): Der unruhige Planet. Der Mensch und die Naturgewalten. Darmstadt: Wissenschaftliche Buchgesellschaft.

Geipel, Robert, Härta, Rainer and Jürgen Pohl (1997): Risiken im Mittelrheinischen Becken: Bericht über ein von der Deutschen Forschungsgemeinschaft gefördertes Projekt. Deutsches IDNDR-Komitee für Katastrophenvorbeugung: Deutsche IDNDR-Reihe Nr. 4.

Graf Hoyos, Carl (1987): Einstellung zu und Akzeptanz von unsicheren Situationen: Die Sicht der Psychologie. In: Bayerische Rückversicherung (Hrsg.): Gesellschaft und Unsicherheit. Karlsruhe: Verlag Versicherungswirtschaft, pp. 49-65.

Green, Colin (2004): The evaluation of vulnerability to flooding. In: Disaster Prevention and Management 13, 4, pp. 323 – 329.

Haltermann, Ingo (2011): Vom Alltagsrisiko zur Katastrophe – Die Veränderung von Naturrisiken und deren Wahrnehmung am Beispiel Accra/ Ghana. In: SWS-Rundschau 51, 3, pp. 349-366.

Hamm, Bernd and Ingo Neumann (1996): Siedlungs-, Umwelt- und Planungssoziologie. Opladen: Leske + Buderich.

Hellbrück, Jürgen and Manfred Fischer (1999): Umweltpsychologie: ein Lehrbuch. Göttingen, Bern, Toronto, Seattle: Hogrefe Verlag für Psychologie.

Hewitt, Kenneth (1997): Regions of Risk. A Geographical Introduction to Disasters. Harlow: Longman.

Hubig, Christoph (1994): Das Risiko des Risikos. Das Nicht-Gewußte und das Nicht-Wißbare. In: Universitas. Zeitschrift für interdisziplinäre Wissenschaft 49, 4, pp. 310-318.

Ittelson, William H., Proshansky, Harold M., Rivlin, Leanne G. and Gary H. Winkel (1977): Einführung in die Umweltpsychologie. Stuttgart: Klett-Cotta.

——— (1974): An Introduction to Environmental Psychology. New York: Holt, Rinehart and Winston.

Jungermann, Helmut and Paul Slovic (1993): Charakteristika individueller Risikowahrnehmung. In: Krohn, Wolfgang and Georg Krücken (eds.): Riskante Technologien: Reflexion und Regulation. Einführung in die sozialwissenschaftliche Risikoforschung. Frankfurt am Main: Suhrkamp, pp. 79-100.

Kates, Robert W. (1976): Experiencing the Environment as Hazard. In: Wapner, Seymour, Cohen, Saul and Bernard Kaplan (eds.): Experiencing the Environment. New York: Plenum Press, pp. 133-156.

Löw, Martina (2001): Raumsoziologie. Frankfurt am Main: Suhrkamp Verlag.

Luhmann, Niklas (1993): Risiko und Gefahr. In: Krohn, Wolfgang and Georg Krücken (eds.): Riskante Technologien: Reflexion und Regulation. Einführung in die sozialwissenschaftliche Risikoforschung. Frankfurt am Main: Suhrkamp, pp. 138-185.

Macamo, Elísio and Dieter Neubert (2009): Die soziale Deutung von Hochwasserkatastrophen und ihre Bewältigung. In: Meyer, Günter, Steiner, Christian and Andreas Thimm (eds.): Katastrophen in der Dritten Welt – Soziale, wirtschaftliche und politische Folgen. Veröffentlichungen des Interdisziplinären Arbeitskreises Dritte Welt, Band 19. Mainz: Johannes Gutenberg Universität, Mainz, pp. 47-66.

——— (2008): Erwartung an Sicherheit – Subjektive Katastrophenwahrnehmungen und Bedingungen der Bewältigung am Beispiel Mosambiks und Deutschlands. In: Rehberg, Karl-Siegbert (ed.): Die Natur der Gesellschaft. Verhandlungen des 33. Kongresses der Deutschen Gesellschaft für Soziologie in Kassel 2006, Teil 2. Frankfurt, New York: Campus Verlag, pp. 858-874.

Müller-Mahn, Detlef and Simone Rettberg (2007): Weizen oder Waffen? Umgang mit Risiken bei den Afar-Nomaden in Äthiopien. In: Geographische Rundschau 59, 10, pp. 40-47.

Pelling, Mark (2003): The Vulnerability of Cities. Natural Disasters and Social Resilience. London, Sterling VA: Earthscan.

Pelling, Mark and Ben Wisner (eds.) (2009): Disaster Risk Reduction: Cases from Urban Africa. London, Sterling VA: Earthscan.

Renn, Ortwin, Schweizer, Pia-Johanna, Dreyer, Marion and Andreas Klinke (2007): Risiko. Über den gesellschaftlichen Umgang mit Unsicherheit. München: Oekom.

Rohland, Eleonora, Böcker, Maike, Cullmann, Gitte, Haltermann, Ingo and Franz Mauelshagen (2011): Why People Don't Leave: Attachment to Place in the Aftermath of Disaster – Perspectives from Four Continents. In: Cave, Mark and Stephen Sloan (eds.): Oral History and Crisis. New York: Oxford University Press, in preparation.

Rowe, William D. (1993): Ansätze und Methoden der Risikoforschung. In: Krohn, Wolfgang and Georg Krücken (eds.): Riskante Technologien: Reflexion und Regulation. Einführung in die sozialwissenschaftliche Risikoforschung. Frankfurt am Main: Suhrkamp, pp. 45-78.

Stengel, Martin (1999): Ökologische Psychologie. München: Oldenbourg.

Weber, Max (1985): Gesammelte Aufsätze zur Wissenschaftslehre. Tübingen: UTB für Wissenschaft.

Weichselgartner, Jürgen (2001): Naturgefahren als soziale Konstruktion. Eine geographische Beobachtung der gesellschaftlichen Auseinandersetzung mit Naturrisiken. Bonn. http://hss.ulb.uni-bonn.de/2001/0175/0175.pdf (accessed 09.10.2011).

II.2 Vulnerability and resilience

"Flood disasters". A sociological analysis of local perception and management of extreme events based on examples from Mozambique, Germany, and the USA[1]

Elísio Macamo and Dieter Neubert[2]

1. Introduction

As extreme events disasters represent very deep gashes into human life. They appear to be universal phenomena to which both developing countries and industrial societies are exposed. At the same time disaster discourse often tends to stress the particular "vulnerability", often combined with a limited capacity for resilience, of deprived sections of the population as well as that of people living in the developing world (Wisner et al. 2004). We share the sympathy and compassion for the victims of disasters expressed by this statement. However, it seems to entail a simple and as we think, misleading relationship: The more deprived and poor people are the more their livelihood is at risk and the lesser are their chances to recover. Before making assumptions about the nature and effects of disasters it seems appropriate to compare empirically the local perceptions of such extreme events in different countries and the action taken to overcome the damage caused by a disaster. Indeed, our aim in this contribution is to introduce the notion of an "ordinary management expectation" to serve as an analytical caveat against an undifferentiated notion of disaster and the blanket use of vulnerability in descriptive accounts of the effects of hazardous events. Our interest focuses group, community or society assessments and reactions i.e. the question concerning how these social entities deal with disasters. In other words, we will not be looking at coping processes on an individual level.

For this purpose we have chosen the example of floods. Floods are a worldwide phenomenon threatening peoples' life and material means in developing countries as well as in highly industrialized countries. This allows for a comparison between different parts of the world. The starting point is a flood in the year 2000 in the Limpopo Valley in Southern Mozambique, an event that at the time received wide media coverage internationally (Christie and Hanlon 2001).

---

1   This paper is based on an earlier version translated by Anja Löbert and Timothy Wise.
2   We are most grateful to Dag Schumann for his support in conducting the interviews upon which the material on the case studies in Germany and the USA is based.

During the field study in Mozambique we noticed the ubiquity of metaphysical accounts of disaster prevention. For this reason, we have chosen to add further cases from other settings in order to put these metaphysical accounts into perspective. Accordingly, we have selected the 1997 flood in the Odra region in the Eastern part of Germany, which was at least a news item in Europe. In this region religion plays a minor role in everyday life. We thought that we could safely assume that metaphysical accounts would play a minor role there. The last case we chose were the 2003 and 2004 floods in the Tennessee Valley – a place situated in the highly religious and so-called "bible belt" region of the USA. The significance of these rather localized floods went beyond the local level in that they represented the concrete consequences of weather phenomena which affected various states and which, both as atomized events and as a sum total, gained public interest.

Before introducing and comparing the individual cases we need to develop adequate definitions which should also yield important criteria for comparison (2. Definitions). This will be followed by the description of the empirical basis for the comparison of the progression of three cases (3. Three Floods). Next, the actual comparison will be undertaken (4. Comparing emergency management strategies and assessments). All of this will be rounded off by reflections on the contextualization of the analysis as well as on the applicability of the concepts of vulnerability and resilience (5. Contextualizing vulnerability and resilience).

2. Definitions

Popular notions of disasters generally include severe natural phenomena such as earthquakes, storms, floods, or tsunamis. Aside from those, accidents of a more technical nature are referred to as disasters. Among these can be explosions in a chemistry factory (Bhopal) or a nuclear meltdown (Chernobyl, Fukushima). Even terror attacks such as "9/11" can be referred to as disasters. The class of phenomena described by the notion of disaster is mostly characterized by a certain amount of suddenness and potential ferocity with a large number of fatalities or considerable material damage.

These popular notions are extended in two ways in common parlance. Firstly, a disaster is stretched out significantly on the temporal axis. Already most floods (with the exception of flash floods) start with rising water levels usually for some days at least some hours. Global warming is presented as disaster and we talk about a "disastrous drought" and thus refer to the effects of long-term processes lasting for a month, year or even decades. The term "disaster" in this case refers to the ferocity of the effects and – in contrast to suddenness – the permanent and continuing effects, as the notion in its etymology refers to a dramatic turning point. Secondly, smaller phenomena are often tagged with the media-effective label of "disaster". When compared with the aforementioned examples, this seems inappropriate, for example, in the case of a train crash with a large number of fatalities. The term "accident" seems more fitting here. An index derived from the amount of damage and the number of fatalities in the sense of a "catastrophe rating" also appears problematic. So where can we draw a sensible and analytically practicable boundary?[3]

Good sociological definitions should be precise, analytically serviceable, and at the same time adaptable to the social reality to which they refer. Beside all differences in the definition of disaster it is generally agreed that disasters are thus appreciated as social phenomena. We suggest a variation of the definition which has been in use in the sociology of disaster for a number of years (Clausen and Dombrowsky 1990; Japp 2000). We briefly define disasters as the "breakdown of emergency management capabilities in extreme events and in cases of threat and crisis".[4] Thus, we, too, combine ferocity and social consequences. In looking at emergency management capabilities we are referring to empirical behavior which simplifies the application of the definition and allows us adequate sociological access.

Ultimately, however, nearly all consequences of disasters are successfully dealt with somehow and at some point. The volcanic eruption of 1983 in Krakatau and 1906 San Francisco earthquake are now history. Even in Aceh, which was most severely hit by the 2004 Tsunami, surviving casualties had been medically treated after several weeks and, subsequently, the reconstruction stage was already well on the way (Schröter 2009). The breakdown of normal life and ordinary assistance and support structures thus usually refer to a certain window of time which is normally limited to the period right after the sudden unexpected event. At the same time even the most perfectly organized aid needs a while to take effect. In a more severe case aid may be absent even long after the locally tolerable times of intervention. Put more simply, "ordinarily expected relief" stays away. Not only does the expected time pass before the arrival of relief, but also the content of what can be expected is not absolutely determined. Rather, for every social context there is an "ordinarily expected relief".

This holds also in the case of the destruction such as the one wrought by the 1970s Sahel drought. Spittler (1989) has shown this quite perceptively as far as the Tuareg are concerned, especially in drawing our attention to the extent to which people relied for years on their coping mechanisms. Only when the local management capacity failed and the famine became uncontrollable did the

---

3 We do not attempt to present a general definition of disaster, which is hardly possible (Quarantelli 1998), but a definition suitable for a sociological analysis of local perception and management of extreme events.

4 "An extreme collective crisis is to be understood as the public perception of a suddenly emerging severe development or manifestation of a problem which cannot be dealt with by means of conventional problem-solving techniques and which is in need of adaptation at all levels." (translated, Geenen 2003: 5).

drought become an extraordinary event. In this way, local expectations of normality are made the centre of analysis and definition. It is safe to assume that disasters are, at least in this sense, socially determined and their threshold values vary depending on the locality.

Our definition of disaster is therefore contextualized in a double sense. Firstly, the expected outcome of relief varies in each case (safeguard from death, freedom from bodily harm, protection of material goods) – this is what we describe as the "ordinarily expected management". Secondly, crisis management capabilities differ depending on the respective strategies, the kind of assistance, the extent, and the capabilities (measures of protection in the construction of buildings, early warning, escape routes, medical care, and regulations regarding compensation) as well as organizational complexity; put simply, the way emergency services and reconstruction are organized.

"Ordinarily expected management" in the case of floods may, for instance, refer to protection from drowning. This can be done through seeking refuge in elevated terrain or through helicopter rescue. In other cases, management includes notions of protecting material goods by means of a dam. As far as organizational complexity is concerned, management can, in some cases, refer to the scope of action of a community or village, as, for instance, in many parts of the developing world. In other cases, "ordinary" management is based on a multi-staged, locally, regionally, and nationally organized network of specific institutions of disaster prevention and control, such as can be found in advanced industrial societies. In other words, disasters are defined according to local relief management capabilities and the associated notions of "ordinarily expected management". As we will show later on, this has an impact on the definition of vulnerability and resilience.

The definition introduced here foregoes the criterion of "suddenness". Thus, creeping developments, too, such as slow-onset droughts, can lead to disasters when crisis management measures fail. The process by which such slow-onset events originate, as well as the surprise effect, diverge from sudden disasters; the local effects, however, are similar.

This disaster definition of breakdown or large failure of crisis management measures is narrower than the one used in everyday language. Only this selectivity offers the possibility of applying the term "disaster" analytically. However, it would be an unnecessary limitation for sociological analysis to look only at those phenomena which qualify as disasters according to the definition. The management of disasters becomes especially apparent when crisis management measures do take effect or when disasters are merely looming.

In this way we would like to suggest a graded conceptual system.[5] The starting point is an **extreme event**, whether a natural phenomenon (as in natural disasters), for instance an earthquake, a volcanic eruption, hard rainfalls with rising river gauges, a storm, protracted droughts and processes of desertification, or a technical accident, such as an explosion, etc. Even profound ecological transformations are initially neutral occurrences that are referred to as "deviation" in disturbance ecology. In this sense, there are no "good" or "bad" ecological processes.

When this occurrence becomes the subject of a significantly negative social evaluation and the fear arises that it might become a problem for the people affected by it, we can refer to this as a **threat**. Threats are part of everyday life. There are strategies to respond to them within the framework of normality.[6]

Only when ordinary strategies fail to have the desired and usual effect, and when the management of the event or its ramifications become uncertain is normality placed at risk. We refer to this as a **crisis**. In the case of a crisis management strategies are put to the test. The usual management strategies fail to take effect, and particular emergency measures are applied which set the occurrence outside of normality. Crises are not normal, but as extraordinary situations they are exceptions to the rule of normality that are generally to be expected or to be feared.

The mastery of a crisis occurs when the consequences of an event are successfully managed. Only when emergency measures fail to take effect and their exercise itself is threatened by the event, is it legitimate to talk about a **disaster**. In this case normality is permanently destroyed. This is the exceptional case that disaster prevention endeavors to avoid and which reveals that the prevention, management, and emergency measures have failed. These events reach beyond what was thought to be possible and what was therefore included in plans for emergency management. In this case relief has to come from outside, e.g. the international community. A further intensification sets in when the categories

---

5 These reflections are inspired by a paper written by Gerd Spittler on the famine in Sahel (Spittler 1989). Meanwhile such ideal systems have become part of the equipment of disaster sociology. For other terminology see: Geenen 2003; Krüger and Macamo 2003: in the livelihood approach the crisis is already a situation that requires relief assistance from outside.

6 Sociology of risk points out that anticipated dangers can be estimated with regard to their effects and accordingly precautions can be taken. Thus, the effects of the initially unspecific dangers can be minimized by appropriate precautions and the remaining/residual threats can be accepted by assessing the likelihood of their occurrence and the feared consequences. The dangers are thus reduced and reflected and transferred to a "risk" that is knowingly taken (cf. Luhmann 1991; Japp 2000). This important analytical approach, however, is not of tremendous significance here. The suggested graded conceptual system refers mainly to empirical behavior. Moreover, our categories apply regardless of whether a danger was recognized and then transferred to a risk.

and expectations of normality themselves are undermined, i.e. when not only emergency measures fail to take effect, but also when both social life in its part-nered nature and the ability to evaluate events can no longer be taken for granted. That is when the stage is reached which we could legitimately describe as the collapse of the social order. (This might be called a **"catastrophe"**). On an abstract level there is a further stage, namely the complete destruction of hu-man life. If this case ever occurred, however, sociological analysis would be im-possible and obsolete.

3. Three floods

In the following we are going to describe three cases of floods in Mozambique, Germany, and the USA in order to analyze the relation between local perception of extreme events and "ordinary emergency management expectation". At the core of this description is the empirical research conducted over an accumulated number of months amounting to nine between 2001 and 2006 in the Limpopo Valley in Southern Mozambique (mainly by Elísio Macamo[7]). The examples in the Odra region and Tennessee represent smaller comparative studies based on in depth interviews conducted by Dag Schumann in 2001 and 2005.[8]

3.1 The Limpopo Valley in southern Mozambique: the 2000 floods
In 2000 Mozambique fell victim to a devastating flood. According to official reports around 700 people died. Moreover 544,000 lost their homes and 700,000 became dependent on emergency services. According to the UN, the floods de-stroyed 10% of the cultivated patches of land and nearly 90% of the irrigation infrastructure. More than 600 primary schools were either destroyed or heavily damaged. The World Bank estimated the immediate economic damage to be 273 million USD and indirect damages to a further 247 million USD. Added to-gether the damage exceeded Mozambique's export earnings from the previous year (Christie and Hanlon 2001; Africa Recovery 2000).

Extreme events are no rarity in this region of the world. During the last cen-tury three out of eight documented floods were part of the local collective memory, as well as three (out of four) droughts, two out four wars. None of the documented heavy storms is part of the local memory. Due to the precarious economic situation in Mozambique, especially in the rural regions, everyday life

---

7   For an extended version of this study see Macamo 2003.

8   The project was developed and conducted jointly by the authors who not only devised the main conceptual lines of the project (including the leading interview questions) but also jointly undertook the comparative analysis. It was part of the special collaborative re-search project "Local action in Africa in the context of global influences" run by the Uni-versity of Bayreuth. We would like to thank the German Research Foundation for the funding of this project.

is a continuous struggle for survival. This results in a kind of permanent precariousness. As a norm, people fall back on family, personal, and group-specific (church, region of origin) relationships for assistance. Government assistance or benefits from other institutions do not feature in the establishment of security.

The flood in 2000 took place in the context of this permanent situation of precariousness. On the local horizon the occurrence of a flood goes beyond the actual time of flooding. It comprises the time following it, i.e. when the water has already receded, and includes both the period of sowing and harvesting. The course of such an event is accompanied by tried-and-tested measures applied by the farmers. For this reason, an exact depiction of the flood requires a tri-partite periodization: (1) prior to the flood, (2) during the flood, and (3) after the flood. Seasonally, floods in the Limpopo Valley occur after the often fierce rains from January through March. The time before that is characterized by the farmers trying to estimate whether the high water is going to exceed normal dimensions. For this purpose they measure the water level of the river by poking canes into the ground. They examine the color of the water based on the belief that certain tones indicate the ferocity of the flood. Furthermore, the farmers look out for the increased presence of a certain insect and the early flowering of a local fruit variety. Informants unanimously agreed that as far as the 2000 Floods were concerned the signs did not point to particularly high water. Hence, no precautions were taken to retreat to higher reaches. Surprisingly for the local population the authorities cautioned against the imminent deluge of water and urged the residents of the valley to go find shelter over loudspeakers. Because this warning contradicted the local perception of the flood no-one followed their instruction.

The mass of water came in a deluge and hardly gave the farmers any time to employ the established strategies. One reason for this was that the flood, as a consequence of intense rain in South Africa, threatened the stability of the South African (reservoir) dams. As an emergency measure the dams were opened (at a late stage) which caused a flood wave. The pillars that the huts were built upon were too low for the uncommonly high water, and even where they escaped the floodwaters, farmers had to share their refuge with undesired company: the dreaded snakes. Others quickly climbed up trees and anxiously awaited rescue through helicopters and boats. Then the intervention of the governmental emergency service and international emergency aid apparatus brought relief.

In contrast to the farmers the latter relied on the interpretation of meteorological data, the dam capacity, the height and the stability of the levees, and the intensity of the predicted flood waves. Although several dozens of people died, most could be rescued and taken out of harm's way. One village located securely away from the flooded territory was quickly enhanced by swiftly erected makeshift housing that the flood victims could retreat to.

According to local experience flood water does not stay longer than one week. In the case of the 2000 Floods, however, the water did not recede until four weeks had passed. Along with it went the emergency services and helpers, who assumed that their presence only made sense as long as the area was flooded. Usually, after the flood the soil is loosened, there are fewer vermin, and the farmers can look forward to a good harvest. This time, however, the water remained until the beginning of the dry period and the soil became hard very rapidly. It was impossible for the farmers to work according to their conventional techniques. They would have needed tractors, which were not available. The consequence was a crop shortfall, starvation, and considerable problems in social life. It was only against this background that the flood, which deviated so drastically from the normal experience, was retrospectively classified as a disaster.

3.2 The Odra flood 1997 in Germany

The Odra flood has, aside other floods, remained in the collective memory of Germany. This can be attributed on the one hand to the severity of the phenomenon, but on the other hand to the intense media coverage of this event. Heavy precipitation in the drainage area of the Odra in July 1997 led to water levels rising by four meter, more than twice the normal level. As a result of breaches in the levees and high water across the border in Poland and the Czech Republic over 100 lives were claimed. In the border region of the Odra the raised water levels lingered over three weeks. As a measure of precaution on the German side 10,000 people were evacuated. Besides fire brigades and the *Technisches Hilfswerk* (German disaster management agency, mobilizing especially trained volunteers) altogether 30,000 soldiers of the German armed forces, out of these 10,000 simultaneously, were employed during the massive disaster operation. In the end, 600 houses were damaged or destroyed in the lowland of *Ziltendorf* and there were 400 additional reported losses. The provincial government estimated the property damage at 650 million DM (360 million USD, at that time). There were no fatalities in Germany (Ministerium des Innern des Landes Brandenburg 1997)

The periodic winter and summer floods in the Odra region are no exceptions. Parts of the region's altitude is below the ordinary high water level and relies on the protection of levees and faces high risk when extreme floods occur. From the multitude of these likely events several individual memories stand out. Thus, the devastating winter flood of 1947 was often mentioned as a reference point during the events 50 years on. Other historic extreme situations that are remembered locally include the fierce battles preceding the Berlin attack of the Red Army in 1945 at the end of the II World War and, relevant for a large faction of the population in the Odra region, the flight experience from Silesia. Accompanying the structural transformation of the area since reunification in 1989 is the cutback of jobs which makes this area to one of the poorer parts of Germany and represents a great source of uncertainty to the population and is hence described as a present-day threat.

The news from the neighboring countries as well as the official notifications regarding the seriousness of the situation were initially met with a certain cool-headedness by the border residents on the German side. While the water was rising and higher water levels were reported upstream and people expected higher water levels over the following days and more water in their basements than usual, the critical situation was not appreciated. Still, at home and at work ordinary precautionary measures were taken. The assessment of the situation and the decision-making were based upon local criteria informed by local experience with extreme events of this nature.

This notion of autonomy of action was maintained even when the intensification of the situation revealed that the potential for one's own action was exhausted. In these cases people relied on their knowledge of an institution that could be summoned to assist them in an emergency. The specially established crisis committee was responsible for the coordination of those involved in disaster prevention. On the basis of the newly identified water levels and the potential for danger it continually updated the warnings to the population. Once alert level IV applied in a particular region, this implied the redistribution of responsibilities. For instance, a certain territory was fenced off by the police, and access was allowed only to those holding a special permit. High-risk communities were completely evacuated.

Simultaneously the local population was involved more fully in the management of the disaster. The principals of the towns and villages relayed information and orders coming from the crisis committee to the population. Central places were equipped with stations for filling sandbags and everybody lent a helping hand. Levee wardens worked overtime. Residents provided soldiers stationed nearby with homemade cakes.

In the course of the intensification of the situation further institutional resources were mobilized. The fire brigades and local committees of the *Technisches Hilfswerk* requested staff and technical devices from the neighboring counties as well as the national level. The initially deployed 200 engineers of the armed forces were gradually increased by the Minister of Defense up to a number of 9,500 soldiers working simultaneously. Thousands of air hours were flown by helicopters of the armed forces.

After two breaches of the levee the lowland of *Ziltendorf* filled up with water. At this point it became apparent that even with the measures at hand not all dangers could be controlled. There was a further dramatic development near the small settlement of *Hohenwutzen* which made another levee breach very likely. Flooding of the more densely populated *Oderbruch* could be prevented once the damaged areas had been filled with pieces of concrete by helicopters, divers had installed tarpaulins at the danger spots, and special drainage systems had been employed to drain the levee basin. Finally, after four weeks the water level decreased at the beginning of August and the situation returned to normal. There were no fatalities in Germany. Extensive revenue from donations facilitated a relatively rapid clearing of the damage to public infrastructure and private property. For the people affected by the flood the event only came to a conclusion when around Christmas they could relocate from their provisional accommodation to their renovated or entirely new homes and when there was a more or less satisfying prospect of compensation and insurance payments.

3.3 The Tennessee Valley in the USA: floods 2003 and 2004

In spring 2003 and twice again in the following year, heavy rainfall in East Tennessee led to the flash floods of various tributaries of the Tennessee River. In the two cases we studied some parts of the towns were completely flooded. Residents were forced to leave their homes. Several inhabited mobile homes were destroyed or carried away. Cars were swept away. The foundations of access roads were washed away and bridges were destroyed.

For the residents in East Tennessee extreme natural phenomena are not unusual. Storms (especially tornados, winter storms, and hurricane troughs) feature prominently. While seasons can be limited and defined to a certain time of year, storms are perceived as a special threat, because the possibility of forecast, as well as prevention, is considered limited. Another threat, though not from natural phenomena, is presented by two nuclear plants operating in the region. They are the subject of ambivalent attitudes indeed. On the one hand they are an important stimulus for economic growth in the region. The power plant and its owner, the "Tennessee Valley Authority" (TVA), offer lucrative employment for both technical and administrative staff. The danger, on the other hand, is for the most part considered to be one that is calculable. The danger of terrorism, which national authorities increasingly warn against, remains rather abstract in the local context. Flood experiences, however, have been very concrete in the region for a long time. Until the beginning of the 20th century the Tennessee was considered one of the most unpredictable and most dangerous rivers of the USA. At regular intervals heavy floods resulted in severe damage and a great many fatalities. Today along the Tennessee there is a dense network of barrages which is controlled in a centralized and computer-assisted fashion. The river is now deemed tamed. It is a different matter, however, with regard to the outlets from the Cumberland Plateau, which rush into the Tennessee in a thoroughly unregulated manner. The geological make up of the mountain causes rainwater to flow away rapidly.

In the course of the heavy rains in spring and summer 2003, as well as after hurricane Ivan in autumn 2004, the rather small mountain streams suddenly and for only a short period of time turned into raging rivers. This made them a serious source of danger to the residential areas in the valley. The extremely heavy rainfall concentrates on small areas in mountains quite a distance away from the settlements. In severe cases the rain causes a flash flood in that particular water catchment, if there are settlements a distance downstream away they will be hit all of a sudden. Whereas the risk of heavy rains and flash floods can be forecasted it is impossible to predict where they will occur. The sudden nature of the event made specific warnings impossible. General warnings concerning the risk of flash floods are usually transmitted via the TV-weather channel. People are on alert, without being able to know whether they will be affected.

In the cases researched people only realized the flash flood when the torrent became audible. Those who live near the stream had only minutes to spare, all that was left to these residents was to move for their own safety. Other parts of settlements in more distance to the stream were flooded in the course of a few hours only. Many of the predominantly lightly constructed buildings were torn from their foundations, damaged, or completely destroyed. Cars and furniture were washed away. For immediate emergency service the local rescue squads, composed of voluntary workers with special training, were available. They alerted those who were asleep, evacuated families in danger, or put up barriers. More costly equipment and technology, such as mobile dam systems, had to be supplied by the county emergency service. Directly after the flood, representatives of relief organizations from churches and the private sector arrived on the scene, gave emergency aid, provided clothing, meals, and accommodation. After a few days traces of what had happened were still visible, but the water had receded.

The state authorities, local authorities, and individuals all saw their potential for action limited where the establishment of normality and the prevention of disaster were concerned. In the cited example an action committee was quickly organized consisting of representatives of various religious denominations. This committee was able to organize efficiently the immediate aid and the distribution of donations. For the clean-up constructions companies based in the area provided heavy machinery without charge. Long-term support in the reestablishment of the status quo prior to the events could only be provided in selected cases. Theoretically available government assistance proved unavailable, either because the affected people were unsuccessful in their dealing with bureaucratic procedures or because they could not meet the formal requirements. Especially those who settled on cheap plots marked as potential flood area due to financial constraints could not claim official compensation.

As for the majority of people, individual protection is not affordable. On all sides the flood problem is generally considered manageable by appropriate hydraulic engineering. However, it seems impossible to agree on the measures. A majority prefers a stabilization of the river banks with concrete and other construction measures to secure easy drainage of flash floods, which contradicts environmental regulations. A smaller group, which includes some experts, opts for a renaturation of the streams to prevent the emergence of extreme flash floods from the beginning. In addition it is disputed who will have to pay for the construction works, the plot owners, the county or the state. From a governmental perspective there is a lack of urgency and cooperation at the local level. Thus, the outcome of the situation remains open for many people even long after the events.

4. Comparing emergency management strategies and assessments

Starting from our definition of disaster the examples introduced can be subjected to a systematic comparison. We will take a look at the components in a double contextualization, as in the following:

1. the particular emergency management strategies
   a. including the type of aid and the strategies of management,
   b. the specific organization and its complexity as well as
2. the "ordinarily expected management" which determines the expected results of (successful) relief assistance.

This comparison is made in a rough manner, as were the brief case studies as well, in order to highlight the essential findings. The main aim is to reveal clear lines of similarities and differences.

4.1 The Limpopo Valley in southern Mozambique: flood 2000
Emergency management is essentially based around individual prevention and escape strategies. The danger is self-assessed by the individuals and the local group based upon locally collected information. If water reaches high levels, people flee to higher terrain or try to save their own lives (and the lives of other members of the community), and their possessions by retreating to buildings on higher ground and, if necessary, climbing onto rooftops. Special technical measures of flood control are beyond the local capabilities and do not, therefore, figure prominently in the local catalogue of demands placed on whomsoever has the responsibility of ensuring individuals' safety.

Emergency management and relief organizations offer only complementary support. They are not linked up with the local level and, along with the involved authorities, are perceived as external agents whose efforts are generally welcome, but which do not count as ordinarily expected flood mitigation or relief. While the authorities' responsibilities are clearly defined at the regional level and incorporated into the state apparatus, international organizations, including nongovernmental ones, became involved in the case of the 2000 flood and decided independently about the type and content of their contribution. Especially the extent and nature of the aid showed that all of this exceeded the ordinarily expected emergency management during the flooding. From a local point of view help was welcome, if somewhat random, and offered new, more spontaneous options for action. To give an example, after their rescue, some affected people returned to their homes and were "rescued" a second time, each time carrying some of their possessions.

To a large degree external relief is uncertain and its arrival unpredictable. It is thus outside of the ordinarily expected assistance and does not have an impact on the commonly assumed management which continues to be primarily determined by local prevention and mastery.[9]

The outcome of the ordinarily expected management is primarily the prevention of high numbers of fatalities and the preservation of the means of minimum subsistence which was at stake in this case since the sowing period was lost due to the late recession of the water. A small number of deaths is of course very unfortunate on the local level, but is not untypical for crises. In addition to that, the flooding of villages is not per se a crisis. Even the destruction of property is counted as an expected flood consequence and the damages and ramifications are not necessarily evidence of a disaster event.

These remarks call for the specific assessment of this flood. It can be considered a crisis. The expected aid measures were partially made ineffective by the unusual sudden nature of the flood wave, and due to the high level of the water many residents had to find shelter on top of their roofs. The extraordinarily high water threatened the emergency management itself, such as the retreat to higher ground with a maximum of possessions and supplies. Initially the situation was stabilized by the unexpected outside aid. It remained a crisis, because neither the number of fatalities nor the material loss made the event a local disaster. It is uncertain whether a disaster was prevented because of the external support or whether it would have remained a crisis regardless.

The late recession of the water in the dry period as well as the loss of the sowing period, however, introduced an entirely new context. A famine loomed ahead. The usual labor migration did not manage to cover the widening gap in supply. The accumulation of essential supplies and the option to eat part of the seeds in case of an emergency were made impossible due to the stock-loss associated with the flood. It is in this sense – flood and missing harvest – that the local emergency management broke down and local livelihood was endangered. From the local perspective this proved to be the real and extraordinary burden, so that the situation as a whole is to be assessed as a disaster. Because the local population was not able to balance out the ramifications of the unanticipated loss

---

9   Nevertheless the nature of the unexpected relief assistance pointed out the potential for assistance. This might have an effect on the definition of ordinarily expected management of future cases.

of the sewing period and the missing yield, and because they were reliant on external relief, the new context was experienced locally as a disaster.

4.2 The Odra flood 1997 in Germany
Aside from the wide potential for individual self-help there are definite expectations focused on institutionalized governmental and legally regulated measures regarding all aspects of security (including social security). Flood control is predominantly realized by technical prevention using hydraulic engineering, especially levee construction and water drawing machines for the drainage of marshland as well as early warning systems. Individual prevention, here, is only a complementary measure (insurance, boarding up of buildings, emptying lower stories or entire buildings). Local information such as the observations of levee wardens is fed into the information pool of institutions in addition to regional, national, and international structures of risk assessment. In this way the local level provides the groundwork for the organization of emergency management.

Technical semi-voluntary structures like the *Technisches Hilfswerk* and local voluntary fire brigades are integrated into communication and command chain of the professional emergency management agencies. Aside from these two hybrid organizations nongovernmental organizations, which play an important role in the social domain, do not feature in flood control.

In keeping with the firmly institutionalized system of security, implementation is based on a complex system of densely networked organizations whose interaction is determined in clearly structured emergency plans. Based upon a schedule of responsibilities these plans also define the interventional obligations of superior levels.

At the center of the local standard of ordinarily expected management is the protection of individuals from death and damage to health. Added to that is the safeguard of material possessions, and in the case of a flood, the protection from flooding and associated damages. The long-term expectation takes into consideration the security of the basics for living and shelter from economic repercussions (preservation of one's employment). Finally, in case of danger or crisis, it is expected that preventive measures are taken in order to avoid a repetition of the event. All these things describe the comprehensive demands upon the management of extreme occurrences. Already a small number of deaths or severe health impairment takes the situation out of the ordinary. Moreover, extensive material damage (without fatalities) can be regarded as an indicator for the absence of ordinarily expected management.

This particular case can be called at least a crisis. Flooding a specially selected basin was a normal measure taken to avoid an emergency. However, water levels were extremely high and the standard structures of crisis management were no longer sufficient to handle flood control. The involvement of the national army in the fight against the flood extended measures foreseen in emergency plans. The co-ordination of local, regional, national actors as well as volunteers who offered help was realized no longer simply by the crisis management group alone but included ad-hoc communication between the agencies, local authorities and other actors active on the spot. Formal rules of communication and decision making were sometimes ignored to adapt to the special situation. The partial breach in the levee and the risk of further breaches fostered fears of a disaster. At the same time the structures of emergency management remained capable of action and, aside from the restrictions mentioned, worked effectively. The disaster for the whole region was averted; the situation during the flood remained in general that of a severe crisis.

However, the people in the lowland of *Ziltendorf*, whose houses, workshops and farms were flooded, were confronted with great losses of material wealth. Their situation looked as if it would become a disaster. The extensive media coverage led to extraordinary effort to support the region and the victims. Apart from insurance payments, government funds and a large sum of private donations were available. This facilitated a rapid cleanup of the damage and a nearly full compensation of all losses in between a few months. Aside from some minor quarrels about the distribution of funds the result was a swift overcoming of the crisis and a return to the high standard of normality usually expected. In fact, in this case the ordinary management expectation was even excelled in some cases.

4.3 The Tennessee Valley in the USA: floods 2003 and 2004
Similar to the situation along the Odra emergency management comprises a broad variety of technical measures and institutional emergency services. However, these are not as strictly interlocked in implementation as in the Odra case. Rather, prevention at a private and community level takes priority. This includes private measures, voluntary initiatives in the community, regional institutions, and national nongovernmental organizations. Although the nongovernmental initiatives are voluntary, they feature among the ordinarily expected relief in emergencies. In addition to that the national level makes available financial and technical resources as well as a complex institutional apparatus in case of need. Although their responsibilities and interaction are based upon a legal foundation, their actual involvement appears for the people in need as a negotiation with uncertain outcomes.[10] Thus, what we have here is an organized basic prevention and management complemented by NGOs and individual initiative.

Similarly, the local expectations towards ordinary management are extensive, as in Germany. At the center is the protection of individuals from death and impairment of health. Added to that is the safeguarding of material possessions,

---

10 The organizational shortcomings of this system became very apparent in 2005 after hurricane Katrina had caused devastating flooding. (Committee on homeland security 2006).

and in the case of a flood, the protection from flooding and associated damages. The long-term expectation considers the security of the basics for living, while indeed a certain degree of individual initiative in reconstruction is regarded as justified. But important is the prevention of a repetition of a crisis or disaster by adjusted preventive measures.

The flood, which was less drastic in contrast to those in the Limpopo Valley and Odra, was not perceived as a disaster. The people including the victims referred to damages by the hurricane Ivan in Florida as disaster. Whereas, the poorer people living in the trailer parks situated nearby the streams, were hit hard by the material damage the flash floods were initially seen as a reasonably well-managed crisis. The emergency service worked quite well, so that no fatalities had to be mourned. With respect to the medium and long-term management there is a noticeable difference from the Odra which becomes apparent in the relationship between the emergency management and the expectations regarding the restoration of normality. The flood victims are dissatisfied with the material aid and feel neglected by the authorities. They would not expect a complete compensation (except for insurance payments to those insured), but they bemoaned at least the lack of material help in resettlement or reconstruction. The small loan scheme offered by FEMA (Federal Emergency Management Agency) is not considered as helpful because it is perceived to be too bureaucratic and poorer people fear not to be able to repay the loan.

While people see a potential for prevention, such as through deepening the riverbed or relocations, the involved institutions handicap one another through their conflicting interests (e.g. ecological and financial). This makes for a status quo in which even in the course of 2005 (the time of the survey) the precarious situation still prevailed according to the local population. It is yet undecided whether the crisis will lead to the reestablishment of normality or to a disaster.

4.4 Relativity of disasters and crises and crises explanations

In our comparison of floods with varying effects, the application of a contextualized conceptual system of threat, crisis, and disaster leads in each case to a similar classification. To put it differently, during flooding itself, only in the Limpopo Valley do we find extensive flooding of buildings, wide-ranging damage and destruction and even fatalities. An event whose consequences from a German or US American vantage point doubtlessly would have been a disaster is seen in the local assessment only as a crisis. The floods on the Odra or the Tennessee with their comparatively minor consequences would not have even been classified as a crisis in the Limpopo Valley.

What stands out most, in our opinion, are the diverging expectations of normality. There appear to be two basic types of expectations of normality. There is on the one hand a type that places emphasis on maximum security and is highly intolerant of any material and human losses. This type, which is more pronounced in Germany and, up to a point, in the Tennessee Valley case studies, is also more likely to have a lower threshold concerning the evaluation of a disaster. The other type, on the other hand, which is more pronounced in the Limpopo Valley case study, but also up to a point also in the Tennessee Valley case study, is less intolerant of material and human losses, which it is more likely to see as part of the normal consequences of disaster events. This type is more likely to apply a higher threshold to the evaluation of such events.

Together with these different expectations of normality we find different expectations of disaster prevention and mitigation as well as completely different ways of prevention. Interestingly, the particular explanations of the crisis or disaster correspond with these differences. We distinguish two general types of explanation.[11] First of all, there are the technical-scientific "how-explanations" that focus on the exact origin of the extreme event and construct a detailed chain of functional cause-effect relations, explained in scientific-technical and administrative or logistic terms, leading to the final impact of the disaster. These explanations concern "how" the flood came about, what its effects were, and how well the management measures took effect. Secondly, there are "why-explanations". They ask "why" it rained so much in this particular place, why was this or that village or this or that family so badly affected. These explanations refer to the metaphysical causes of this misfortune and construct a general relation between certain behavior of or evil in the community struck by the disaster at one side and a metaphysical power on the other side eliciting the disaster without a detailed explanation how the metaphysical power acts. The focus is on the question what triggered the action of the metaphysical power leading to the disaster or why the metaphysical power acted.

In each of the three regions we can find respective "how-explanations". This is also valid for the Limpopo Valley in which the correlations between intense rainfalls in South Africa, the far too late opening of the South African dams, and the force that caused the flood are part of the common knowledge of the people in the region. In addition to these "how-explanations" there are important "why-explanations" in the Limpopo Valley. These refer to disunity in the village, insufficient sacrifices to ancestors, and witchcraft. "How-explanations" appear to have been of no significant consequence to everyday life. No attempts were made, nor even considered to influence the water management of the South African dams. The "How-explanations" seemed to be useless for preventive action. "Why-explanations", by contrast, were the subject of intense discussion. People made a direct link between the elimination of evil to the prevention of further threats. The only question was how to eliminate the evil in order to avoid another devastating flood.

---

11 For a more comprehensive presentation of these types of explanation see Macamo and Neubert 2004.

The Odra region represents the opposite extreme, as it were. Even after intensive questioning among, for instance, members of the church, or a pastor no "Why-explanations" were uncovered. The respondents often stated clearly that there is definitely no metaphysical explanation at all. And respondents often were surprised about this kind of question. All explanation presented were "How-explanations" concerning the water management often including very detailed technical arguments. Nearly all these "How-explanations" were directly related to the technical-administrative complex of disaster management. All respondents were convinced that flood control is possible once the right measures are implemented.[12]

We selected the Tennessee area because of its location in the so-called "bible belt" of the United States, a region with significant fundamentalist Christian groups with an important influence on the life of the communities and on public debates. Indeed, the question about metaphysical "Why-explanations" was usually taken up. One third of our respondents reacted positively and referred to explanations like a "sinful life" or an ordeal imposed by god. Even for those who rejected this kind of explanation it was a familiar concept.

The issues concerning management and prevention, however, were exclusively, as along the Odra, explained by the technical-administrative complex of disaster control. While the church services of intercession around the time of the flood were taken very seriously, they were not regarded as a suitable measure by which to manage or prevent disasters.

Our claim with regard to the relative weight of types of explanations is that their presence corresponds with the organization and capacity of the emergency and prevention measures. In places where management and prevention are undertaken by an agency deemed competent to master natural forces – and statutorily charged with the task – the explanations relevant for action are based on scientific-technical "how-explanations". "Why-explanations", however, tend to coincide with situations of very limited technical means for management and prevention. Material and physical impairment seem to be accepted as typical of crises. Our claim should be understood as a caveat against an undifferentiated definition of the notion of disaster. Indeed, attention to the degrees of ferocity eliciting a differentiated local appreciation is central to understanding how individuals and communities account for and address severe events.

5. Contextualizing vulnerability and resilience

The context-related variation of the extent and capabilities of "ordinary", expected management does not remain without implications for the conceptual dichotomy vulnerability and resilience often used with regard to the effects of

---

12 However the priorities differed. Interestingly, none of our respondents proposed to resettle people into less flood-prone areas (see also Felgentreff 2003).

threats, crises, and disasters. In development research and, particularly, in development policy the assumption is often made to the effect that people from developing countries and especially the poor or poorest are the most vulnerable. Once hit by the consequences of an event, they struggle particularly hard to reestablish a state of normality. The assumption is that they have little resilience at their disposal. According to this idea, events which in industrial societies present at most a threat will in the developing world quickly become crises and climax in a disaster.

What at first sight seems plausible neglects the connection between the contextualization of the concept of disaster and poverty. Indeed, there is no lack of examples where events with a disastrous outcome in the developing world had a milder effect in the industrial countries. Landslides in squatter areas built in the mountainous town outskirts of a developing country, for instance, repeatedly lead to disasters with a high number of casualties because the poor are forced to build homes in areas not suitable for habitation. Still, however, we deem a contextualization of the concept of vulnerability necessary, especially when the determination of crises and disasters is also contextualized.

As the three floods compared here clearly show, the severe nature and the human and material toll of a flood do not immediately yield an extreme evaluation of the hazardous event itself. In the Limpopo Valley the flooding of dwellings is to begin with a danger met by flight to rooftops or to elevated terrain. Only when the water rises so high that the roofs are covered does it become a crisis. A state of normality is established when people can return to their homes after the water recedes. Similar events at Odra or in the Tennessee Valley elicit much more dramatic evaluations.

The limited material damage and even the lamentable fatalities that accompany a flood in the Limpopo Valley do not exceed what would ordinarily be expected in such a situation. It is against this background of ordinary expectations that flight to rooftops, rescue of enough individual property and avoidance of too many deaths can be seen as successful disaster management and, hence, crisis mastery. The idea of failure intervened rather in connection with the extraordinarily long duration of the flood and the resulting hindrance of the annual sowing. The ensuing famine which affected the entire local population provided the warrant for a retroactive evaluation of the flood as a disaster[13].

On the individual level the consequences of floods can be (and indeed were) highly problematic and devastating. We appreciate that, for instance, the loss of a single cow is capable of severely worsening the livelihood of a family, so that in the foreseeable future the reestablishment of the previous life situation will be impossible. Thus, those families have little resilience. Two things, however, need to be pointed out. Firstly, in order for a flood to become a disaster, it needs

---

13 See Macamo 2003 for a more detailed analysis of the local evaluation of the floods.

to present a problem for the whole of the community or local population. This is simply because ordinary expectations concerning the management of a hazardous event refer to a concrete local historical experience which provides individuals with a common normative vocabulary to make discursive sense of events. Hazardous events challenge primarily the cogency of such normative frameworks. Secondly, while profound individual setbacks such as the loss of a cow are real enough and offer individuals a perspective from which to evaluate the force of hazardous events, they are also part of a larger local descriptive framework. This framework lists such individual losses as inevitable consequences of extreme events. Furthermore, given the primacy of "why-explanations" individual losses are not necessarily seen in connection with individual failure, but rather with circumstances beyond individual control that can only be adequately tackled from a communal perspective.

The analytical specification of disaster that we present suggests that the perception of an experience as a disaster is closely related to both the expectation of security and the methods of management developed for this purpose. These have a dialectical relation to one another because the expectation of security is fed by the confidence in emergency management and disaster prevention as much as the latter depends upon the expectation of security. Resilience and vulnerability gain their descriptive value only from this dialectical relation.[14] This argument is reminiscent of a thesis about the poor made by Georg Simmel: poor is he who is the object of poverty assistance (Simmel 1999). To put it simply, the vulnerability and resilience of individuals and groups need to be determined by looking at their expectation of security and their methods of management.

The judgment of whether the Odra or Tennessee flood was a crisis or a disaster is not made by looking at the population's fall into a survival crisis or otherwise. Rather it is asked whether the preservation of life and of the physical well-being of the entire population is guaranteed and whether possessions can be safeguarded from flood damage. Even a temporary flooding of buildings ruins, at least, part of the extensive (compared to Mozambique) material wealth. Single fatalities combined with noteworthy material destruction, indeed, are enough to classify the events as potential or manifest disasters[15]. After the recession of the water major clean-up and restoration work is needed to reestablish normality. Resilience is measured by the speed in which this is done and how fast prevention measures against a repetition are taken. In this case this could imply considerable efforts. In other cases external aid which is not part of the common

---

14 In a somewhat similar way Aguirre 2007 shows the dialectics between vulnerability and resilience. He argues that higher risk-taking needs an adaption of disaster mitigation to keep resilience working. Our additional point is that societies do not only take higher risks because they trust their ability to protect, but because their expectation towards security rises at the same time, too.

15 A single fatality alone would be connoted as accident.

measures of management can be arranged, such as an emergency fund and the private donations for the Odra flood victims. In this respect resilience can be limited even in materially comparatively affluent societies. In Tennessee, for instance, this is indicated during the survey period by the obvious uncertainty among the population as to whether or not the reconstruction assistance and prevention measures will be forthcoming.

From this vantage point, then, materially better off societies can be said to be particularly vulnerable to natural phenomena and, for this reason, they pursue highly elaborate disaster prevention efforts. The potentially greater threat by natural events goes along with massively increased prevention and management efforts. Ultimately, the actual vulnerability is determined empirically according to the balance of increased threat on one side and improved prevention and security on the other. In addition improved methods of prevention offer to revise risk calculations. Effective flood control offers new opportunities for settlements in areas which used to be unsuitable for permanent inhabitation, like in the Odra case. The ability to control nuclear fusion led to nuclear power plants. Highly industrialized societies take high risks based on the trust[16] in their ability to master nature due to their technological development (Beck 1987).

Depending on the case this can lead either to an increase or to a decrease of vulnerability. This is also accompanied by expectations of normality with an enhanced claim to survival, freedom from harm, and preservation of material standards. While single fatalities in the course of extreme events are within the scope of normality in developing countries, the demands in industrial countries for protection from material and bodily harm are growing massively.

Thus the approach of livelihood analysis (Carney 1998) is to be preferred to the still widely used concept of vulnerability. The livelihood analysis once developed to improve poverty reduction in the considering ecological consequences too has the potential for a more detailed analysis of how people formulate their expectations against the background of specific technical, cultural, and social possibilities. Once this approach is consequently applied it relates the strategies of livelihood to the local and context-specific normality.

The argumentation introduced here may be suspected of being cynical. After all, the reduced vulnerability and enhanced resilience detected in a case study from the developing world is simply a result of low expectations of normality born out of hardship and poverty. In fact, what we are criticizing is the normative charge inherent in the ordinary use of the concept of vulnerability in development research where the expectations of normality are not defined in relation to the context, but according to a universal category of dignified human existence. The merit of this usage of the term vulnerability is not so much in its contribution to the sociological analysis of crisis management, as in its ethical and

---

16 For the conceptualization of trust see Luhmann 1988.

normative implications. In contrast to the livelihood approach this concept of vulnerability is suitable for the objectives of development policy. This does not, however, disavow this particular use of the term. It limits, in our view, its analytical usefulness. This, ultimately, calls for the clarification of what is concerned in each case: the most precise analysis of a country's dealing with extreme events or a development-political debate about developmental goals and standards of a dignified human life.

6. Bibliography

Aguirre, Benigno E. (2007): Dialectics of Vulnerability and Resilience. In: Georgetown Journal of Poverty Law & Policy, 14, 1, pp. 39-59.

Beck, Ulrich (1987): Risikogesellschaft. Frankfurt a.M.: Suhrkamp.

Carney, Diana (ed.) (1998): Sustainable Rural Livelihoods. What Contribution Can We Make? London: Russell Press, Ltd.

Christie, Frances and Joseph Hanlon (2001): Mozambique and the Great Flood of 2000. Bloomington: Indiana University Press.

Clausen, Lars and Wolf R. Dombrowsky, Wolf, R. (1990): Zur Akzeptanz staatlicher Informationspolitik bei technischen Grobunfällen und Katastrophen. Bonn: Bundesamt für Zivilschutz.

Committee on homeland security and governmental affaires United States senate (ed.) (2006): The hurricane Katrina. A nation still unprepared. Special report of the committee on homeland security and the governmental affaires United States senate. Washington: Superintendent of Documents.

Felgentreff, Carsten (2003): Post-Disaster Situations as "Windows of Opportunity? Post-flood perceptions and changes in the German Odra river region after the 1997 flood. In: Die Erde, 134, 2, pp. 163-180.

Geenen, Elke M. (2003): Kollektive Krisen. Katastrophe, Terror, Revolution – Gemeinsamkeiten und Unterschiede. In: Clausen, Lars, Elke M. Geenen and Elisio Macamo (eds.): Entsetzliche soziale Prozesse. Theorie und Empirie der Katastrophe. Münster: Lit Verlag, pp. 5-23.

Japp, Klaus Peter (2000): Risiko. Bielefeld: Transcript-Verlag.

Krüger, Fred and Elísio Macamo (2003): Existenzsicherung unter Risikobedingungen. Sozialwissenschaftliche Analyseansätze zum Umgang mit Krisen, Konflikten und Katastrophen. In: Geographica Helvetica, 58, 1, pp. 47-55.

Luhmann, Niklas (1988). Familiarity, Confidence, Trust. In: Gambetta, Diego (ed.): Trust. Making and Breaking Cooperative Relations. Oxford: Oxford University Press, pp. 94-107.

—————— (1991): Soziologie des Risikos. Berlin: de Gruyter.

Macamo, Elísio (2003): Nach der Katastrophe ist die Katastrophe. Die 2000er Überschwemmung in der dörflichen Wahrnehmung in Mosambik. In: Clausen, Lars, Elke M. Geenen and Elisio Macamo (eds.): Entsetzliche soziale Prozesse. Theorie und Empirie der Katastrophe. Münster: Lit Verlag, pp. 167-184.

Macamo, Elísio and Dieter Neubert (2004): Die Flut in Mosambik. Zur unterschiedlichen Deutung von Krisen und Katastrophen durch Bauern und Nothilfeapparat. In: Schareika, Nikolaus and Thomas Bierschenk (eds.): Lokales Wissen – Sozialwissenschaftliche Perspektiven. Münster: Lit, pp. 185-209.

Ministeriums des Innern des Landes Brandenburg (1997): Hochwasserkatastrophe Juli, August 1997 an der Oder. Erfahrungsbericht des Ministeriums des Innern des Landes Brandenburg, Referat Brand- und Katastrophenschutz. Potsdam: Ministerium des Innern des Landes Brandenburg, Referat Brand- und Katastrophenschutz.

Quarantelli, E. L. (ed.) (1998): What is a Disaster? Perspectives on the Question. London and New York: Routledge.

Schröter, Susanne (2009): Katastrophe als Prozess oder als Zäsur? Aceh zwei Jahre nach dem Tsunami. In: Meyer, Günter, Christian Steiner and Andreas Thimm (eds.): Katastrophen in der Dritten Welt - Soziale, wirtschaftliche und politische Folgen, Mainz: Interdisziplinärer Arbeitskreis Dritte Welt, pp. 85-105.

Simmel, Georg (1999): Soziologie. Untersuchungen über die Formen der Vergesellschaftung. Frankfurt am Main: Suhrkamp.

Spittler, Gerd (1989): Handeln in einer Hungerkrise. Tuareg in der Dürre 1984/85. Opladen: Westdeutscher Verlag.

Wisner, Ben, Piers M. Blaikie and Terry Cannon (2004): At Risk. Second Edition. Natural Hazards, Peoples' Vulnerability, and Disasters. London: Routledge.

Internet Source

Africa Recovery, 2000, Vol.14: Country in Focus: Mozambique. United Nations. http://www.un.org/ecosocdev/geninfo/afrec/vol14no3/mozamb1.htm (accessed 22.08.2010).

Squatters on a shrinking coast: environmental hazards, memory and social resilience in the Ganges Delta

Arne Harms

1. Introduction

"The place where our house was is rotting in the middle of the Ganges. When it was gone, we moved on to this embankment again", lamented elderly Binodini Das, as we sat on one of the countless embankments in the Ganges delta of coastal Bengal.[1] To be more precise, we were sitting on one of the fragile embankments built to protect the tiny islet of Ghoramara from the brackish waters of the vast delta.[2] Sitting there, a panoramic view of the muddy river engulfing the island and its ripe paddy fields unfolded. So low-lying is the delta landscape that the view is blocked only by trees, thatched roofs and the ubiquitous embankments, which form an integral part of these delta tracts. Houses of illegal squatters, lined up like pearls on a string on the various outer and inner embankments, catch the eye, as in the vicinity of all those settled areas or whole islands, which are frequently described as being in the mouth of the river, as having been abandoned by the government or, in scientific terms, as being subject to unchecked and threatening coastal erosion. While only in some places directly neighbouring each other, the houses of those who have lost their land to the eroding waves, powerful tidal currents or dangerous high tides and subsequently shifted to only marginally safer spaces on the embankments are always at a short walking distance from one another. Thus populating the embankments of densely populated islands, the squatters as a group might be described as be-

---

1   This paper is an outcome of my on-going research on environmental relations and natural hazards in the Indian part of the Ganges delta, made possible through funds granted by the German Academic Exchange Service (DAAD) and the Hans-Böckler-Foundation. Critical comments by Prof. Ute Luig and by commentators in Bremen's Research Centre for Sustainability Studies (artec), where I presented an early version of this paper, greatly helped to clarify this paper. While I acknowledge these contributions, all the limitations of this paper remain mine.
2   Throughout this paper I use the term Ganges delta as a shorthand for the various more or less accurate designations (e.g. Ganges-Meghna-Brahmaputra delta or Ganges-Brahmaputra delta) for the delta region that comprises the largest part of Bengal, that is, contemporary Bangladesh and the eastern Indian state of West Bengal.

ing enmeshed in two, quite distinct types of collectivity. On the one hand, as villagers they maintain relations with former and contemporary neighbouring villagers, as well as with other villages. On the other hand, as landless squatters they form a dispersed collectivity marked by distinct spatial arrangements and environmental relations. While the latter are clearly marked by intensified social vulnerabilities and chronic crises, the dispersed group of squatters nevertheless also manifests specific understandings, capacities and practices, which, taken together, engender social resilience.

In the present chapter, I shall try to analyse social resilience in highly vulnerable circumstances, not only as geographically or spatially situated, but also as related to culturally mediated interpretations of the present. Stepping beyond solely ecological approaches to resilience, which tend either to be silent regarding cultural constructions or to treat them as a function of ecological processes, I will emphasise the importance of processes of social remembering for a more nuanced understanding of social resilience. In respect to both a critical present that has to be dealt with and to expected futures, remembered events facilitate or seem to suggest what I understand as an ethic of endurance precisely in conditions of persistent environmental and social crises. After a brief overview of recent theoretical developments in these fields, I will therefore outline the distinct environmental, historical and social conditions that have culminated in a chronically critical present in large parts of the Indian Ganges delta.[3] Drawing on my own fieldwork on two of its most rapidly shrinking islands, Sagar and Ghoramara, I will then offer a contribution to the recent rethinking of resilience by substantiating the claims made so far.

2. Resilience, crises and memory

In recent years and indeed decades, the notion of resilience has attracted increasing interest. From initially being coined and reflected upon in parallel debates within developmental psychology and ecosystems, the notion has today found wide resonance in an array of disciplines across the sciences and humanities. Early conceptualizations of resilience – as, either, a personal capacity to psychologically develop under trying circumstances (Werner, Bierman and French 1971) or as an (eco-)systemic capacity to absorb external shocks and still

---

3   While an impressive body of scholarship on the (social) consequences of riverine or river bank erosion in Bangladesh emerged some two decades ago (see, for contradictory accounts, Schmuck-Widmann 2001; Zaman 1989) and may have stimulated the discussion of similar processes on the Indian side (see e.g. Lahiri-Dutt and Gopa 2007), coastal erosions in either Bangladesh or India have hardly attracted any social scientific attention so far.

persist (Holling 1973) – have given way to a plethora of definitions.[4] While debates about so-called natural disasters and anthropogenic climate change may be understood as the academic and political fields from which the term rose to its contemporary popularity, this route cemented a widespread eco-systemic understanding of resilience. But even though resilience's conceptual counterpart – vulnerability – has been considerably clarified and specified by, among others things, sustained efforts to embrace its social and historical constituents (e.g. Wisner et al. 2004; Bohle, Downing and Watts 1994), parallel developments pointing to the social dimensions of resilience still seem to be in their infancy (see Hastrup 2009). To contribute to these theoretical developments – that is, to a theoretical move beyond a narrow ecological or politico-economic analysis of coping strategies and continuity – I consider two approaches to social resilience in highly vulnerable life worlds to be particularly interesting. Drawing on my ethnographic material, first I stress the chronicity of specific environmental hazards, that is, the chronic crises they contribute to, which serves as a condition under, not against which social resilience may unfold. Secondly, I draw attention to the relevance of remembered pasts for such social resilience, not simply as a remembered coping strategy, but rather as a culturally mediated version of the past, which in itself helps to endure even chronically critical conditions. But since both crisis and memory are frequently addressed in relevant debates, albeit in a different manner, I will now try to clarify how they may help us think about social resilience.

While a large number of natural disasters, or rather, natural hazards that have turned into social disasters, may accurately be understood as punctual, delimited events that bring large-scale devastation and may well be contrasted with a peaceful normality, specific social conditions and/or types of environmental hazards clearly point beyond the underlying logic of the inconspicuous periods and intense interruptions that are more prevalent. Thus, in questioning the concept and experience of normality with respect to environmental hazards, process may again have to be distinguished from condition. Especially in studies of social vulnerability and disasters in the Global South, the processual character of disasters – that is, their dependence on structural conditions, and potentially their long periods of gestation and their ensuing influence on these very social conditions – have been powerfully demonstrated. Beyond these causal explanations, and precisely in addressing the experience of environmental hazards in marginalised contexts in the Global South, it may be argued that specific environmental hazards not only figure as processual interruptions of the normal, but blend into various detrimental processes, which, taken together, figure as a *nor-*

---

4   A comprehensive overview of the more influential recent definitions has yet to be published. But see Kelman (2008) for a critical engagement with some of the most important ecosystemic conceptualizations.

malised condition. This may hold true for a range of socially marginalized or vulnerable populations and their interpretations of and dealings with various environmental hazards. In shanty towns, slums or on embankments, environmental hazards hardly form isolated threats to well-being, livelihoods or survival, but effect groups or individuals only in combination with economic relations, political conditions and the like (see Auyero and Swistun 2009).

Nevertheless, slow and hazardous environmental changes, such as a rise in sea level in low-lying, but highly populated coastal zones, may serve as conditions for social practices, theoretically almost by themselves. That is, erosions of the land, the encroachment of sea-water, the saliniziation of drinking water and so on do not have the effect of an intensively devastating interruption of normality, but are better understood as a slowly destructive, but contingently present normality, or, in the words of the anthropologist Henrik Vigh (2009), as chronic crises. With respect to social resilience, this characterisation implies, on the one hand, a distinct set of environmental relations, individual or social capacities and situated practices with which any given society can act or navigate through these conditions. On the other hand, precisely because of their slow, predictable and rather uneventful impacts, these conditions hardly figure as prominently remembered, disastrous events. Rather, their persistent and looming presence must to be understood as interpreted and endured by and through, among others, remembered pasts. Arguing in this way for an understanding of resilience through the analysis of social memory, my approach differs from the linkages between remembering and resilience as established by current environmental theory. I shall therefore briefly turn to these now.

It has become something of a commonplace (e.g. Berliner 2005) to admit that the whole range of scientific inquiry is, in differing degrees, involved in the vast boom of studies into the conditions and workings of social memories. The idea that memories are constructed in complex interplay and are continuously being reshaped form a common basis illuminating diverse topics such as neurobiological formations, philosophically framed subjectivities, counter-narratives, cultural geographies and identity politics (Assmann 1988; Halbwachs 1992). It is therefore remarkable that, in most of the literature on social resilience and climate change, social memory, while occasionally quite broadly defined,[5] is actually treated either as a process of uncovering so-called local knowledge (Adger et al. 2005) or as a form of cultural capital constituted by past experiences (Berkes, Colding and Folk 2003). Leaving well-founded criticisms of associations of certain populations with certain places and certain types of knowl-

---

5   Drawing on the anthropologist Roderick McIntosh, Fikret Berkes and his colleagues define social memory as "the arena, in which captured experience with change and successful adaptations, embedded in a deeper level of values, is actualized through community debate and decision-making processes into appropriate strategies for dealing with ongoing change..." (Berkes, Colding and Folk 2003: 21).

edge aside for now,[6] this approach to social memory appears to be severely limited. Again, tensions, varieties, hierarchies and intertextual borrowings and entanglements in historical relations are lost when mnemonic practices are treated only as a retrieval of 'traditional' knowledge (in the singular) or as means of intergenerational learning.[7] In contrast, a variety of scholars, situated on the sometimes fuzzy boundary between environmental history and environmental anthropology, have decisively taken up social remembering as a tool with which to understand environmental relations in times of disaster and beyond. Analysing politically charged, orally or performatively established relations with a distinct version of the past, these authors mainly demonstrate the relevance of environmental events, processes or 'states of nature' for processes of identity formation (see e.g. Gold and Gujar 2002). Repeatedly referred to in everyday speech or commemorative rituals, as well as via patterned interpretations of artefacts or landscapes, environmental events or states have proved once again not to be beyond the cultural, but as intimately related to it. Conversely, it was demonstrated that remembered experiences or conditions are relevant for present environmental relations, especially the culturally mediated sense-making of these relations (Agrawal 2005). The fact that these and related analytical patterns are actually hardly clearly separable from one another only underlines the intimate relationship between past and present, and, by extension, the future.

Connecting the present to a specific past and to possible future processes of social memory shapes, among other things, environmental practices and distinct fragments of social resilience. On the embankments and shores of the Ganges delta, it does this, I argue, less in establishing a relationship with traditional knowledge and more in burdening the present with histories of marginalisation and loss.

3. The view from the embankment: materialities of an amphibious landscape

In the delta's archipelago, bound together in a dialectic relationship between reclamation and loss, embankments are central markers, if not an emblematic condensation of an ever-threatened, ever fragile life world. As such they are not only highly charged objects of negotiation and conflict (Mukhopadhyay 2009; Danda 2007), but are, in their very materiality, articulations of environmental relations and related subjectivities (Cosgrove and Petts 1990; D'Souza 2006). It

---

6   In the wake of postcolonial interventions, a number of anthropologists have criticised theoretical moves that treated marginal populations as being incarcerated in localities (e.g. Tsing 2005; Appadurai 1988).

7   That is, when it is reduced to a homogenising capacity. The same applies, as I have shown above, to an understanding of social resilience as capacity and may be avoided, I argue, by an understanding of social resilience as a situated practice.

is to this social relevance of embankments, which serve as a window into environmental relations, that I turn now.

Viewed from a coastal embankment, the gaze wanders far into a space at the dividing line of land and water, a space that is both landscape and waterscape. Standing here, it is of little wonder that dikes and coastlines are often thought of in analogy to front lines (Schmitt 1954; Carson 1999). Both are, in a sense, directed landscapes (Lewin 2006); they mark an end but, at least for land-based societies, only rarely a beginning.[8] In the Ganges delta, as in other low-lying landscapes, man-made embankments figure as a boundary between what are understood respectively as an inhabitable landmass and a hostile sea (Allemeyer 2006; Nienhuis 2008). Expelling the tides into an outer space and condensing broad transmission zones into one or a few parallel lines, embankments enable sustained and fundamental changes in landscapes that were, until their erection, amphibious. Embankments are, furthermore, not only important means of intervention but also striking memorials to them, since they have made the landscapes what they are: dry and slightly salty soils, farmland, village lands, urban spaces. But in the historical, technical and philosophical fabric of western modernity, two classes of hazardous event or process were largely forgotten, but could nonetheless be actualised under the conditions of anthropogenic climate changes and thus continue to prevail in eastern India, with which I am concerned here: first, a repeated failure of embankments; and secondly, chronic retreats of embankments.[9]

Sagar, the island that marks the geographical point of reference of my work, is located on the western fringes of the world's largest delta, that of the Ganges. Depending on one's point of view, here begins or ends a vast landscape of water, mud and salt that covers not only large parts of the Indian state of West

---

8 From the anthropological, historical and philosophical perspectives, this distinction is, of course, far from universal. In this vein, marine anthropologists have demonstrated the spatial orderings of and discursive elaborations (Hoeppe 2007) on seascapes by fishermen, historians have analysed historical entanglements across oceans (Gilroy 1993), and philosophers have reflected on the relevance of voyages and ships as spatial figurations (Foucault 1986). While, taken together, these various approaches prove that seashores do not simply figure as an end, but could better be understood as access points, the fact remains that many land-based societies imagine coasts as endings and borders (Helmreich 2011). While, therefore, rivers and journeys on them have a strong hold on the Bengali imagination – articulated, for example, in frequent use of this theme in Bengali Hindu mythological stories and modern Bengali prose alike (see Pokrant, Reeves and McGuire 1998) – the sea and sea-bound journeys figure much less prominently and, when they are alluded to, it is in a strikingly more dangerous terminology.

9 Where coastlines retreat in contemporary Europe, this rather seems to be a calculated step (either because it is economically viable or due to engineering practices that are considered environmentally friendly), rather than being a condition for a local society that sees itself as helpless.

Bengal, but also the greatest part of neighbouring Bangladesh. Brought down by some of Asia's mightiest rivers, thousands of tons of sediment are washed ashore annually, deposited or amassed to create muddy formations in an intricate interplay of river currents, wave activities and geological features. Some of these accumulations give rise to islands, some raise riverbeds for the time being, and some disappear into the muddy waters again after months, years or decades. In the active parts of the delta, however, even after the deposits may have grown into islands, these islands mark ambivalent spaces between land and water. Until they grow into being a part of the more elevated regions of the inactive delta, they lie beneath the high-tide levels and are therefore subject to both the incoming and outgoing tides. According to geomorphological configuration and to time of day and year, whole archipelagos are either flooded or rise out of the waters. These cyclical, rapid changes and ambiguities, then, have not only given rise to distinct ecosystems and forest types, they have also precipitated ambivalent perceptions and interpretations of these landscapes. As islands that were not such for hours at a time, they were ascribed amphibious traits. But as a territory where, because of its amphibious nature, only mangroves and other 'swampy creatures' could flourish, it was a wasteland, and in a sense both an aesthetic and a fiscal challenge to the British (Greenough 1998).

Up to the middle of the nineteenth century, the largest part of the active delta was sparsely populated. Only pirates, wandering groups of honey-collectors and woodcutters, and scattered peasants braved the uncertainties of the huge archipelago, its tides, tigers and cyclones. After the consolidation of British rule, the colonial rulers seriously turned their attention to the huge and swampy jungle immediately to the south of their then capital, Calcutta. Following the then double British strategy of relentless exploitation of specific areas and the simultaneous protection of other areas (Grove 1997), colonial bureaucrats marked thousands of hectares in the south-central parts of the delta as protected areas or nature reserves and promoted the complete settlement of all other parts (Sarkar 2010).

In the effort to populate and settle the unprotected areas of the delta, fiscal interests combined with political and aesthetic concerns. This dangerous, tiger-infested wasteland was, in the words of one British writer, to be transformed into "a seat of plenty" (Huggins 1824: 3). At the same time, this would end the wanderings of collectors and woodcutters and transform them to settled subjects of the Empire. Sedentariness marked simultaneously an important step in the transformation of the commons into parcels of private property. Drawing on and enforcing a distinction between forest dwellers and peasants (see Agrawal and Sivaramakrishnan 2000), the settled life-style of the peasants was favoured by colonial officers as the best way of combining the reclamation of the forests with an increase in fiscal revenues. This was not least based on the understanding that peasants would safeguard the private use rights of their taxable land and thus enforce individualised property relations against collective usages.[10]

Fundamental to these profound changes in ecological and social relations were embankments, that is, the diking of whole islands, or at least of cultivated plots, because only these measures could ensure what one might call the landness of the land. Only by delimiting embankments could the land itself be made stable and, as a consequence, stable settlements and a rice-based agrarian system be assured. In other words, given an area that vanishes twice a day into brackish waters and continuously changes, meaningful tenures and leaseholders from the point of view of the state (see Scott 1998) cannot be clearly mapped, nor can they be made agricultural. Embankments, dikes and seawalls made this possible by separating the water from the land, facilitating stable allocations (homesteads, fields, roads) and finally introducing an agricultural regime that depends on sweet water and is organised into fields.

Thus, within the last 150 years, and in a postcolonial continuation of colonial practices, almost all non-protected parts of the delta were gradually embanked, drained, cleared and transformed into agricultural land of generally inferior quality. Clearances and settlement operations proved to be more difficult, and setbacks as more intense, than the British had imagined. But they were fuelled by the widespread landlessness and endemic poverty of the neighbouring districts, where, beyond the active parts of the delta, feudal property relations dovetailed with the clearance of all cultivable land, as well as with insufficient industrialisation and mounting population pressures, which became intense from the 1920s onwards (Bose 1993). Against the backdrop of Bengal's marginalisation in the colonial and postcolonial political landscape, rural histories are therefore marked not only by poverty but by multiple losses, which were again intensified in the aftermath of the Great Bengal Famine (1942-1944) and the Midnapur Cyclone (1942). When narrated, such losses condense in the form of lost or scarce land. When, therefore, older residents remember and narrate their or their fathers' relocation to the deltaic islands, it is almost always poverty, loss of prestige and inner family conflicts that are invoked, precisely as a result of scarcity in land. Fragmented, sold or lacking, land figures prominently in these narratives, the promise of its availability in the active parts of the delta luring thousands of migrants to settle permanently in these dangerous regions, where they had to endure man-eating tigers, poor drinking water, recurring storms, devastating storm surges and failing embankments.

While storms and storm surges have played a decisive and devastating role in the colonial history of Sagar, processual coastal erosions have intensified in

---

10 Although actual social and environmental relations undermined these ideal orderings and re-orderings, the role of the peasants both as subject to individualised property relations (D'Souza 2006) and as disseminating agents of these changes have been documented (Brara 2006).

recent decades, with deepening effects on the islands' populations, which increasingly depended on the immediate coastal stretches for both agricultural and dwelling purposes. The flexible coastlines that characterise deltaic landscapes themselves became hazardous in an interplay of social vulnerabilities and the intensification of erosive processes. While, as stated above, erosions form an integral part of deltaic landscapes, gradual geomorphological changes in the upper reaches of the Ganges, which have been documented since the sixteenth century, caused a sharp decrease in the river water received by the western parts of the whole delta (Chakrabarti 2001). Since this meant that fewer and fewer sediments were brought here, erosions by tidal currents and sea waves now heavily outweigh accretion (Ghosh, Bhandari and Hazra 2003). These geomorphological changes were intensified, as my interlocutors argued, by economically driven interventions in the river channels and rises in sea level (Gopinath 2010), amounting cumulatively to higher pressure on the embankments. On Sagar and adjoining islands, where mangroves have almost entirely been cut down and where accretions of soil hardly occur now at all (Bandyopadhyay 1997), embankments are therefore the single most important, if ever imperfect means of defence against the powerful currents, high tides, swelling of waters in the monsoon and, finally, rises in sea level. However, these very embankments – made of mud and only rarely reinforced by wood, bamboo or sandbags – are in themselves constantly at risk, as well as being a risk to the society living behind them because of their neglected state. In other words, in contrast to what might be called the balanced flexibilities of other delta regions, where accretion may even outweigh erosion, here flexibility causes shrinking landscapes and therefore life worlds marked by what Henrik Vigh (2008) has called chronic crises, that is, by neither normality nor a sudden interruption of this very normality by disastrous events, but by "slow processes of deterioration, [social] erosion and negative change" (Vigh 2008: 9), which persist as conditions. Beyond the analysis of processually increasing or decreasing vulnerabilities (see Wisner et al. 2004), to which his approach might be related, Vigh emphasizes the need to analyse actions and agency not despite critical situations, but precisely as navigations in and through these conditions (Vigh 2008: 10f.).

When, therefore, on the now completely settled islands year after year, in every rainy season and during the monthly high tides (*kotal*), embankments collapse at endangered locations, this not only entails the partial or total devastation of vital crops, it also marks the advent of coastal erosion at these sites and hence of a hazardous environmental relationship that, from now on, acts as a condition. While it marks, in other words, the impact of interlinked environmental processes on such plots, which in many cases lead ultimately to the land vanishing into the muddy waters, as a chronic crisis it requires to be engaged with it as an enduring presence, even when the erosive processes have turned marginal farmers into landless squatters. Once they reach this stage, not only do they remain permanently in critical conditions as vulnerable squatters, the further encroachment of the material landscape reworks and changes the social landscape and with it the very conditions of social resilience.

4. A shrinking world

Seen from an embankment, the local world therefore becomes accessible as a shrinking landscape. Processual erosions, cyclical collapses, partial flooding and the destruction of large areas during or after tropical storms undercut and erode not only the land, but also the social world. Year after year, strips of land are lost and leave entire families landless. Although these families only occasionally lose immobile possessions[11] – except for land itself – and only very rarely are lives lost, given that the land is a central economic resource, in this agricultural society a central marker of prestige is lost with it. Year after year, those families, fields and homesteads that are located on the immediate coastline that is, immediately behind or on the embankments, are affected by erosion. Quietly and unspectacularly, their immovable property (land, shrines, cultivated areas) gradually shrinks and finally dissolves.

Sitting in teashops or conversing at night on the embankments, local residents know very well which embankment will collapse in the next rainy season, which field will probably again fail to produce harvestable paddy and, finally, where the remaining lands adjoining the next line of inner embankments will soon have to be abandoned. It seems clear to them which family will have to join the ranks of the landless, if not this year, then soon, and who will afterwards have to live in the outermost and most endangered zones – who will, as it were, soon be caught by the encroaching river.

In contrast to other delta regions, where erosion is successfully checked by technological solutions, coastal erosion here predominantly entails a chronic deterioration of both the material and social landscapes. Slowly but relentlessly, the coast draws closer to one's own field, to one's own house, harvests decline until they disappear altogether, and finally, when everything is gone, a new place to live has to be found. Some of the squatters told me that they had had to move their houses twice or three times on their own land, before finally everything was lost and they had to find somewhere else to live. Some were able to buy a new piece of land, which they later lost again, while most of those I spoke to settled immediately after the loss of all their land on one of the inner embankments in the vicinity. There they squatted on a space big enough to rebuilt huts, activated aid from among pre-existing relationships, looked for possible incomes and hoped for government rehabilitation. Squatting, as a spatial practice, thus highlights the importance of embankments from another angle.

---

11 Houses themselves belong to the category of moveable property.

As I have tried to show above, embankments are a central embodiment or materialization of the threat their residents and adjoining settlements are under. They are prone to collapse, and their proximity means that coastal erosion or the intrusion of saline waters into one's own homestead is a very imminent threat. Due to colonial policies and their uninterrupted continuity in the Indian post-colony, embankments are not managed locally or collectively, but by the state department of irrigation. In the economic and bureaucratic conditions of present-day India, with its emphasis on industrialisation and a remarkable neglect of the interests of farmers, this, sadly, explains much of the neglect and the bad shape the embankments are in. But in the absence and passivity of this locally not at all loved department (Mukhopadhyay 2009), as well as in the history of squatting in the Indian state of West Bengal (see e.g. Sanyal 2009), the embankments in themselves offer a place to stay when everything else is gone. As such they cannot be seen as part of the commons, but more as a space regulated by an absent authority and increasingly also by the lowest administrative level of the Indian government, the *panchayat*. Therefore, the landless squatters currently depend on the permission of sorts granted by the *panchayat* or its members to stay on the embankments. I never came across a case where this permission was not granted when newly landless people squat there, their plight being known in the immediate vicinity. Moving further inland or into a slightly less vulnerable area seems to involve a greater amount of political loyalty to the relevant local politicians and incorporates the squatters more fully into the structure of patronage and the pervasive struggles for political power. In conclusion here, while the absence of an institution to take systematic care of the embankments explains their fragility, it is precisely this absence that opens them up as a place to stay, a space in which to live.

However, in conversations with local leaders, representatives of NGOs, journalists and anthropologists, squatters on the embankments greatly emphasise their victimisation. Here, on the coastal periphery of one of the poorest Indian states, structural disadvantages, political neglect and largely absent development measures dovetail with processual, encroaching natural hazards.[12] Taken together, these factors or conditions deepen existing social vulnerabilities to a host of so-called external shocks and thus contribute to persistent and chronic crises. While these vulnerabilities and crises are experienced and articulated, among other things, as a loss, decay or the workings of dirty politics (Ruud 2000), local actors are nevertheless not deprived of strategies or an underlying agency. On the one hand, then, a variety of economic strategies whereby local squatters deal with the intensified poverty that coastal erosion locally entails can be demon-

---

12 Stepping beyond the important insight that disasters are generally better understood as a process, rather than an insulated event, for the sake of this paper I treat coastal erosion as a hazardous process in itself.

strated ethnographically.[13] Because these form an integral part of a locally enacted social resilience, I will now turn to them.

5. Fragments of social resilience (I): economic strategies

On-shore fishing, especially of tiger-prawn fry or *meen*, seems to be of crucial importance for the squatters. Caught by men, women or children operating alone or in small groups almost throughout the year with cheap nets built of wooden frames and widely available mosquito nettings, *meen* offer a reliable, if very low income. Immediately sold to contractors, the fry are brought to controversial aquacultural farms, fed, raised and finally sold mainly to Europe and East Asia. The fishing of fry, being the first step in a long value chain, has taken place almost daily and on a large scale all over the Bengal delta since the 1980s (Jalais 2010). Although *meen* is subject to seasonal price fluctuations and an overall drop in prices, this fishing practice is nevertheless crucial for day-to-day survival and social resilience in an ecologically hazardous and economically marginal context. The nets are cheap and widely available, connections to regional markets via contractors are well established, the fishing is done in creeks, rivers or the sea – that is, in an effectively unregulated commons – and an income, however low, seems guaranteed. For estuarine and marine ecosystems, this activity is nevertheless potentially calamitous: with these fine-meshed nets even the tiniest fish and fry are caught, thereby interrupting reproductive cycles and food chains from the bottom up. The ecological consequences of the *meen* boom of the last three decades are already perceptible according to marine biologists. On another analytical level, then, this dependence on a socially resilient practice which nonetheless has potentially devastating consequences for wider temporal or systemic scales of the socio-ecological system illustrates the need to ask: whose resilience and whose sustainability?

While catching *meen*, and occasionally small fish for consumption, relies heavily on female labour, the second most important strategy of survival under hazardous conditions, temporary labour migration, involves mainly men. Hardly ever undertaken on an international scale, predominantly young and unmarried men travel from Sagar far to the south, west and northwest of India. They are recruited mainly as unskilled workers by contractors, either alone or in groups, work for a couple of months at a stretch at construction sites, brick mills or

---

13 Leasing plots or to enter sharecropping relations, which seem to have been common strategies in past times of distress, are only very rarely found at the present day. In the context of population pressure, which only underlined the scarcity of land as a resource, as well as in the political context of land reform and the empowerment of sharecroppers (Mallick 1993), which made sharecropping for the landowner a risky strategy in itself, land is at present worked as much as possible by family labour. Where sufficient family labour is lacking, neighbours or relatives are called in as day labourers.

poultry farms and return home with their earnings. Temporary labour migration of women, on the other hand, seems to be much less common, at least among the people I am concerned with. When women do migrate temporarily, they do so only very rarely in the company of groups of men, preferring other circuits and types of work, mainly as domestic workers in the urban areas of Kolkata.

In the summer months, moreover, a large number of men are engaged on boats in seasonal deep-sea fishing. Recruited by ship-owners or pilots (*majhi*), these men only seldom work on a daily basis, and more often receive a share of the catch in payment. But since the economically viable season coincides with the storm-prone months of the rainy season, these activities are considered very dangerous. Beyond that, deep-sea fishing is also economically and politically risky, since, on the one hand, actual incomes are related to the catches and may be very low in bad seasons. To gain access to the better fishing grounds off Bangladesh or simply because they get lost, local fishermen also cross the international border and may end up in jail or having to pay heavy bribes to Bangladeshi coastguards (Gupta and Sharma 2009).

Another important source of income for landless squatters relates to the important pilgrimage destination of Gangasagar on the southern coast of Sagar Island. Associated with the mythical descent of the holy Ganges River down to earth, the beach and its temple draw a steady influx of pilgrims from the northern, western and eastern parts of India. Furthermore, an annual pilgrim festival (Gangasagar Mela) is held in January, attracting several hundreds of thousands of pilgrims each year. Transporting, catering for and accommodating the pilgrims, especially before and after the annual festival, therefore contributes greatly to economic survival.

Even though an analysis of economic strategies may reveal important aspects of social resilience, of agency in critical circumstances, and may allow us to situate these fragmentarily on larger social and ecological scales, local perceptions, interpretations and their relevance to social resilience still remain hidden. Earlier works have demonstrated how technological or political debates, as well as religious cosmologies, play a significant role in social articulations of environmental relations and as discourses that shape subjectivities and practices (see, for example, Agrawal 2005) and, by extension, resilient social worlds (Crane 2010). While these important approaches could and should be further related to analyses of social resilience, in this paper I intend to emphasise social memory as crucially important to understanding social resilience. To demonstrate this, I now turn to three aspects or workings of social memory: first, the moment of remembered multiple losses; secondly, the establishment of a community of loss; and thirdly, related imaginations of the future.

6. Fragments of social resilience (II): the workings of social memory

As mentioned earlier, from the nineteenth century onwards, Sagar and adjoining islands were systematically cleared, settled and subjected to an agrarian regime. Realising the overall hazardousness of the local environment and the very fragility of their settlements, the early settlers tried to secure their holdings behind mangroves and, as much as possible, out of reach of estuarine waters. Therefore, most of the early settled areas are to be found in the inner regions of the island, and only a very limited number border directly on the banks of a river or the sea. Hence, over most of the island the immediate coastal stretches were left untouched and were only settled quite recently. Not only do old maps prove this, so do the local oral histories I collected. Unanimously, all the squatters I spoke to emphasised the recent immigration of either themselves or their fathers and grandfathers on to these islands.

Furthermore, all my interlocutors pointed to their common origin in the adjoining mainland directly on the other side of the river in the west, that is, in the district of East Medinipur, or, to a much lesser extent, from the northern mainland in the district of South 24 Parganas. Despite having left for good, the vast majority of the islanders still entertain relations with the mainland. Building, among other things, on pre-existing relations, caste membership and dialects, cultural formations as diverse as marriage relations, religious pilgrimages and political loyalties crisscross the muddy rivers between the mainland and the islands. While these formations and processes greatly contribute to constantly negotiated and reworked social continuities and collective identities – that is, to particular social memories– a shared burden of remembered pasts (Lambek 2003) forms another fragmentary, but integral set or facet of these very identities. Utterances, narratives or speeches indicating the islanders' shared history point to distinct losses as experienced on the mainland beyond the rivers. Oral histories and related practices of social remembering are, in other words, marked by narratives of loss, which, as I will show, inform not only social identities, but also distinct fragments of social resilience. Poverty, density, disenfranchisement and conflict – all culminating in the loss or absence of land – serve as structural elements for remembered histories of destitution and migration.

Heuristically, then, two narrative motifs involving why and how it became necessary to leave, relocate and remake a life in environmentally hazardous conditions can be distinguished. Interestingly, both motifs are narrated in the same terminology of gradual destitution and landlessness, which in itself attests rather to conditions marked by chronic crises than to single disastrous events. On the one hand, then, migration is remembered as having become necessary because of landlessness, poverty and intra-familial conflict caused by general conditions in the Indian hinterland. On the other hand, migration became necessary because of landlessness, poverty and intra-familial conflict in the wake of disastrous events. Narratives of the first type refer to the landlessness of second-, third- or fourth-born sons and their families, to inadequately small agricultural plots, to the absence or insufficiency of the means to earn a living, and to intra-familial conflicts rooted in these economic conditions. These detrimental conditions forced mainly young men to migrate, reclaim and cultivate land or at least acquire reclaimed land on one of the islands, establish themselves there, bring their families and enjoy a nostalgically valorised phase of prosperity, that is, of agricultural abundance.

The second type of narrative ties loss and impoverishment to the main social disasters that haunted the south-western part of Bengal in the twentieth century: the devastating storm surge in the wake of the Midnapur Cyclone (1942), which is remembered as the 'Red Flood' (*lal bonya*), as well as the Great Bengal Famine (1942-1944). While the latter was caused by a fatal million-fold increase in food prices that had been slowly accumulating related mainly to the war economy (Sen 1981; Alamgir 1980), the Midnapur Cyclone and the ensuing flood are remembered as a sudden event that caused the immediate devastation of large coastal areas (see also Greenough 1982). It is precisely the temporally delimited but nonetheless drastic features of a very tacit devastation brought about by the deluge that probably accounts for the strength of its hold on contemporary figurations of social memory. Nevertheless, both cyclone and famine intensified widespread poverty marked again by material destruction, mortality, related diseases or, indirectly, distress sales, indebtedness and so on.

If, therefore, both these narratively constructed histories of migration are narrated in the same idiom, remembered pasts are constructed with recourse either to normalised destitution or to worsening disastrous processes. In their narrative structure, as already noted, both nevertheless point in varying degrees to non-disastrous, even normalised conditions of scarcity or absence of land, poverty and conflict. Therefore, contemporary landlessness and the related intensification of poverty constitute an actualisation of rather recent and well-remembered events, not a single, previously unknown or forgotten experience. Seen and experienced through these very present histories, the hazards of coastal erosion definitely mark painful losses, but at the same time they also mark losses that can be related to historically. In addition, this actualisation of remembered pasts in and through environmental hazards may play an important role in the evolving of collective identities across the asynchronous effects of these hazards. Let me elaborate.

Writing on disastrous events or chains of events that have affected large populations at roughly the same time, it has been shown that communities of survivors may be considerably levelled and marked by heightened feelings of fellowship. In the face of adverse conditions, a 'brotherhood of pain' (Oliver-Smith 1999) may replace rigid hierarchies, at least for certain periods of time, and may later be remembered as a truly beautiful aspect of a world that has fallen apart (Hoffman 1999). While these earlier considerations clearly relate to collective experiences of often single, but at least temporally well-defined events that affect entire populations, I argue that they also have implications for the subject of this paper. Even if the group is not affected all at the same time, nor in the same way by coastal erosion, specific collective experiences and identities are both, actualised and endowed in an interplay of actualised pasts, recent losses and expected futures. In other words, when erosions affect only individualised families and clearly definable strips of land at any given moment, but precisely because of this push the coast inland and on to adjoining fields and homesteads, then collective identities are constituted not in the experience of a disaster in itself, but in necessarily remembered and individualised events. This is reflected in patterned social memories, which bridge diverse courses and experiences of ever asynchronous erosions by strongly emphasising losses. Relevant narratives are therefore dominated to such an extent by lost plots, lost securities and a lost prosperity that experiences of waters rushing in, of devastated harvests or of futile attempts to check erosions tend to be marginalised, if not forgotten.

Beyond a further engagement with forgetting itself, I now want to deal with layers or fragments of a collective identity that are enabled by this kind of forgetting. For this purpose, one of Judith Butler's reflections on what could be termed a community of loss seems to be particularly helpful. She writes (Butler 2003: 468, italics in original): "Loss becomes condition and necessity for a certain sense of community, where community does not overcome the loss, where community cannot overcome the loss without losing the very sense of itself as community." While this certainly does not speak in favour of resurrecting those much-criticised 'cultures of poverty', it nevertheless helps us understand the everyday aspects of social resilience, as well as the continuous dwelling in conditions of heightened vulnerability. That is, in sharing histories of this kind, which pervade the everyday and resurface again and again in lively recollections of conditions beyond the deltaic islands, the society on and behind the embankments qualifies as a community of loss. Being characterised, even haunted, by losses, squatters are therefore certainly not traumatised – if it is meaningful at all to apply this notion to collective phenomena – but build social resilience on it. While recent or contemporary losses – losses due to eroding coasts or the flexibility of the coastlines – actualise past experiences, mutual assistance and even solidarity among the inhabitants are rooted in this community and are in themselves crucial in creating and maintaining the material or economic layers of social resilience. Therefore, not only are relations among kin or neighbours activated to ensure a livelihood when everything is gone, but a community is addressed, its shared history related to, in order to evade or smooth over the conflicts that commonly arise when the natural resources on which local communities highly depend become even scarcer.

Beyond that, squatters on embankments rely on the community as a kind of last resort and as a political instrument to ensure fragile futures. Burdened by histories of social and environmental disenfranchisement, which are articulated as losses, on the one hand it seems to be a rational strategy to rely on the community, that is, to stay back, not to migrate permanently and to cling to a life on the embankments, even if this means living under conditions of heightened social vulnerability to environmental hazards. Here, squatters can be sure, they are hardly likely to be driven away and can make a space for themselves, because, I argue, the community is marked by losses to such an extent that everybody can relate to them.

On the other hand, the community of loss is situated and acts in the political context of a regional state (Sivaramakrishnan and Agrawal 2003), which is experienced predominantly through its lowest administrative level, the *panchayat*. Here I wish to stress not the importance of institutional adaptation to environmental change, but rather the social agency of hopes and expectations of governmental assistance or rehabilitation that rely on a burdened community as a political society (Chatterjee 2006), demanding and allocating allotments. Living on embankments, squatters try to exert political pressure on the local and regional governments, which presupposes a shared interest with local politicians that is rooted mainly in electoral politics, but also in the community of loss already mentioned. The importance of such liaisons may be illustrated by a locally powerful distinction between the squatters on embankments and the 'ordinary' roadside squatters, the *rasta lok*.[14] While the former are included in the political societies just mentioned, the latter are not. Being seen as very recent immigrants, the roadside squatters are mainly classified in morally derogatory terms (as thieves, illegitimate couples and so on) and excluded from a community of loss precisely in the denial of a shared history.

Finally, it has been argued not only that social remembering plays a part in shaping present conditions and reworks imagined pasts, but also that processes, narratives and enactments of remembered pasts are embraced in intimate relationships with the future (Stewart and Strathern 2003). Remembering is therefore not an act of retrospection or engagement with the present alone, but is better understood as powerfully moulding specific futures by investing imaginations with a sense of plausibility, giving rise to expectations or being the staple of (strategic) claims. Reaching into near or distant futures, social remembering

---

14 In an exclusively low-lying landscape like the deltaic islands, the typical brick roads are built on the inner embankments themselves. Hence many of the squatters I am concerned with in this paper do actually live on the road. The distinction from what I call 'ordinary' roadside squatters therefore relates to a social demarcation grounded and enacted in "narratives of the past, by which they may alternately question, support, and undermine the claims of others while projecting their own" (Gottschalk 2001: 71).

is again relevant to the present and plays a dialectical role in its very constitution and experience.

Seen in the light of remembered experiences – that is, of memories of poverty, deprivation and conflicts in areas beyond the active delta – these very areas are, on the embankments, hardly imagined as a place to go or return to. The squatters I spoke to clearly preferred staying on and enduring a hazardous environment, intensified poverty and a bleak future to permanent out-migration. Such migration, they feared, would bring them back to the conditions they or their forefathers had once escaped. Again, they would have to make a living under harsh circumstances, would have to struggle to make a space for themselves. Most likely, they would ultimately be turned into 'ordinary' roadside squatters in foreign areas, where they would have to manage with only very limited influence and beyond a shared past.

Set against futures which are imagined through a dialectical interplay of past and present, of present experiences and social memories, to stay on even under extremely vulnerable conditions emerges as a rational strategy. Socially vulnerable as these living conditions are, they engender the continuity of neighbourly relations, of a social environment of mutual help, which the squatters clearly prefer to the deprivations they can expect as foreigners living in foreign areas.[15] In other words, in the light of these interpretations it appears safer to live on an embankment that may collapse, be flooded, is haunted by intense storms or cleared by politically driven evictions than to relocate to uncertain distant areas lacking social relations.

Grounded in a shared past, contemporary experiences and relations, as well as in future expectations, which are all intimately related, an ethics of endurance therefore emerges. Not to be confused with fatalism, as I have tried to show this ethical stance engenders social resilience.

7. Conclusion

Reflecting on the highly vulnerable, yet socially resilient spatial and social arrangements of squatting in coastal Bengal, in this paper I have proposed a twofold shift in the analysis of social resilience. To improve our understanding of its multifarious contributions, genealogies and limitations, I have first of all argued for an ultimately historicising approach to social resilience. Building on that, I have argued for the importance that socially remembered pasts may have – as a shared burden and as culturally mediated expectations of the future – for localised forms of social resilience.

---

15 The designation of foreign areas (*bidesh*) may, according to speech situations, apply to all places beyond the immediate locality, to places beyond one's island or to regions beyond Bengal.

The underlying intention of this paper has therefore been to problematize the predominant understanding of resilience as either a personal, social or systemic capacity. While social resilience may be partially understood as a culturally mediated capacity to endure, to be flexible or to adapt, it is certainly also a situated practice. In the latter sense, it therefore draws on and is enacted through a range of material relations, economic practices, social articulations and cultural interpretations. Furthermore, processes and forms of social remembering might be of crucial importance to facets of social resilience that are only distinguishable analytically.

On the one hand, therefore, social memories might – perhaps as countermemory – be a means "to summon aesthetics, emotion, and imagination to inspire a swell of pride and a sense of possibility" (Nazarea 2006: 329), thus reinforcing social capacities, ethical stances or emotive attachments which amount to important aspects of social resilience. On the other hand, social resilience as a situated practice unfolds in given ecological, historical and political contexts, which are read and made sense of by, among other things, processes of social remembering. Possibly strengthening social capacities, as well as framing situated practices, social memory might therefore be crucial to resilient life worlds.

While social resilience has therefore to be understood as constituted or enacted in and through diverse localizations, correlations or borrowings, precisely these complicated involvements could serve as a reminder to engage critically with associations of groups and places or landscapes, instead of taking them for granted (Appadurai 1988; Raffles 1999). When, therefore, groups are necessarily involved in social-ecological systems, the way they relate to wider systems is not a function of this embeddedness, nor is it to be understood as a necessary contribution to the system's overall robustness. While this is certainly a truism with respect to western societies, recent re-appraisals of local knowledge and resilience in the Global South tend not to problematize but to further what has recently been called the eco-incarceration of distinct populations (Shah 2011). This, of course, is not meant to deny the importance of local knowledge in itself as a means to learn about or engage with hazardous environments, but is rather meant to problematize the neo-orientalist figure of thought that postulates an end to environmental crises by and through an activation of local knowledge alone. Beyond the homogenising blurring of hierarchical relations and its inability to solve the problems at hand alone, this figure again, at least implicitly, inscribes certain populations into certain places and negates agency beyond the environmental.

To ask for the multifarious constitutions of social resilience serves, I think, as one important step in this direction, as does another reformulation of two quintessential postcolonial questions: Whose (social) resilience are we talking about? And how does this relate to an overall resilience across scales?

8. Bibliography

Adger, W. Neil, Hughes, Terry, P. Folke, Carl, Carpenter, Stephen R. and Johan Rockström (2005): Social-Ecological Resilience to Coastal Disasters. In: Science 309, 5737, pp. 1036-1039.

Agrawal, Arun (2005): Environmentality: Technology of Government and the Making of Subjects. Durham: Duke University Press.

Agrawal, Arun and K. Sivaramakrishnan (eds.) (2000): Agrarian Environments: Resources, Representation, and Rule in India. Durham and London: Duke University Press.

Alamgir, Mohiuddin (1980): Famine in South Asia: Political Economy of Mass Starvation. Cambridge: Oelschlager, Gunn & Hain.

Allemeyer, Marie Luise (2006) 'Kein Land ohne Deich .!' Lebenswelten einer Küstengesellschaft in der Frühen Neuzeit. Göttingen: Veröffentlichungen des Max-Planck-Instituts für Geschichte 222.

Appadurai, Arjun (1988): Putting Hierarchy in Its Place. In: Cultural Anthropology 3, 1, pp. 36-49.

Assmann, Jan (1988): Kollektives Gedächtnis und kulturelle Identität. In: Assmann, Jan and Tonio Hölscher (eds.): Kultur und Gedächtnis. Frankfurt am Main: Fischer, pp.9-19.

Auyero, Javier and Debora Alejandra Swistun (2009): Flammable: Environmental Suffering in an Argentine Shantytown. New York: Oxford University Press.

Bandyopadhyay, Sunando (1997): Natural Environmental Hazards and Their Management: A Case Study of Sagar Island, India. In: Singapore Journal of Tropical Geography 18, 1, pp. 20-45.

Berkes, Fikret, Colding, Johan and Carl Folk (eds) (2003): Introduction. In: Navigation Social-Ecological Systems: Building Resilience for Complexity and Change. Cambridge: Cambridge University Press, pp. 1-30.

Berliner, David (2005): The Abuses of Memory: Reflections on the Memory Boom in Anthropology. In: Anthropological Quarterly 78, 1, pp. 197-211.

Bohle, Hans-G., Downing, T.E. and Michael J. Watts (1994): Climate Change and Social Vulnerability: Towards a Sociology and Geography of Food Insecurity. In: Global Environmental Change 4, pp. 37-48.

Bose, Sugata (1993): Peasant Labour and Colonial Capital: Rural Bengal since 1770. Cambridge: Cambridge University Press.

Brara, Rita (2006): Shifting Landscapes: The Making and Remaking of Village Commons in India. New Delhi: Oxford University Press.

Butler, Judith (2003): Afterword: After Loss, What Then? In: Eng, David L. and David Kazanjian (eds.): Loss: The Politics of Mourning, edited by. Berkeley: University of California Press, pp. 467.474.

Carson, Rachel (1999[1973]): The Edge of the Sea. London: Penguin.

Chakrabarti, Dilip K (2001): Archeological Geography of the Ganga Plain: The Lower and the Middle Ganga. New Delhi: Permanent Black.

Chatterjee, Partha (2006): The Politics of the Governed: Reflections on Popular Politics in Most of the World. New York: Columbia University Press.

Cosgrove, Denis E. and Geoffrey Petts (eds.) (1990): Water, Engineering and Landscape: Water Control and Landscape Transformation in the Modern Period. London: Belhaven.

Crane, Todd A. (2010): Of Models and Meanings: Cultural Resilience in Social-Ecological Systems. In: Ecology and Society 15, 4, 19. http://www.ecologyandsociety.org/vol15/iss4/art19/.

D'Souza, Rohan (2006): Drowned and Damned: Colonial Capitalism and Flood Control in Eastern India. New Delhi: Oxford University Press.

Danda, Anurag (2007): Surviving the Sundarbans: Threats and Responses. An Analytical Description of Life in an Indian Riparian Commons. University of Twente: Twente.

Foucault, Michel (1986): Of Other Spaces. In: Diacritics 16, 1, pp. 22-27.

Ghosh, Tuhin, Bhandari, Gupinath and Sugata Hazra (2003): Application of 'Bio-engineering' Technique to Protect Ghoramara Island (Bay of Bengal) from Severe Erosion. In: Journal of Coastal Research 9, pp. 171-178.

Gilroy, Paul (1993): The Black Atlantic: Modernity and Double Consciousness. London: Verso.

Gold, Ann Grodzins and Bhoju Ram Gujar (2002): In the Time of Trees and Sorrows: Nature, Power, and Memory in Rajasthan. Durham: Duke University Press.

Gottschalk, Peter (2001): Beyond Hindu and Muslim: Multiple Identity in Narratives from Village India. New Delhi: Oxford University Press.

Gopinath, Girish (2010): Critical Coastal Issues of Sagar Island, East Coast of India. In: Environmental Monitoring and Assessment 160, pp. 555-561.

Greenough, Paul (1982): Prosperity and Misery in Modern Bengal: The Famine of 1943-1944. New Delhi: Oxford University Press.

————— (1998): Hunter's Drowned Land: An Environmental Fantasy of the Victorian Sundarbans. In: Grove, Richard R., Damodaran, Vinita and Satpal Sangwan (eds.): Nature and the Orient: Environmental History of South and Southeast Asia. New Delhi: Oxford University Press, pp. 237-269.

Grove, Richard H (1997): Ecology, Climate and Empire: Colonialism and Global Environmental History, 1400-1940. Cambridge: The White Horse Press.

Gupta, Charu and Mukul Sharma (2009): Contested Coastlines: Fisherfolk, Nations and Borders in South Asia. New Delhi: Routledge.

Halbwachs, Maurice (1992[1925]): On Collective Memory. Chicago: The University of Chicago Press.

Hastrup, Kirsten (ed.) (2009): The Question of Resilience: Social Responses to Climate Change. Copenhagen: The Royal Danish Academy of Sciences and Letters.

Helmreich, Stefan (2011): Nature/Culture/Seawater. In: American Anthropologist 113, 1, pp. 132-144.

Hoeppe, Götz (2007): Conversions on the Beach: Fishermen's Knowledge, Metaphor and Environmental Change in South India. New York: Berghahn.

Hoffman, Susanna M. (1999): The Worst of Times, the Best of Times: Toward a Model of Cultural Response to Disaster. In: Oliver-Smith, Anthony and Susanna M. Hoffman (eds.): The Angry Earth: Disaster in Anthropological Perspective. New York: Routledge.

Hollig, C.S. (1973): Resilience and Stability of Ecological Systems. In: Annual Review of Ecology and Systematics 4, pp. 1-23.

Huggins, William (1824): Sketches in India. London: John Letts.

Jalais, Annu (2010): Forest of Tigers: People, Politics and Environment in the Sundarbans. New Delhi: Routledge.

Kelman Ilan (2008): Critique of Some Vulnerability and and Resilience Papers. Version 2, 17 November 2008. Downloaded from http://www.islandvulnerability.org/docs/vulnrescritique.pdf

Lahiri-Dutt, Kuntala and Samanta Gopa (2007): 'Like the Drifting Grains of Sand': Vulnerability, Security and Adjustment by Communities in the Charlands of the Damodar Delta. In: South Asia: Journal of South Asia Studies 32, 2, pp. 320-357.

Lambek, Michael (2003): The Weight of the Past: Living with History in Mahajanga, Madagascar. New York: Palgrave Macmillan.

Lewin, Kurt (2006[1917]): Kriegslandschaft. In: Dünne, Jörg and Stephan Günzel (eds.): Raumtheorie: Grundlagentexte aus Philosophie und Kulturwissenschaften. Frankfurt am Main: Suhrkamp, pp. 129-140.

Mallik, Ross (1993): Development Policy of a Communist Government: West Bengal since 1977. Cambridge: Cambridge University Press.

Mukhopadhyay, Amites (2009): On the Wrong Side of the Fence: Embankment, People and Social Justice in the Sundarbans. In: Bose, Pradip Kumar and Samir Kumar Das (eds.): Social Justice and Enlightenment: West Bengal. New Delhi: Sage Publications, pp.118-152.

Nazarea, Virginia (2006): Local Knowledge and Memory in Biodiversity Conservation. In: Annual Review of Anthropology 35, pp. 317-335.

Nienhuis, Piet H. (2008): Environmental History of the Rhine-Meuse Delta: An Ecological Story on Evolving Human-Environmental Relations Coping with Climate Change and Sea-level Rise. Dordrecht: Springer.

Oliver-Smith, Anthony (1999): The Brotherhood of Pain: Theoretical and Applied Perspectives on Post-Disaster Solidarity. In: Oliver-Smith, Anthony and Susanna M. Hoffman (eds.): The Angry Earth: Disaster in Anthropological Perspective. New York: Routledge, pp. 156-172.

Pokrant, Bob, Reeves, Peter and John McGuire (1998): The Novelist's Image of South Asian Fishers: Exploring the Work of Manik Bandopadhyaya, Advaita Malla Barman and Thakazi Sivasankara Pillai. In: South Asia: Journal of South Asia Studies XXI, 1, pp. 123-138.

Raffles, Hugh (1999): 'Local Theory': Nature and the Making of an Amazonian Place. In: Cultural Anthropology 14, 3, pp. 323-360.

Ruud, Arild Engelsen (2000): Talking Dirty about Politics: A View from a Bengali village. In: Benei, Veronique and C.J. Fuller (eds.): The Everyday State and Society in India. New Delhi: Social Science Press, pp. 115-136.

Sanyal, Romola (2009): Contesting Refugeehood: Squatting as Survival in Post-partition Calcutta. In: Social Identities 15, 1, pp. 67-84.

Sarkar, Sutapa Chatterjee (2010): The Sundarbans: Folk Deities, Monsters and Mortals. New Delhi: Social Science Press.

Schmitt, Carl (1954): Land und Meer: eine weltgeschichtliche Betrachtung. Stuttgart: Reclam Verlag.

Schmuck-Widmann, Hanna (2001): Facing the Jamuna River: Indigenous and Engineering Knowledge in Bangladesh. Dhaka: Bangladesh Resource Center for Indigenous Knowledge.

Scott, James (1998): Seeing Like a State: How Certain Schemes to Improve the Human Condition Have Failed. New Haven: Yale University Press.

Sen, Amartya (1981): Poverty and Famines: An Essay on Entitlement and Deprivation. New Delhi: Oxford University Press.

Shah, Alpa (2011): In the Shadow of the State: Indigenous Politics, Environmentalism, and Insurgency in Jharkhand, India. New Delhi: Oxford University Press.

Sivaramakrishnan, K. and Arun Agrawal (eds.) (2003): Regional Modernities: The Cultural Politics of Development in India. Stanford: Stanford University Press.

Stewart, Pamela J. and Andrew Strathern (2003): Introduction. In: Stewart, Pamela J. and Andrew Strathern (eds.): Landscape, Memory and History: Anthropological Perspectives. London: Pluto Press, pp. 1-15.

Tsing, Anna L. (2005): Friction: An Ethnography of Global Connection. Princeton: Princeton University Press.

Vigh, Henrik (2008): Crisis and Chronicity: Anthropological Perspectives on Continuous Conflict and Decline. In: Ethnos 73, 1, pp. 5-24.

Werner, Emmy E, Bierman, Jessie M. and Fern E. French (1971): The Children of Kauai: A Longitudinal Study from the Prenatal Period to Age Ten. Honolulu: University of Hawaii Press.

Wisner, Ben, Blaikie, Piers, Cannon, Terry and Ian Davis (2004): At Risk: Natural Hazards, People's Vulnerablity and Disasters. New York: Routledge.

Zaman, M. Q (1989): The Social and Political Context of Adjustment to Riverbank Erosion Hazard and Population Resettlement in Bangladesh. In: Human Organization 48, 3, pp. 196-205.

II.3 Reflecting on methods:
how can we make sense of disasters?

Researching coping mechanisms in response to natural disasters: the earthquake in Java, Indonesia (2006)

Manfred Zaumseil and Johana Prawitasari-Hadiyono

1. Introduction

Natural disasters are a ubiquitous and historically persistent phenomenon, one that is apparently growing in importance due to human-generated impacts on the environment. What is new is the speed at which information about these events spreads around the globe, and the understanding that these disasters and their consequences are not produced by a nature that stands apart from humankind as a separate entity, a given. Instead, various research approaches concerning natural disasters have attempted to embrace the notion of complex interaction between humans and the environment. Based upon this assumption, such disasters should be handled as much as possible through anticipatory "risk management." Also new is the worldwide coordination of disaster aid that is determined by both the dynamics of mostly special-interest organizations collecting donations in rich countries, and the policies that rich "donor nations" follow in allocating their aid. In the places in which disasters that qualify as "natural disasters" occur, regional and national conditions not infrequently channel, re-shape and counteract the international relief effort, the result being a more or less successful coordination with the relief effort that originates in national and regional bodies. Less often examined in the published literature or by NGOs are the more unofficial self-help efforts and long-term coping with the disaster within the networks of meaning in the populations affected. Often no attention is paid to the fact that the people involved are concerned with making sense of the disaster itself, as well as putting the external aid into a context in which local culture, history, shared interpretations and strong emotions are important. As a result, the technical side of handling disasters often encounters constraints and resistance that can lead to large-scale frictional loss in the long-term.

In this paper, we use a research example to show how research approaches in cultural science can help unlock the structure of conditions, meanings, practices and psychosocial dispositions following a natural disaster in the place it occurs, and how this knowledge can be turned to use in coping with disaster. This report therefore deals with knowledge production in very disparate fields. The goal is to demonstrate, using a practical example, how the risk and vulnerability approach in disaster research, research into psychological coping and knowledge of local and cultural specificity in cultural anthropology interrelate with each other. We also discuss the question, whether and how the exchange and transfer of knowledge can be organized with the people who have experienced and suffered from the disaster and have had to cope with it in various ways.

What is under consideration, therefore, is the question of the production of knowledge, the employment of theoretical building blocks from diverse scientific areas in the analysis of data and reflections on their use. We begin with a short description of our research project, which we use as a case study. We will also discuss various disciplinary and methodological approaches that are relevant for coping with disasters. As examples, we will then present some results of our study, focusing in particular on the psycho-spiritual dimensions of coping with disaster, in order to demonstrate our initial process of developing categories and their relations with one another. We will then illustrate the further process of a theory-reflecting analysis by projecting our data on to some of the theoretical models introduced earlier. In doing so, we wish to point out how the combination of multiple disciplinary approaches, as well as dialogue with the people affected, can advance our knowledge of how people cope with disaster.

2. The research example

The title of our ongoing project (February 2011) is "Individual and collective ways of coping long-term with extreme suffering and external help after natural disasters: meanings and emotions".[1] For the purposes of this research, the disaster which took place in 2006 in the region (DIY) of Yogyakarta (Bantul) and Central Java (Klaten) in Indonesia provides an example with which to explore the intertwined issues of coping and aid. The earthquake that occurred in the early morning of May 27 resulted in the destruction of 280,000 houses and the deaths of almost 7,000 people. Three villages, which vary in distance from the epicentre of the tremor, have been chosen as research sites.

The aim of the project is to find out how people who have experienced extreme suffering and misfortune continue to cope in the long-term. We are also

---

1   The research was funded by the Fritz Thyssen Stiftung from November 2008 to October 2011. Gavin Sullivan PhD, Monash University, Melbourne, was a co- applicant and senior researcher in the first research period. Research assistants were Silke Schwarz (psychology), Mechthild von Vacano (cultural anthropology); also associated with the project were Dr Jeane Indradjaja (Psychology). Close cooperation occurred with the Faculty of Psychology, UGM (Dean: Prof. Dr. Faturochman) and the Yogyakarta-based NGO ICBC (Institute for Community Behavioral Change; Director: Edward Theodorus). Cooperating researchers from ICBC were Nindyah Rengganis, Lucia P. Novianti, Tiara R. Widiastuti and Yohanes K. Herdiyanto.

interested in how this suffering is handled and interpreted in connection with disaster aid, as well as the ability of the local population to deal with enduring threats. As a result, we aim to find starting points for the creation of a more culturally sensitive approach to disaster management and to find out how cultural meanings and practices, especially in the psychosocial and spiritual spheres, mediate subsequent resilience and disaster preparedness.

2.1 Method

To collect the data, we conducted narrative and guideline-supported interviews with villagers and aid workers, focus-group interviews and field observations during multiple short field trips of seven to fourteen days. In addition, opportunities for feedback on the village and regional levels provided a different set of data. The data were collected in either the local language (Javanese) or the national language (Bahasa Indonesian). Data collection and analysis were conducted in close cooperation with the university (UGM) and the employees of the local NGO "ICBC" (Institute for Community Behavioral Change). In the last phase of the research (2011), we made more frequent use of methods of participatory research with co-researchers from the three villages. The responsible chief investigator for the research in Java, Prof. Dr. Prawitasari-Hadiyono, as well as members of "ICBC" have well-established field contacts which were created earlier during a relief programme and a participatory research project using traditional forms of theatre. Prof. Dr. Zaumseil has enjoyed longstanding research cooperation with the local university (UGM) for thirty years.

3. Overview of the relevant approaches to coping with disaster

We treat the processing of the disaster as our central issue, and see coping as process of social and individual construction. The event partitioned time into a before and an after, and many changes that occurred afterwards are attributed to it. Personal and collective perceptions, and the processing of what happened and what followed, have many facets, of which this report will concentrate on just one: the retrospective psycho-spiritual processing of the direct consequences of the earthquake two to five years afterwards. Thus we are not dealing with the whole range of project topics such as the practice of coping with the disaster, the handling of the earthquake relief effort or the fairness debate: the social changes that can be traced back to the earthquake, the traumatization attributable to the event and coping with disabilities are also of great concern. In order to be able to assess adequately the long-term coping process after a disaster in the local context, it is necessary to relate to at least three disparate scientific areas, which themselves are characterized in turn by a great degree of heterogeneity. And finally, we aim to integrate possible methods of participatory knowledge production, conducted in cooperation with the village people (see the diagram below).

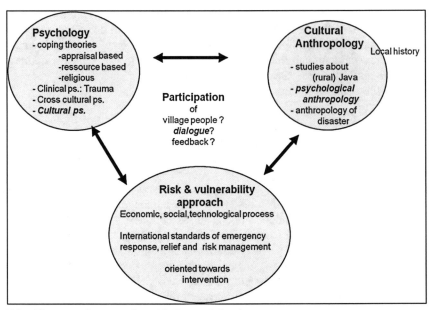

Scientific approaches to coping with disaster in local context.

3.1 The risk and vulnerability approach

Wisner, Blaikie et al.'s influential publications on the risk and vulnerability approach (Blaikie et al. 1994; Wisner, Blaikie and Cannon 2004) reflect concerted international efforts to reduce the impact of disasters and to shift the focus from disaster management to disaster prevention. Theirs is the most influential contribution to the framework concerning funding of international disaster relief and prevention and the establishment of corresponding standards (see Hyogo Framework 2005; International Strategy for Disaster Reduction (ISDR) of the UNO 2005; World Bank 2006, 2010), including the collective and individual interpretation and processing of the event.

The risk and vulnerability approach is concerned with appraising the state of a social entity in a region. Technological, geographical, economic and social measurements are used to examine and assess, where possible, the vulnerability or resilience of a social entity to the dangers posed by an environmental hazard. A risk is the product of vulnerability and endangerment (Wisner, Blaikie and Cannon 2004: 51). A natural hazard can be partly overcome using this technological approach. A disaster with huge losses and many victims will occur only when a pre-existing vulnerability and a lack of preparedness, as well as the social causes that form the basis of these deficiencies, are also present. In the framework of this approach, interventions are aimed at controlling the risk through, for instance, early-warning systems, disaster plans, earthquake-ready building techniques, etc., and at increasing resilience through, for instance, community-based risk-management programs, as well as minimizing vulnerability to the risk. The ability to cope is thus of interest as a more or less already available and trainable skill: together with access to resources, it belongs to the notion of resilience as an individual or collective skill and psychological readiness. This static model (the so-called Crunch model) is expanded by Wisner, Blaikie and Cannon (2004) through an additional model.

The further progression of this situation into a disaster (including the coping mechanisms) is developed in the related, dynamically designed "Access Model." "The Access Model sets out to explain the establishment and trajectory of vulnerability and its variation among individuals and households on the micro-level. It deals with the impact of a disaster as it unfolds, the role and agency of the people involved, what the impacts are on them, how they cope, develop recovery strategies and interact with other actors (e.g. humanitarian aid agencies, the police, the landlord and so on)" (Wisner, Blaikie and Cannon 2004: 88).

The authors themselves describe the model as an economic, structural and implicitly quantitative application (Wisner, Blaikie and Cannon 2004: 97). They write: "It is important to complement the economistic and quantitative aspects of our access model with an understanding of the ways in which the disaster event was experienced by different people, and how it altered their sense of well-being and their strategies to reconstitute that well-being in a new, post-disaster world" (ibid.: 110). They consider that part of the Access Model that is less liable to explanation in economic or political-economic terms, that is, the psychological coping process, as describable "only" in qualitative terms. "It is very difficult to model, predict or find regularities in agency or inventiveness" (ibid.: 97). Taking up this point, we suggest that it is not necessary to capitulate before this centrally important question, and that methods of social research and knowledge from the fields of psychology and cultural and social anthropology have much to offer in coming to a deeper understanding and better possibilities to support the answer to natural disasters as they unfold under specific local socio-cultural conditions.

3.2 Psychology
3.2.1 Appraisal-based approaches to coping
Psychological coping models address, in various ways, what is lacking in the risk and vulnerability approach. The dominant tendency is to consider coping primarily as an individual answer to stresses and challenges. Subjective appraisals are here considered particularly relevant to the progression of stress. There is the subjective appraisal of the threat on the one hand and, on the other, one's own potential for dealing with it (Lazarus and Folkman 1984). This is often accompanied by strong emotions. "Emotions continue to be integral to the coping

process throughout a stressful encounter as an outcome of coping, as a response to new information, and as a result of reappraisals of the status of the encounter" (Folkman and Moskowitz 2004: 747). Compared to Lazarus and Folkman's model (1984), the understanding of coping has broadened (see Folkman and Moskowitz 2004). Narrative approaches revealed the importance of meaning-making and causal attribution in stressful situations. But there is disagreement regarding the different forms of coping and how to categorize them. Within the more than one hundred category systems and coping scales examined by Skinner et al. (2003), no two included the same set of categories. But the suggested categories claim to cover the phenomenon in a universalistic manner.

Lazarus and Folkman's well-known distinction between problem-focused coping (actively changing the situation) and emotion-focused coping (controlling negative feelings if the situation cannot be changed) seems to be problematic because these activities are often mixed. Skinner et al. (2003) prefer the distinction between engagement coping, which brings the individual into closer contact with the stressful situation, and disengagement coping, which allows the individual to withdraw. Social coping is done by seeking social support. The individual handling of natural events occurs in relation to people's values and social resources. This "embedding" can be viewed within the model of "sense of community" (Sarason 1974) and the "social capital" framework (Putnam 2000). The latter concept in particular is often used in the context of development aid work (Grootaert and Bastelaer 2004).

Of special interest to this contribution is a form of coping within appraisal-based approaches which has been called "meaning-centred coping". This is seen as positive cognitive restructuring, examining beliefs and values, reordering life priorities, infusing new meanings and finding benefits in adversity, especially in cases of chronic stress that may not be amenable to problem-focused efforts (Folkman and Moskowitz 2004). Later a future-oriented preventive coping sometimes called proactive coping was added.

The array of coping strategies originating in psychology has thus far defied generalizing categorization and systemization with the aim of creating definitive classifications valid for all cultures. Concealed behind this goal is a view, likely stemming from a more Euro-American academic context, of coping as a universal psychological mechanism inherent in every single person, rather than something that one summons forth in cooperation with other people in social communities and that is, therefore, differently constructed and experienced in different communities. According to the dominant tendency of psychology, one tries to elicit cross-cultural generalizations by limiting one's analysis to cognitive inquiries, conceptualizing emotions on a rather cognitive level and de-emphasizing the context of coping whenever possible. In this research strategy, meaning-making must also be conceptualized so that one records the fact of meaning-making (and, if need be, whether it is "positive or negative" meaning-making), but largely refrains from considering the content. In this way, one tries to manage difficult interrelationships that are very obviously connected with coping, such as the religious and spiritual spheres and, more generally, the socio-cultural integration of coping and the variety of coping strategies that grow out of it.

3.2.2 The psychology of religion: religious coping

Religious psychology (see, for example, Emmons and Paloutzian 2003) and a psychology of religious coping (see Pargament 1997; Pargament, Koenig and Perez 2000) thus came into being, through which generalizations and methods of measurement were developed that were intended to be universally valid. In their concept of forms of religious coping, Pargament and his working groups focused strongly on cognitions (appraisals) that are capable of being studied, and brought the search for meaning to the forefront of their model of religious coping. According to this view, the search for meaning filters and influences the interpretation of the situation. The system of religious beliefs, which is in turn a part of the general person-specific belief system, mediates between the situation and the answer to it. In addition to the search for meaning, other psychological "functions" emerge that presumably play a role in religious coping, such as acquiring control over the situation, obtaining comfort and closeness to God and gaining intimacy with others. Pargament portrays this accepted religious belief system (insofar as it is related to coping) as a relatively static array of options that are shared to variable and measurable degrees by the adherents of that system. Certain religious convictions or options (for example, the concept of a punishing God) are less conducive than others to the well-being and health of believers, again to a measurable degree.

3.2.3 Resource-oriented coping approaches

Hobfoll (1998, 2001) follows a more social-psychological approach and devotes greater attention to the socio-cultural context in which coping occurs. His coping/stress model is thus located between the universalizing individual-related models and those that search for culturally specific forms of coping in a social context.

In his theory of resource conservation (COR), Hobfoll presents an alternative to the appraisal-based stress/coping theories, because the COR theory focuses on the objective as well as culturally constructed character of the environment. Hobfoll (1998, 2001) and Hobfoll and Buchwald (2004) put access to resources instead of subjective appraisals in the centre of their stress/coping theory, in which loss of resources is the deciding ingredient in the stress process. In this process, resources will also be used to prevent a loss of resources. As more and more resources are lost, people become increasingly vulnerable to the effects of stress and fall into a spiral of loss, with serious consequences. Though the "objective" loss of resources is the focus of the theory of Hobfoll et al. (2003), as in the appraisal-based model resource change is assessed by questioning individual subjects, either by using a questionnaire or by conducting interviews.

Hobfoll criticizes appraisal-based stress theories for providing too little insight into how assessments of a situation as threatening are reached, how they are produced as a result of learned or overly internalized interpretive rules, and to what extent they represent the shared cultural scripts of a community. When one excludes the surrounding social and cultural context and explains the assessments only based on the views of individuals, one creates, according to Hobfoll, a significantly faulty construction.

3.2.4 Cultural Psychology

Cultural psychology adheres to a research strategy, by which one assumes that, in addition to what people have in common, embedded socio-cultural differences and peculiarities exist that reach deeply into psychological mechanisms and ways of functioning.[2] Interestingly, this age-old conflict between nature and nurture has taken on new relevance through biology, as neuro-imaging has helped provide support for the cultural specificity of psychological mechanisms. Kitayama and Uskul (2011) thus consider that the debate between the cross-cultural and cultural psychologists has been decided in favour of the latter, and write: "It is thus reasonable to hypothesize that recurrent, active, and long-term engagement in scripted behavioral sequences (what we call cultural practices or tasks) can powerfully shape and modify brain pathways" (Kitayama and Uskul 2011: 421). In their model of "neuro-culture interaction", psychology is no longer cognitive. The influence of cultural behaviours on the brain, then, is unmediated by any symbolic or cognitive representations, and cultural influence is mainly exerted by doing and practising what is relevant in the cultural context (Kitayama and Uskul 2011). This sounds almost like Bourdieu's concept of the habitus, which is likewise not conveyed through cognitive representations, but rather conceives of the body as the embodied social (Bourdieu and Wacquant 1996; Bourdieu 2002).

The search of cultural psychologists for psychological particularities in specific socio-cultural contexts was propelled forward above all by developmental psychology (see, for example, Greenfield et al. 2003). Anthropologists developed a similar perspective in psychological anthropology (LeVine 2010). Sometimes the term ethno-psychology is also used. In the cultural psychology of religion (see Belzen 2010), one presupposes the cultural and contextual specificity of religious phenomena. "Psychology of religion should try to detect how a spe-

---

2 In contrast the school of so-called "cross-cultural psychology" developed universal dimensions and their corresponding measurement instruments (e.g. regarding the distinction between an interdependent and an independent self), within which psychic structures and functions are supposed to vary (see Zaumseil, 2006).

cific religious form of life constitutes, involves, and regulates the psychic functioning of the persons involved" (Belzen 2010: 344-5).

In sum, it can be said that psychological research and theory-building with regard of the psychology of religious coping offers a rich potential stimulus for the treatment of natural disasters in particular cultural contexts. In the psychology of coping and its subspecialties, an array of research approaches exist that are more universalized and more related to the individual and his or her cognitively influenced appraisals. These strategies focus on inner, mostly cognitive processes within the person and deemphasize the role of context. So far there has been no cultural psychology in the special field of coping with disaster. The book edited by Marsella et al. (2007), which promises ethno-cultural perspectives on disaster and trauma, is written by experts on individual and collective trauma experience, posttraumatic stress and related syndromes, as well as emergency and crisis intervention. It is focused on intervention and follows a somewhat essentialist minority approach. Following a special scheme in each chapter, emotional, psychological and social needs, as well as the communal strengths and coping skills that emerge in disasters, are documented for major minority groups in the United States. The authors offer generalizations for African Americans, Native Americans, Arab Americans, Asian Indians, Chinese Americans, Caribbean Americans, Latin Americans, Native Hawaiians and Vietnamese Americans. This is rather different from our understanding of cultural specificity (see below).

3.2.5 Trauma and "disaster psychology"

The psychiatric or clinical category of traumatization due to extreme psychological stress occupies an exceptional position in association with pressure situations. Like the coping model, scientific models and findings with regard to the conceptualization of extreme suffering and the processing of such suffering contain cross-cultural, generalizable statements mixed with interpretations largely based in Euro-American culture (see Young 1995; Argenti-Pillen 2000). The discourse on trauma and the practices associated with it are clearly demonstrated examples of how the global discourse is linked with local beliefs. The western concept of trauma is a particular way (individualizing and pathologizing) of understanding terror and extreme suffering and of treating it medically ("therapeutically"). As such, it emerges as a mostly foreign assessment and aid technique in international disaster management. In our examination, we were not concerned with the diagnosis of traumatizations in the sense of a clinical definition of PTSD (posttraumatic stress disorder). The psychiatric classifications tend to make ontological implications about cumulative classifications of disease. In the Euro-American cultural context, it is only relatively recently that these phenomena have been seen as symptoms of a disease (posttraumatic stress disorder or PTSD) that is amenable to analysis and treatment by psychologists and psychiatrists ("officially" in the American Diagnostic and Statistical Manual of Psychological Disorders or DSM III for the last thirty years). In the beginning "disaster psychology" (see Reyes and Jacobs 2006) was identical to disaster mental health and trauma psychology. Since then it has become intervention-oriented and follows a broader psychosocial community approach instead.

3.3 Cultural anthropology

3.3.1 Concepts of culture

Coping is a process that takes place within a local cultural setting and a local history. Cultural anthropology, which addresses and conceptualizes the socio-cultural and the locally specific scientifically, is itself diverse and in a process of change. As psychologists, in trying to orient ourselves within the broader scope of the social sciences, we find ourselves confronted with various conceptions of culture. Among these we rely on a "more recent" understanding of culture (see below), being aware that we may oversimplify matters when we dissociate ourselves from a "traditional" essentialist concept, according to which culture follows a sort of container and is regarded as a secluded, original entity which is constant, homogenous and given by fate. Instead of this outdated essentialism of culture, Hörning and Reuter (2004) conceptualize culture as practice, as something which is done. They reject conceptions which confine culture in a cognitivist manner to mentality, to text, to web of meaning or to a system of values.

Schlehe (2006a) follows this conception according to which culture is interwoven with other phenomena of the social order because it permeates all spheres of (practical) social life. She shows the difference between the essentialist and the new conception by contrasting an understanding of culture as complex, dynamic and hybrid with the outdated essentialism of culture. According to her, cultural complexity means diversity within culture, while polyphony and multivocality are permanently negotiated. Cultural dynamics understand culture as process with a historical dimension. Cultural hybridity means that culture and cultural identities exchange with other cultures and are not secluded, so through exchange there is a constant formation of hybrid forms.

3.3.2 Conceptions of culture in studies about Indonesia

If one approaches ethnographic findings on Indonesia or Java with this conception of culture in mind, critical distancing is needed on the one hand; while on the other there is an opportunity to reflect critically on the essentialization of Java and its embeddedness in power structures. Java and also Bali were important settings for instrumentalization, as well as for the attempts to shrug off essentialism. This went hand in hand with discontinuities, infighting and a shift in the paradigmatic and methodological preconditions, mirrored in the formulation sketched out above concerning how we want to approach cultural phenomena. These critical approaches have the goal of debunking the changing purposes of anthropological research, with its roots in the colonial period (see Pemberton 1994; Boellstorff 2002; Antlöv and Hellman 2005).

According to Antlöv and Hellman (2005), the image of "Java" was part of Orientalism – a refined, mystical "Other" concocted to maintain the self-image of a rational Europe. The Leiden school of former Dutch colonial anthropology created sophisticated tools for Dutch rule and laid down their construction of "ethnolocality" (see Boellstorff 2002) between 1910 and 1955 in 46 Volumes of *adat* (traditional law and custom). This strategy focused on the local and the ethnic. According to Antlöv and Hellman (2005), this form of cultural engineering was continued and reinforced during the rule of Suharto in the big "Proyek Inventarisasi dan Dokumentasi Kebudayaan Daerah (Project to inventory and document the culture of the country)". It looks quite ironic that, in the present period of democratization and decentralization in Indonesia, there is again a revival of *adat* (Henley and Davidson 2008), now in the name not of colonial rule but of self-determination and/or the reestablishment of local elites. In this way the editors, Davidson und Henley (2007) – the latter again from the Royal Netherlands School at Leiden– offer a self-reflexive trajectory from colonialism to indigenism.

3.3.3 Javanese specificity?

Until now, we have demonstrated how culture can be understood and the aims to which the construction of a specific Javanese culture or any other cultural specificity can be put. In what follows, we will review the particularity of Javanese culture and how this specificity can be related to coping strategies following the earthquake. This problem is the same for Java as for any other cultural context, such as the recent disaster in Japan (March 2011), when speculations abound concerning the Japanese mentality, which is held responsible for the specific way of dealing with this catastrophe.

For geopolitical reasons there has been increasing American involvement in Southeast Asia since the 1950s. Clifford Geertz (1960, 1973, 1985, 1996) became the "founding father" of post-colonial anthropology on Java and introduced the notion of "culture" as a symbolic system with a focus on cultural expression. The goal of this interpretive approach was to reveal not peoples' everyday practices and concerns but the most enduring cultural features of Java and a generalized view of the typical Javanese. There was a tendency to speak in broad outlines and cultural formulas and to reduce human experiences analytically to expressions of culture (Braten 2005). We devote further attention to the views of Javanese experts (see Hildred Geertz 1961; Magnis-Suseno 1981; Koentjaraningrat 1985/89; Mulder 1978, 1990, 1994; Stange 1984) in the text below.

Possibly the "Javanese" have been over-generalized and over-typified in the interpretative approach of Geertz and many of his followers. Expressive forms in art, theatre, ritual and even social life were read and interpreted as "texts" from which to distil the typical "Javanese". Through this essentialization, the cultural dynamics, inner diversity and external influences were outside the scientific endeavour. This picture of Javanese culture often came from urban male middle-class informants (e.g. in the popular books of Mulder). Anthropologists like Schlehe (2006a, 2008, 2010) or Beatty (1999) offered a more dynamic theorization of culture.

Beatty (1999) presented a differentiated analysis of the religious life of a village in East Java (near Banjuwangi) and found, at the beginning of the 1990s, a great variety within what he called "Javanism" as well as "practical Islam" in the countryside. Javanese mysticism was embedded in the currents of popular religions, instead of being limited to urban or aristocratic elites. He found that "folk beliefs, orthodox Islam and mysticism came to flourish alongside each other in a single setting among rice farmers of broadly similar background" (ibid.: 35). Ricklefs (1979), among others, has suggested revisions of the standard categories introduced by Geertz (1960)[3] by finding multiple shades of grey between the polar types. But according to Beatty, the complexity of Javanese civilization resides not just in plurality but in interrelation, in the dynamics of religious adaption and change.

In sum, one can say that, for a differentiated observation of religiously or spiritually inspired coping in Java, it is necessary to take into account very different forms of this inspiration. Geertz is to be commended in that he employed a highly inclusive approach that went beyond narrow definitions of religiosity. Many authors concur with the juxtaposition of Javanism and Islam. Woodward (1989) goes so far as to consider Javanism, in its popular as well as its mystical forms, as being ultimately rooted in Sufism, the old Muslim form of mysticism, and therefore perceives it as embodying the local form of Islam itself. Beatty states that this view can be supported, at most, for the court cultures of Yogyakarta and Solo (for a discussion, see Beatty 1999: 29, 35).

Spirituality is seen as a broader understanding of religiosity which includes the subjective feelings, thoughts and behaviours that arise from a search for the sacred. The sacred may be a divine being, divine object, ultimate reality or ultimate truth perceived by the individual (Hill et al. 2000: 68). Prawitasari (2008: 329) stresses the diversity from an Indonesian perspective: "We have plenty to offer in this approach since we do have spirituality and mysticism in each part of this country. Each ethnicity has its own local wisdoms as to how to behave with others, with the divine, and with nature." Mysticism is thus a special form of

---

3   Geertz distinguishes three categories: pious rural Muslims (*santris*), farmers with indigenous Javanese traditions (*abangan*), and members of a traditional, more urban aristocracy, who are nominally Muslim but practice a form of mysticism (*priyayi*).

experience and insight which can be included within the phenomenon of spirituality.

Beatty (1999: 158) specifies the difference between Islam and Javanism in the following way: "Where Islam promises heaven through ritual compliance and devotion to the Koran, Javanism (*kejawen*) takes the everyday world as its key text and the body as its holy book." "The world is permeated with symbolism, and it is through symbols that one meditates on the human condition and communicates with the divine" (ibid.: 160). Learning therefore means the pursuit of self-knowledge "...in the sense realizing one's intrinsic universality. There is the conception (or experience) of God as dwelling in man...and the experiencing self or some animating principle within the self is in some sense divine" (ibid.: 164). And further:

> The basic schema of Islamic ethics – the categories of obligatory, recommended, neutral, disapproved and forbidden – is at odds with the way moral issues are framed in Javanist thinking. For Javanists obligations come from outside and are therefore considered inferior to internally motivated acts. What comes from oneself and is freely given is superior to what is done to order. (Beatty 1999: 182)

Particularly relevant for our study are the various publications of Schlehe (2006b, 2008, 2010), in which she studied various aspects of the same earthquake in central Java with which we are concerned. The publications also represent the newer anthropological approach to disaster research in that she attempts to link the political, ecological and technological aspects (Schlehe 2006b). As with Frömming (2006), her aim is especially to pursue the question of continuity and change, and to show that nature represents a cultural construct that is subject to constant changes, new definitions and manipulations (Schlehe 2008). In the publications from 2008 and 2010, she emphasizes a religious anthropological approach to disaster research. She tries to identify discourses about the 2006 earthquake near Yogyakarta. In her interpretive generalization, she identifies a tension between what people see as tradition and as modernity (Schlehe 2010).

Using the example of the earthquake, she cautiously believes she can identify a new "tendency."

> Obviously, many Javanese do not harmonize anymore, but rather nowadays tend to polarize, thereby using globalised idioms and characterizations: they create binary categories for what they call traditional, local values on the one hand – and on the other hand they position moral decline, exploitation of nature, materialism, Western democracy and what they define as modernization. It remains to be further investigated in other contexts and over a longer period of time whether this observation holds true for a general new tendency in Javanese society. (Schlehe 2010: 119)

This finding, documented shortly after the earthquake, would certainly be interesting, and would also relate to psychological modes of processing and the ways in which the people in the village interact with one another. With regard to the possibilities for generalization, she puts forward the view that:

> Religious ideas are explored here to reveal what they can tell us about the ways in which Javanese construct and conceive natural hazards.... A culturally specific and site-based reading of the discourse on a natural disaster exemplifies that disaster-linked cognitive coping strategies are always unique and contingent formations in response to local culture and politics. (ibid.: 113)

In so doing, she expressly incorporates the Javanese tradition and local religious beliefs as sources of meaning attribution, as well as taking into account transcultural dynamics: "the culturalization of natural events brings both cultural and transcultural dynamics to light"(ibid.: 112).

If the knowledge and reflexive potential of cultural anthropology is translated on to other scientific domains, it seems desirable to remember the very construction of the "Javanese" and of tradition in order not to overstress their specificity but to use them as a potential background for the interpretation of data. Many findings suggest the great internal diversity and dynamics of Javanese inwardness, spirituality, and the attitudes linked to them.

3.4 The participation of village people and local structures

The production of knowledge does not limit itself to questions of access and application with regard to disciplines, but can and should be integrated into the field of research. Empirical methods of theorising, like Glaser and Strauss's (1967) grounded theory, do not produce any universally valid theory. According to Strübing (2004) they are based on a conceptualization of reality following Peirce and Dewey, which is never finished, in the sense that their theoretical formulations have to be tested by acting on reality in such a way as to develop that reality further.

Thus we must develop a position for how we deal with our research results, deciding to what extent we make them available to those we have researched and, through this process of feedback, to what extent we participate in the villagers' process of coping with the ongoing threat, or deciding whether we want to abstain from this. In other words, we become, through our research, a part of the coping process and therefore a part of what we are researching. All the difficulties involved in gaining access for research purposes, the local authorization procedures through the district governor and our integration into the villages make us a part of a local political process. The co-author, as a representative of a large local university (Gadjah Mada University or UGM), is responsible for the research project in Java and the local colleagues and cooperative structures there, and ensures that the project is embedded in a particular way. Reflection upon this process is, in our opinion, essential.

The co-author and her colleagues had already acquired experience of participatory research before the project started and were thus able to make it fruitful for our research project. With her team she developed a new method of performative social research ("Happy Stage") by using the tradition of the local theatre (*Srandul*). This project (Hadiyono-Prawitasari et al. 2009) took place in the same village where we did the research and was a very valuable experience. Exactly one year after the earthquake, that is, between July and December 2007, the Happy Stage method was performed as a cycle of action research. Voluntary groups of five neighbourhood units performed specific topics. These included the time before and directly after the earthquake, the moment of the earthquake, hidden social conflicts due to perceived injustices over the distribution of material aid and the ongoing programme of reconstruction at the time of the performances. It was interesting to see that conflicts of power concerning the performance reflected the attempt to exercise influence on critical scenes and thus became a topic in itself.

In later interviews and discussion groups, an attempt was made to evaluate the impact of the whole event. The method was considered quite useful "in order to generate a process of reflection in the villages, to revitalise local knowledge and to manage or mediate perceived social conflicts. It is questionable whether the performance of village life by the members of the community will endure because the elites still have the power to determine all processes of decision" (Hadiyono-Prawitasari, 2009: 287).

In the last phase of our research (2011 at the time of writing), it is planned to do more participatory research in order to deal with the future as a further topic, not in the sense of risk management by the village people, but with the aim of open results which will include the perspective of another perceived possible threat.

4. Coping psycho-spiritually with the earthquake

This report is intended to show how the local psycho-spiritual coping process following the earthquake can be understood, and which generalizations can be drawn from it. The project as a whole encompasses a broader spectrum of inquiries and conclusions, which are, however, also linked to psycho-spiritual coping.

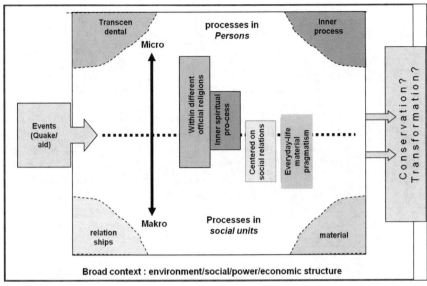

Framing the results: different ways / axes of coping & change.

Drawing on a broad conception of coping, we conceive of the occurrence as a process that highlights the earthquake and the subsequent aid as incidents that may lead to transformation or preservation. Transformation or preservation of what exists can be observed on the micro-level of persons and families or the macro-level of larger social units. In interviews, focus-group discussions and field observations in the three villages that were studied a variety of ways of processing emerged, not any unified process. We have given structure to this diversity in the form of axes or dimensions, within which diverse forms of coping and focal points can be distinguished from one another. Psycho-spiritual coping plays a large role in the statements of many interview partners and is sometimes expressed and described very emphatically. It plays itself out largely in the framework of the official religions, above all in Islam but also in Christianity, or else in the framework of spirituality and inner life. Overlaps between Javanese beliefs, which are not recognized as a separate set of beliefs, and the official religions are partially dealt with below the surface. Other interview partners, in describing their experiences during and after the disaster, concentrate more on the axis of social relationships and the social support that they experienced or lacked, and less on the psycho-spiritual dimension. Still others addressed the issues of daily life and the material dimension of their efforts to cope after the disaster. The axes mentioned and the terms, attitudes and beliefs associated with them came to the fore in almost every interview, and in the beginning we tended to assume relatively unified beliefs. However, related traditional, apparently Javanese concepts and terms have different meanings for different people and seem to describe only a unified framework, as Beatty (1999) similarly demonstrated with regard to the ritual practice of *slametan* (the common celebratory meal) in eastern Java.[4]

This common framework represents a particularly local collection of orientations that is sometimes termed "local wisdom". Perhaps it is better to see it as manifold local wisdoms with an inner heterogeneity. It comprises interrelated webs of beliefs and embodied stances coloured by feelings, values and practical attitudes, as well as by practices with particular virtues and "philosophies". As the elements are interrelated, they are difficult to separate from one another: an analytical dissection would rob them of their meaning. Thus the situation is indeed interpreted through their description and denomination, and it is met with specific attitudes that have strong emotional overtones and virtues. The villagers act in relation to one another, negotiating with one another as well as about what they feel and perceive (such as the almighty power of God) and on what is to be done. We understand "coping" as this interdependent, composite entity and have therefore not separated the explanation and interpretation of the earthquake from the answer to it and the attitudes and virtues to be realized from it. We call this web of beliefs and practices "Javanese" with reservations, as in doing so we run the danger of regarding a part of local existence as a cultural phenomenon. We believe that what we describe locally is summoned forth by many influences, of which what can be classified as Javanese tradition represents only one aspect. The notion of what is Javanese is a construction, and also a work in progress, in which many interests are invested and many stakeholders involved.

4.1 Methodological section: descriptive analysis and theory production

In qualitative social research,[5] computer programs are available to support data evaluation and theory building. We use the program ATLAS (see Muhr 2004). Using coding, one searches from the very beginning for theoretical concepts for the phenomena being researched, thus identifying unities of meaning. Against this background – and always maintaining close contact with empirical reality

---

4 In addition to this coexistence of diversity, Beatty found that much effort was devoted (above all in the collectively practiced ritual meal, the *slametan*) to finding a collective language and to maintaining a collective ritual practice: "the unifying factor is a willingness to make concessions in order to maintain social harmony (*rukun*) in the neighborhood" (Beatty 1999: 156). This seemed to be important for survival. The bloody history of the massacres of 1965/1966 and the violent outbreaks that occurred again and again in different areas of Indonesia on the boundaries of this process force the Javanese almost to exaggerate *rukun* in order to maintain this diversity –which is always threatening to disintegrate –together through the tolerance of many voices and through inclusive ritual practices such as *slametan*.

5 See Glaser and Strauss's (1967) influential method of grounded theory.

and the data – one searches systematically for new instances that unlock more knowledge, single phenomena and their contexts, in order to work out larger unities of meaning, categories and interrelations within the categories with the aim of developing a theory suited to the subject at hand. At decisive points in the research process and in constant contact with the reality of the data, the various research styles that are grouped under the label of "grounded theory" engage in a creative re-configuration and re-discovery of theory by using a "theoretical sensitivity" (see Glaser 1978). This is an awareness of specialized scientific theories and concepts that is not to be understood in the sense of the acceptance and use of these models. One should have an array of theoretical tools available, in which sundry critically deconstructed parts lie ready for creative new combinations (see Zaumseil 2007).

For the implementation of practical research, Clarke (2005) has developed a fruitful approach to the social-scientific analysis of complex situations. Basing this approach upon and exceeding the boundaries of grounded theory, she creates links among the disciplines and various scientific viewpoints.

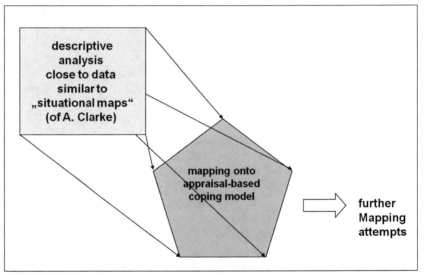

Playing with theory – using coping as an example.

Accordingly, we first develop "situational maps" closely based on the data. These "lay out the major human, nonhuman, discursive, historical, symbolic, political and other elements in the research situation of concern and provoke analysis of relations among them. These maps are intended to capture and discuss the messy complexities of the situation in their dense relations and permutations" (Clarke 2005: xxxv). Modifying Clarke's approach in the next step of theory reflection and production (see below), we use one of the family of stress/coping models (see below, the example of appraisal-based models) and attempt to build on what we have discovered in the process of descriptive analysis by playing with the models, as Clarke does. This helps us advance, in that one sees what one must change in order to be compatible with the data, and where the model does not fit. In our visualization (above), the square would be somehow projected upon the pentagon.

Next, one can try to map the results of the descriptive analysis onto other theoretical models such as the resource-oriented or salutogenetic models, or one can attempt to employ the vulnerability model of disaster researchers and see what is missing and what needs to be fleshed out. In this way, we try to bring various insular disciplinary approaches into contact with one another.

4.2 Some samples from the descriptive analysis of psycho-spiritual coping
The inquiry into psycho-spiritual coping is work in progress, and there is not enough space here to elaborate on the findings so far. We would therefore like to present some examples of descriptive analysis and theory work below.

4.2.1 Attribution of meaning and explanation of the earthquake
In most of the interviews, some interrelated concepts for the explanation of the earthquake are used that delineate the framework for various connected ideas. The event is understood
- as a warning or reminder (*peringatan*),
- as text (*cobaan, ujian*)

It is understood less often as an expression of the wrath of Allah (*murka Allah*).

These interpretations are most often employed in the service of a greater intimacy and connection with God. They are often used in parallel with scientific explanations and are, for example, linked with the knowledge that the region in which one lives is located along a fault line. The standard formulas being used appear to be an umbrella for an extensive array of personal meanings and spiritual ideas.

The exact nature of the meaning attributed to the earthquake cannot be separated from the coping process. It is first developed partly as a personal meaning during the coping process, and the decoding of this initially hidden meaning (*hikmah*) is an important aspect of spiritual development, described to us as moral maturation through the earthquake.

4.2.2 The coping process
One easily runs the risk of imposing conceptual and terminological dualisms from the Euro-American understanding of the world and people onto local beliefs, whose inner logic does not admit such separations. We use the term *Haltung* ("stance" as a more embodied term compared to "attitude") in order to ex-

press complex orientations that are contained in the concepts of *nrimo* (acceptance), *pasrah* (surrender) and *syukur* (gratitude when face to face with a higher power). These stances delineate the framework of the coping process.

The translations mentioned are always incomplete and rudimentary, exactly because these stances contain the unity of a positioning toward God and to the spiritual and social worlds in the sense of moral order and to the cosmos that one has to realize physically, corporeally (also as posture), emotionally and as a bearer of meaning (including the spiritual dimension). In this respect, the emotions, for example, or emotional management in the sense of coping cannot be separated from these stances because the realization of the stances relies on emotions in unity with the situation at hand (see Beatty 2005a, 2005b). Values and virtues often described as Javanese are similarly closely linked to these stances. The moral orders, including these stances and virtues, rely upon particular notions of God or a higher power, and the deeper meanings (*hikmah*) of the earthquake can be deduced via divine agency or spiritual sources. The meaning of active coping (*usaha*) with regard to the earthquake can also be deduced from this web of references.

What experts on Java in the 1980s described as Javanese wisdom, ethics and spiritual inner life (see Magnis-Suseno 1981; Koentjaraningrat 1985/89; Mulder 1978, 1990, 1994; Stange 1984), does not seem to us to contain such a high degree of cultural specificity, but rather uses the holistic, broad interconnections mentioned above, within which new configurations again emerge out of new and old elements.

4.2.2.1 The stances: Nrimo/Pasrah/Syukur/Usaha[6]

In a first step we tried to obtain clarity about these stances. This implies that we distanced ourselves from cognitive dissections, like separating the emotional, cognitive or action-centred dimensions in order to do justice to the complex contextualization and interrelatedness of the terms. The stances which interest us are perceived as diverse figures of configuration which are used to confront the events and initiate a process of orientation.

Nrimo/Usaha

Nrimo means to accept or receive what happens (non-acceptance means further suffering and sickness). The outer occurrence is predetermined (*ditakdirkan*) and becomes an assumed (*diterimo*) fate: *nasib*.

A woman from Village B speaks about her situation regarding her foot, which was severely injured in the earthquake:

---

6   Due to lack of space we limit our discussion to these four attitudes and do not consider *semangat* (with dedication), *sabar* (patient), *santai* (serene) or *menata hati* (control of one's inner being).

Yang membesarkan hatinya seumpama sudah nasibku ini begini, ya diterima dengan senang, diterima apa adanya, ya kan, seumpama orang harus begitu ceritanya ya kan, nasibnya, saya ya harus ceritanya umur segini ya diberi ini sama Yang Kuasa, ya apa kakinya begini, ya untuk bekerja ya butuhnya segitu, harus diterima, menghadapi kisahnya.

What was heartening was to take it simply as my fate (*nasib*) like this, yes, to accept it with contentment, to accept it as it is (*diterima*). When such things happen to a person, it is fate (*nasib*). It is like a story (*cerita*) given from the Almighty. My story, my fate (*nasib*), was to have at this age a foot like this. Yes, I need it for working like this. It must be accepted (*diterima*), I face my story (*kisah*). (Bu S, Village B)

She describes the emotional stance out of which she accepts her disability and assumes the position of listener to her own story, which the Almighty has given her. Her acceptance of her fate does not mean being idle. On the contrary, she tries to find an active counterbalance; the search for alternatives prevents despair.

Orang kalau menerima kan kita sudah berusaha Mbak. Kita berusaha tidak putus asa. Nah gitu, jadi terus ayo berusaha. Nggak punya rumah ini ya bekas puing ayo dibikin rumah untuk berteduh. Sampai nanti ada bantuan untuk bikin rumah lagi. Menunggu itu juga sampai 7 bulan di tenda. Nunggunya itu cuman 7 bulan setelah itu kan bikin, bikin, lha dari puing-puing itu kan bisa dipakai.

People who accept (*menerima*) it have already tried (*usaha*) something, Miss. We have tried (*usaha*) not to be desperate (*putus asa*). That's how it is, so let's keep trying (*usaha*). If there is no house, let's make a house out of these ruins to shelter us until there is more aid to make a house. We waited up to seven months in the tent. We waited only seven months. After that, we built houses from the ruins that we could use. (Bu D, Village B)

Pasrah

With far more than simple acceptance, with an emphatic faithfulness (*pasrah*), the people give themselves up to the Almighty, and this emphasis and unrestraint – as our interview partners stressed again and again –simultaneously lends them strength, serenity and peace.

'Sunami-sunami', ada orang begitu, lha terus aku kan ditinggalin to Mbak? Ditinggalin sama orang-orang semua. Saya bawa anak dua, mbopong itu Prastowo, terus (....) dah ditidurin. Haduh ya allah, lha saya kan nggak mungkin kan Mbak, bawa anak lari 2 kan Mbak? Ya udah, ya sudahlah kalau memang ya wis nek ana banyu yo wis aku kudu mati karo anakku gitu kan. Ya nggak mungkin saya bawa satu terus anak satu gini dalam keadaan terkapar gitu. Ya udahlah aku nerima nasib aja.

'Tsunami, tsunami!', people said. Then I was left behind, I was left by everybody. I carried my two children, and carried Prastowo on my back, and then (...) so I leaned him against my back. Oh my God, Allah, you know, I wasn't able to run away and carrying two children right, Miss? So I settled myself (*ya udah, ya sudalah*), so if the water would come, I was OK to die with my children. It's just impossible to run away with one child, whereas the other one was severely injured. So I was settling myself (*ya udalah*), I accepted my fate (*nerima nasib*). (Bu E, Village B)

Bu E discusses here the meaning of *pasrah* in connection with her severely injured son:

Waktu itu kan ya belum Mbak. Cuma ya udah kan ini yang diobati dulu yang ini. Ya udah, sementara yang Prastowo udah tak biarkan kan ya udahlah kalau memang dia memang dah nggak ada ya udah saya relain gitu. Kan yang kalau yang diobatin kan yang kelihatan luka-lukanya dulu Mbak sana. Kalau yang nggak papa tapi cuma gini kan dah dibiarkan, memang gitu. Jadi saya juga harus memang ya udah harus pasrah seandainya dia tu nggak ada. Kan nggak mungkin saya bawa infus terus ini sementara ini anak saya [...] diminumin kan mutah Mbak. Katanya kalau gitu kan gegar otak gitu. Jadinya saya ya udah. Kalau wong suami saya juga begitu, terus anaknya nggak ada jadi saya ya allah ya usah saya pasrah aja. Yang penting yang ini dulu kan. Katanya 'Mbak ini infusnya harus dipegangin lho!' Jadi saya gini, saya gini, nuggu anak yang ini dah diam aja, tak gini-gini tak nganu nggak.. yang harus saya lakukan ya yang kelihatan ini apa lukanya dulu gitu lho Mbak.

At that time, we didn't get any treatment yet, Miss. So it's okay (*ya udah*) that this child was treated first. It's OK (*ya udah*). In the meantime I left Prastowo untreated. I accepted (*ya udah, relain*) the condition in case he would die. The ones who received immediate treatments had open wounds, Miss. But the ones that were like Prastowo were treated later. So I had to accept (*ya udah*) the condition at that time in case he would be dying. Since it's impossible for me to hold the infusion while the other kid was [...]. Since he was vomiting, Miss, when I gave him to drink. They said, that means he's got a brain concussion. So I accepted it (*ya udah*). While my husband was away, and I was going to lose my son, so Oh my God Allah, so I submitted my fate (*pasrah*) to God. So it's better to take care of the other child first. They said, 'Miss, you have to hold the infusion carefully!' So I was doing like this, and waited for my other son. He just kept quiet, I moved him, but he kept quiet. So what I did was to take care of the open wounds first, Miss. (Bu E, Village B)

Gratitude (Syukur)

Being grateful, a semantic field of meaning, is closely related to a complex of patience, relaxation, humbleness, repose and inner peace.

Waktu gempa itu terus ya saya bilang sama apa ibu saya itu 'Yang sabar aja lah buk', gitu. 'Saya udah ngrepotin segala macem (...) Ya saya jalani dengan ya bersyukur lah masih diberi kesempatan untuk hidup. Jadi cuma santai aja lah!'

In the period after the earthquake I said to my mother, 'Hopefully you can deal with this patiently (*sabar*)', like that I said. 'I myself have already had troubles with all of this (...) Anyway, I faced it by being grateful (*syukur*) that I was still given a chance to stay alive. I faced everything in a relaxed (*santai*) way.' (Mbak A, Village B)

Another interviewee:

Itu harus kita syukuri, apapaun walaupun dibandingkan dengan yang lain, itu sedikit katakanlah. Dia dapat tiga atau empat, kita cuma dapat satu, kita harus tetap bersyukur. Itu kuncinya, kalau itu, apapun yang kita dapat itu kita syukuri, pasti tentram dan damai.

Even when we receive less in comparison to others, we have to be grateful (*syukur*) for it. When someone gets three or four things and we only get one, we still have to be thankful (*syukur*). That's the key. Whatever we have received, we should be grateful (*syukur*), and then we will surely experience tranquillity (*tentram*) and peace (*damai*). (Pak B, Village A)

4.2.2.2 Intricate relations: situational maps
Meaning and coping
We used situational maps (Clarke 2005) in order to formulate hypotheses about the connections between elements. One example is the following structuring of relations, which visualises how coping strategies are linked with assumptions about the meaning of the earthquake.

In using the concept of *hikmah*, many interview partners (IP) assume that the Almighty keeps a deeper positive meaning for the catastrophic events that happen. We discovered no simple precise meaning for this concept. In the following passage, we present a "messy map" of meanings that we have drawn together experimentally into a bigger picture using interview statements from various people. In this way, we developed something like a field of meanings, but we obscure the variety. For example, for some interview partners the meaning of *hikmah* boils down more to material gain than to spiritual refinement. Our impression is that it is a mistake to demand an absence of ambiguity from Javanese terms. In this case ambiguity, in the sense of a fuzzy concept or a "boundary object" (see Star and Griesemer 1998), provides a terminological umbrella for diverse interpretations, similar to the various possibilities for bringing diversity (and sometimes contradictions) under the umbrella of collectively practised ritual stances (see Beatty, 1999).

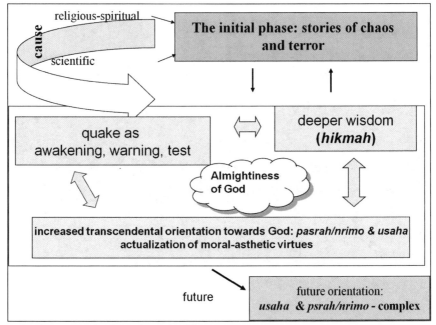

Hikmah: an example of mapping a specific concept.

Thus *hikmah* contains, among other things, something like a representation of inner growth that is embedded in the cultural context of a moral order. Often deciphering *hikmah* is described as an arduous process. At the beginning many report that they had to go through a period of inner protest and bemoaning towards God. Similar to these data, literature on religious psychology (see, for example, Schuchardt 2002) offers different ways of accepting one's fate (passively, actively, bemoaning, etc.). In the diverse manifestations of relationships with the spiritual (with corresponding notions of God), we can perceive a diversity that may reflect the status quo. To what extent this may mirror recent cultural dynamics on Java is difficult to assess so far.

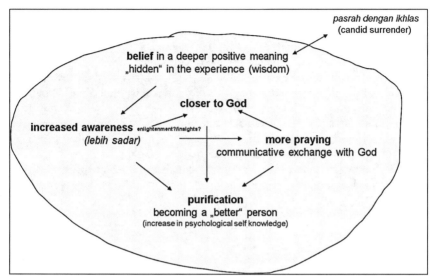

Semantic field of *hikmah*.

4.3 Theory-reflecting/producing analysis

4.3.1 Relationships to the appraisal based coping model

We would like to present and discuss an example drawn from theory-producing analysis of how theoretical building blocks from various disciplines can be played with most productively.

The appraisal-oriented coping model referred to above, based on Folkman and Moskowitz (2004), can be loosely sketched out as shown below:

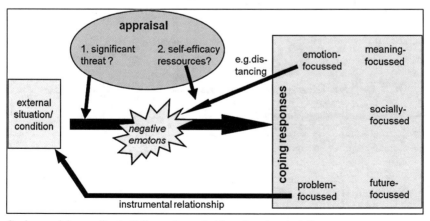

Coping model according to Folkman (appraisal based coping model).

An outside situation is subjectively perceived as a threat or challenge and likewise subjectively assessed in relation to one's own coping resources. The coping process that is thus put into motion is accompanied by negative feelings. The coping response is emotionally oriented, with a focus on negative emotions, and problem-oriented as much as possible, with a focus on the instrumental processing of the catalyzing situation. This basic model has since been expanded through other coping responses, only a few of which are presented here. It is also possible to manipulate the subjective meaning of the event, to mobilize social support (which is linked in turn to resources), and one can also react preventatively and in a future-oriented manner to upcoming threats.

We will now attempt to map the loosely sketched-out results of the descriptive analysis shown above onto the model, which would then look something like this:

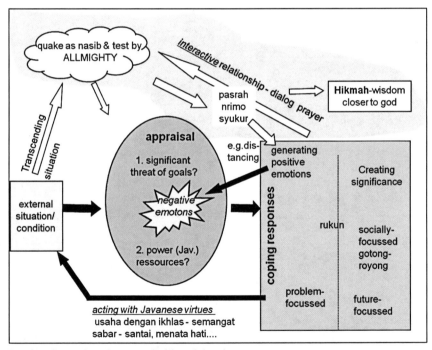

Local coping projected on appraisal-based theories.

We see that the external situation has been transformed or doubled. It is not only the external condition that has to be confronted with the earthquake, it is a test from the Almighty which has to be accepted as one's own destiny (*nasib*). There is not only a situation in the external world which has to be changed by an instrumental relationship. The situation has been transformed into a test from the Almighty within an interactive relationship with God and a dialogue with him. This implies a change in the stance towards the situation characterized by positions like *pasrah*, *nrimo*, *syukur,* etc., with the evoking of positive feelings which are included in this stance and which balance or overcome the negative emotions evoked by the initial appraisal of the threat.

With our scheme we are trying to show that the spiritual dimension we found in the local way of coping is more fundamental than being just a variation of emotional or meaning centred coping. First it is important to understand that the spiritual is experienced as reality. Secondly, it means a transformation of the whole psychological process and calls for a different conceptualization.

Inherent in the psychological coping model is the construction of special dichotomies that are specific to a special cultural view of reality. There is a special separation of person and environment and of external and internal processes which we do not find in the same way in Javanese thinking and feeling. In the western coping model, the individual does something with an external situation (this mirrors the behavioristic stimulus-response-thinking) and manipulates or manages his or her own emotions by a cognitive artifice called "meaning-centred coping" (like wishful thinking, benefit finding or other ways of manipulating reality.

In the coping form we found, the relationship with an external situation is expanded through an interactive relationship with the Almighty or with the spiritual world which encompasses specific meanings, shapes assessments of the event, activates attitudes charged with positive emotions and ultimately represents an aspect of active, communicative coping (for example, through invocations or prayers). The practical, situation-specific action (for instance, helping others, ensuring survival, etc.) is infused with and carried along by the attitude and positive feelings that result from the simultaneous creation of a relationship with the spiritual. At the same time, virtues are realized through this action, such as *dengan ikhlas* (whole-hearted and willing), *semangat* (with dedication), *sabar* (patient), *santai* (serene) and *menata hati* (control of one's inner being).

4.3.2 Relationships to religious coping

We are currently researching to what extent this form of religious coping reflects general patterns as described by Pargament (1997), Schuchardt (2002), Fontana (2003) and others. For Pargament (1997: 90), "coping is a search for significance in times of stress." He still deals with control of the external situation, but God appears in the postulated coping styles. The deferring style is more passive because control of the situation is placed entirely in God's hands. In the collaborative style, one seeks control through a close connection with God. In this way – when one seriously considers the assessment and the subjective construction of the situation – God becomes a conversational partner and a support in the situation, present and ready to be called upon (see quotation example). This changes the situation very significantly because this communicative collaboration with the higher being and its associated stances and feelings attain greater significance and also affect the practical (instrumental) processing of the situation. Here the question emerges of whether it makes sense to classify the particular forms of religious coping among the already acknowledged universal categories, or whether one should extract the specific forms of psycho-spiritual coping as defined by the cultural psychology of religion (Belzen 2010). Closely connected to this question are investigations – beginning with Geertz (1960) and cogently explored in more recent times by Beatty (1999) – that address the issue of the diversity of spiritually and religiously inspired relations to the world. An emphasis on diversity would also be appropriate to the recently observed revival of *adat* (traditional custom) in post-Suharto Indonesia. As Henley and Davidson (2008) assert, this may be related to the process of democratization and decentralization that followed the end of Suharto's authoritarian rule. Modifications would also be conceivable, in keeping with the aim of playing with theoretical models (see above).

When someone like Pargament differentiates among various kinds of relationships with God, one could speak on the one hand of a punishing God who demands submission, and on the other hand of a gracious and benign God, with whom one has a very personal and conversational relationship, which quite possibly has a somewhat cooperative character. It also speaks for the fact that such notions are in flux, and that a shift in the relationships and possibilities for understanding between people finds its counterpart in a shifting relationship with God:

> *Ibaratnya itu kan aku meminta, memohon dan me..curhat, curhat tapi curhatnya langsung kepada Yang Diatas gitu lho. Jadi mungkin ee sedikit berharap agar semoga keluargaku tetep selamat, tetep dijaga, nggak kekurangan apapun. Mungkin seperti itu juga lebih, tetep itu juga membantu. Membantu aku tu kayanya pasrah gitu lho.... Intinya mungkin pasrah, menyerahkans segala sesuatunya kepada Allah, tapi aku tetap berdoa agar keluargaku baik-baik saja.*

> It is like beseeching God, like sharing (*curhat*) my feelings, but sharing them directly with God. So maybe I was hoping a little bit that everyone in my family was safe and got what they needed. Maybe like that. But it did also help me. It helped as I was submitting everything to Him (*pasrah*).... The key was maybe submission (*pasrah*), handing over everything to God (*menyerahkan*). But I still kept on praying so that my family would be fine. (Mbak P, Village B)

In connection with *pasrah*, this interviewee expresses an almost intimate relationship with God. *Curhat = curahan hati* means "to pour out one's heart", an expression that one usually uses among friends.

The interpretative hypothesis presented here, that the psycho-spiritual coping process we have observed expresses itself as a changing relationship with a higher power, is a speculative hypothesis, particularly because no data are available for long time periods.

Continuing to play with theoretical models, we can also interpret our data against the background of older cultural anthropological notions of what is typically "Javanese", as below.

4.3.3 Javanese orientations (*kejawen*)

Against this background, which has been criticized as being too essentialist and outdated, the meaning of the traditional Javanese orientation, with its proximity to Javanese mysticism, has proved hard to assess for the design of the coping process. According to Koentjaraningrat (1960, 1989), Geertz (1973), Mulder (1975, 1978), Magnis-Suseno (1981) and Stange (1984), there is a special Javanese psychology of self. In this orientation the internal–external dichotomy is handled quite differently compared to the concepts which are behind the coping model.

This orientation differentiates between an outer realm of behaviour and appearance (*lahir*) and a pure inner being with an inner order (*batin*). The intuitive, emotional *rasa* affords one the ability to connect oneself with the essence of all existing life and the divine spiritual dimension behind earthly things. The concentration upon one's own deep inner life also reveals, through a mystical notion of communication, the rules of the cosmic course of events. With the practice of *kebatinan* and the refinement of inner resources (*batin* and *rasa*), one attains a noble and refined (*halus*) inner nature.

Among scholars of Javanese culture, there are different views of the balance between *nrimo* (acceptance of fate and what one has been given) and the active, transformative deed (*usaha*), which one can impute to the practical, situation-changing, coping form. Koentjaraningrat (1960) stated that, for the Javanese, every human life is predetermined in its beginning, its possibilities for self-realization and its endpoint, and that no one can avoid this predetermination. Magnis-Suseno (1981) occupies himself in his book with Javanese wisdom and ethics, creating a portrayal in which active participation plays a notably lesser part. In this portrayal, the Javanese person must bring passions (*hawa nepsu*) and egotism (*pamrih*) – both *kasar* or unrefined feelings – under control in order not to seem socially disruptive, provoke conflicts and tensions in the community, or endanger social peace. Being free from self-interest (*sepi ing pamrih*) leads to calm self-ownership, patience and an ability to be humbly accepting of one's assigned role in society. *Nrimo* does not mean, however, to "apathetically give in to everything" (Magnis-Suseno 1981: 123), but rather to react reasonably to disappointments and adversity, neither breaking down nor struggling against them. *Ikhlas* "brings psychological peace through a lack of attachment to the external world" (Koentjaraningrat 1960: 96). The right deed in the world is the fulfilment of each respective task (*ramé ing gawé*), not in any sense "changing the world", as Magnis-Suseno writes. It is the active renunciation of one's own interests in the interests of social harmony. One keeps a knowing distance from the outer world. "Giving oneself over to goals in the outside world, the drive to change this outer world, passionate engagement for the betterment of society, is considered to be misguided against this background" (Magnis-Suseno 1981: 127).[7] *Sepi ing pamrih* expresses fundamental humility about one's place in society and the cosmos, *ramé ing gawé* the fulfilment of the role one has been given in this place. The one thing that has meaning in the Javanese universe is adherence to the right order (and the right manners): any attempt to make one's own self-interested goals from the outside world the basis of one's actions is reprehensible.

Pak P, Java.

---

7   This is very reminiscent of the division of the universe into the profane and the sacred, described by Durkheim (1984). Holiness is thereby impersonal, commands respect, is overwhelming and at the same time uplifting, and elicits enthusiasm. It motivates one to be unselfish and self-disciplined, and allows one to disregard one's own interests. For Durkheim the origins of morality derive from holiness, whereas Habermas sees morality as emerging from communicative action oriented toward the Almighty.

Pak P[8] from Village A appeared to us to be an ambassador for Javanese tradition, someone who seems to embody the old values. He is the custodian of the graveyard, and had fasted and meditated and foreseen a disaster, but nobody believed him. He himself and his family suffered no harm. For him, Islam is un-Javanese and comes from foreign Arab lands:

Lha yang membuat saya itu tu tentang shalat itu sebetulnya cuma potongan dari asing luar negeri. Tadinya kan yang Pak Harto presiden kan mau dibuat negara Isla. Tapi Pak Harto tidak boleh, soalnya sini sudah punya negara sendiri, Indonesia.

What I heard about the Islamic praying (*sholat*) is actually that it is something from foreign countries outside of Indonesia. President Suharto wanted to make an Islamic nation out of this land. But he couldn't do it, because this country has already been established with its own rules. (Pak P, Village A)

Bu L's viewpoint strongly evokes the stances that are identified by Javanese experts as typical.

She is a 31-year-old mother of four children; her ten-year-old daughter was buried under the rubble of her house.Her family is poor:

He-eh, lebih pasrah aja. Gitu lho, mBak. Dan ngga' yang ngongso (harus terpenuhi keinginannya), gitu lho, mBak. Dulu 'kan yang saya bilang nerimo itu ya, seadanya, lah. Apa yang Tuhan kasih. Kalau memang anu, rejeki saya segini, ya segini aja, gitu. Ngga' usah yang ngongso, ngga' usah yang iri dengan tetangga, gitu ngga' usah. Kalau dulu 'kan yang tetangga kaya, iri, atau apa. Akhirnya kalau kena gempa habis juga! Terus kalau cuma ngga' dikasih pemerintah juga ngga bisa buat rumah. Ha, itu 'kan juga anu, hikmah dari semua, ngga' usah nerimo-ngga' usah ngoyo-ngoyo, gitu aja.

Yes, just surrendering myself (*pasrah*) more. Like that, Miss. And don't have to *ngongso* [forcing a will to be fulfilled], like that, Miss. Back then, what I called *nerimo* was like, yeah, referring to whatever we owned. To whatever God had given us. If my fortune was to be like this, yeah, it would be like this, like that. No need to *ngongso*, no need to be envious of the neighbours, no need to. Before that, when a neighbour was rich, we were jealous or something. Finally, when the quake happened, everything was gone, destroyed! And if the government hadn't provided help, they couldn't have reconstructed the houses. Ha, it's also, well, a deeper wisdom (*hikmah*) in all these things, like there is no need to *nrimo*, no need to *ngoyo* (push harder). (Bu L, Village B)

Bu L modestly accepts what she perceives as the place God has prescribed to her in the village (in the sense of *nrimo*) and refuses enviously to pursue (*ngongso*) the wealth her neighbours have. This orientation is important for the

---

8 We thank Pak P for his permission to use his statement and his photograph for this publication.

understanding of coping. Pargament (1997) points to the (Western) cultural roots of the coping concept, which would contradict such an attitude: the coping concept could only arise out of cultural surroundings in which human beings are assumed to have a personal choice about the direction of their lives, and in which they undertake the individualized construction of their own life plans. He suggests that, for cultures "with a compelling prescriptive approach to the world" (ibid.: 74), the coping model is incomprehensible. One must now ask oneself, however, whether the traditional Javanese orientations described here have not themselves undergone a transformation. A closer examination of Bu L's position then reveals a possibly more sophisticated view of subjectivity and individuality in dialogue with God than is presently found in Java (see below).

The process of change in Indonesian worldviews and spiritual orientations is portrayed as extraordinarily dynamic and locally diverse. Schlehe and Rehbein (2008) conducted an analysis of the process of modernization and the reinforcement of tradition, a process that is multi-faceted, interwoven and superimposed upon itself: the traditional is partly reinvented, or a traditional quality is simply added to what is new. At the same time, global discourses influence efforts to reclaim indigenous culture and traditions. Ricklefs (2008), who is deeply knowledgeable about the religious history of Java and above all its Islamic currents, paints a highly contradictory picture of the religious and social dynamic, and almost seems to fear a renewed polarization such as that which originated between 1850 and the violent outbreaks in 1965:

> There are active Islamic forces of a multitude of styles and there is active opposition to them. I believe that I could make a case for the increasing strength and influence of an Islam that is puritan, inflexible, intolerant of other cultures and faiths, rejecting of local culture, opposed to mysticism, hoping to impose its version of Islam from the top down and assertive or even willing to use violence. I think I could just as easily show that the forces opposed to such versions of Islam are increasing in strength: people and organizations that are liberal in their interpretations, supportive of gender equality, supportive of multiculturalism and welcoming of other faiths, valuing local culture, accepting of mysticism, politically disinterested and peaceful in their approach. (Ricklefs 2008: 133)

4.3.4 Kejawen obsolete?

In his second investigation, conducted at the beginning of the 1990s, Mulder (1994) observes an extensive cultural shift – above all, after a good twenty years of the "new order" under Suharto. He observed that:

> the resources of Javanese wisdom are increasingly out of reach, especially for the younger generation that seems to respond with eagerness and spontaneity to the opening up of society.... the best channel for social participation appears to turn to the "modern" monotheistic religions, such as Islam that offers a righteous conviction, congregational life, and modern organized activities. (Mulder 1994: 183)

Javanese mysticism, which Mulder portrays as almost a spiritual individualism[9] (every person forges and finds his own path to the spiritual/heavenly world), would, then, be a trailblazer for a Javanese form of coping in the framework of the great religions, and would suggest a particular use or experience of the spiritual that approximates to a personal relationship with the Almighty.

Nevertheless, he is of the opinion that the traditional Javanese spiritual and moral orientations are simply being given a new façade. Mulder (1994: 184) concludes that "all Javanese whatever their degree of Islamization (we could add Christianization) share in Javanese culture. That culture is not necessarily religiously expressed, but contains a common vision of man, society and the ethical conduct of life."

There is a growing body of literature about the increasing influence of Islam in Indonesia. Fealy and White (2008) trace how noticeably the diverse forms of Islam have infiltrated into daily life, above all in the Javanese middle classes. Many publications address the danger of violence and radical Islam. Serious publications scrutinize the enormous inner diversity and dynamics of Indonesian Islam and demonstrate that the radical-terrorist groups are fringe groups that do not enjoy widespread acceptance.

In his study of a village in the east of Java, Beatty (1999) shows how pluralities in Java co-exist with one another and examines what the connective tissue consists of and what, in this context, the oft-cited Javanese syncretism means. The theoretical outlines of syncretism have been reconceived. According to Steward (1995), this does not at all mean a mixing and melting together in the sense of the elimination of diversity. In support of this, Beatty (1999: 3) declares that it is "a systematic interrelation of elements from diverse traditions, an ordered response to pluralism and cultural difference." He shows that this interplay does not lead to a fusion of diverse cultural influences, and in no way to any sort of product.

We would like to introduce Bu L again,[10] this time at more depth. In her idiosyncratic way of coping, she vividly demonstrates the possible combinations of various elements of world- and self-understanding, and in this context, makes clear what is presently possible in the village:

---

9   He does not use this term.
10 We thank Bu L for her permission to use her experiences and her photograph in this
    publication.

Bu L with children, Java.

She lost her daughter in the earthquake. Afterwards she struggled with her fate and her God. She then pleaded with him for forgiveness for this, because she understood, through a subtle, self-reflexive process of self-analysis, the hidden meaning (*hikmah*) of his "bitter reminder":

Tahu diri itu kalau, sebenarnya kalau saya tahu diri itu 'kan saya ngga' sempurna, ngga' berdaya, gitu lho, semua itu 'kan sudah ditentukan Tuhan, gitu lho, mBak. Kalau saya 'kan sering sok, gimana ya? Sok sombong itu 'kan pasti, itu lho. 'Kan anu, saya sendiri lho, 'kan sombong, angkuh, padahal...ngga' tahu diri 'kan saya itu. Kalau setelah ada musibah 'kan harus tahu diri, gitu lho mBak. Sebenarnya kita-kita ini 'kan cuma manusia, gitu lho. Kalau Tuhan menghendaki kena reruntuhan harus mati, ya, mati juga kita. Lha mbok (walaupun) cantik, lha mbok kaya, lah mbok gimana 'kan bakalan mati juga. [...] Wong anak saya itu sehat, lucu-lucunya, anak saya yang mati itu. Yang saya maksud tahu diri ki (itu), oo, lha, manusia itu cuma segitu. Kalau memang Tuhan menginginkan mati, ya mati. Anak saya itu yang, yang...sering apa ya? Yang ngga' lupa 'kan udah perawan, cantik kalau anak saya yang gede itu. Bener, ngga' sombong tapi bener yang cantik, begitu. 'Kan saya sering kangen...gitu. Jadi itu itulah, saya sok sering, " [...] Anu, anak saya pendiam, rajin, bener, yang rajin, pendiam, nurut sama orangtua. Bener, ngga' pernah ngebantah sekalipun! Ngomong jelek, ngga' pernah. Nakal, ngga' pernah, betul yang nurut sekali. Itu yang...saya yang sering sombong di hati saya 'kan itu, mBak. [...] Tapi saya ngga' pernah ngomong sama orang lain, ngomong sama saya sendiri. [...] Ya setelah Tuhan menginginkan si anak saya [...] Tuhan adil itu 'kan saya bilang gini, apa ya, mBak, istilahnya

begini, mBak, kemaren 'kan ada cobaan begitu. Pahit...banget 'kan itu?
istilahnya gempa itu membuat ujian saya pahit sekali. Kalau (2 Sec.) seandainya
ini. Seandainya ngga' ada gempa, mungkin.

Self-knowledge (*tahu diri*) is like, actually when I examine myself (*tahu diri*), I
see that I am not perfect, I am powerless. Everything was arranged by God. As for
me, I was often haughty, how was it? I was haughty and arrogant, that's for sure. I
was haughty and arrogant because in fact I had not examined myself. Actually we
are only human. When God wants us to be buried under rubble and to die, yes,
then we will be dead. It doesn't matter whether we are beautiful, rich or whatever
else, we too will die... [...] My child was healthy and funny – the one who died.
To know oneself means to see humankind's limits. If God really wants someone
to die, the person will die. My child was often.... She was still a teenager, she was
beautiful, my oldest child. She was not arrogant but really beautiful, I miss her
often.... It was often – yes, I was often proud of her [...] – she was a demure child,
diligent and obedient to her parents. No talking back – never – no nasty talk –
never rude, it is true, I was proud in my heart to have such a daughter. [...], I
never showed my pride outwardly [...]. After that, God took my daughter from
me. [...] God is just, I say it like that, because what... Miss. The earthquake was a
very bitter test (*cobaan*) for me. In case (2 sec.), if there had been no earthquake,
perhaps I wouldn't have been able to repay my debts. (Bu L, Village B)

Bu L perceives that she has been seen through and admonished, and that God
has brought her, through a very personal and private dialogue, onto the right
path of psychological self-knowledge. And she sees him as entirely just and fair,
he has rewarded her now with a new baby and has made it possible for her to
love her four children much more and more deeply, and to appreciate them more
than was possible for her before. She sees both of these things as his act and his
message to her, and here she sees herself in conversation with him.

These are two perspectives on Bu L which demonstrate that, from the dialogue between the data and the different theoretical perspectives of psychology
and cultural anthropology, different results emerge. The view of the Javanists
offers different insights than a perspective which focuses on the psycho-spiritual
self- and world-view being changed by other influences. Due to shortage of
space we have been unable to analyse Bu L and her family's restricted access to
resources. This would have enabled us to understand different aspects which are
discussed on the one hand regarding the person as explicated in the social psychological resource-based coping model of Hobfoll (1998), and on the other
hand regarding the household through the economic-technical resource access
model of Wisner, Blaikie and Cannon (2004). Another dimension which we
have not yet discussed is the Islamic perspective. Bu L discovers her *hikmah* in
the context of a religiously oriented rectification towards Islam in close exchange with her husband who contributes many insights from his Islamic religious schooling in the *madrasa*. Her relationship with her husband, which we

did not analyse either, would open the door into a familial and social dimension which we could not enter.

5. Conclusion

Theories and models each illuminate a part of reality. The corresponding instruments (for example, for the investigation of coping strategies) facilitate comparisons and quantifications, but one must take the limits of the data into account. In our methods of gathering data, we have chosen a relatively open approach.

Combining diverse theoretical models for the same research subject makes richer insights possible. We have tried to show how playing with various theoretical and disciplinary approaches facilitates a more comprehensive understanding of psycho-spiritual coping. The examples of Bu L and others enabled us to demonstrate how different disciplinary perspectives and their theories can be related to our data, thus generating new forms of generalization. We believe that this is a promising path to follow in order to overcome the compartmentalization of disciplinary research and the separation of approaches even within disciplines. Things that Bu-L told us about her coping strategies could be better understood by mapping it onto different models. Through her use of additional thoughts and insights, she showed us how we could develop our understanding further. We are eager to discuss with her again what we believe we have thus far understood, and are curious about what we can learn from the villagers in the participatory part of the research that is now beginning.

6. Bibliography

Antlöv, Hans and Jörgen Hellman (eds.) (2005): The Java that Never Was: Academic Theories and Political Practices. Münster: Lit Verlag.

Argenti-Pillen, Alexandra (2000): The Discourse on Trauma in Non-Western Cultural Contexts: Contributions of an Ethnographic Method. In: Shalevm, Arieh Y., Yehuda, Rachel and Alexander C. McFarlane (eds.): International Handbook of Human Response to Trauma. New York: Kluwer Academic/Plenum Publications, pp. 87-102.

Beatty, Andrew (2005a): Feeling Your Way in Java: An Essay on Society and Emotion. In: Ethnos 70, 1, pp. 53-78.

——— (2005b): Emotions in the Field: What Are We Talking About? In: Journal of the Royal Anthropological Institute 11, pp. 17-37.

——— (1999): Varieties of Javanese Religion: An Anthropological Account. Cambridge Studies in Social and Cultural Anthropology. Cambridge: Cambridge University Press.

Belzen, Jacob A. (2010): Towards Cultural Psychology of Religion: Principles, Approaches and Applications. London: Springer.

Blaikie, Piers, Cannon, Terry, Davis, Ian and Ben Wisner (1994): At Risk: Natural Hazards, People's Vulnerability and Disasters. London: Routledge.

Boellstorff, Tom (2002): Ethnolocality. In: The Asia Pacific Journal of Anthropology 3, 1, pp. 24-48.

Bourdieu, Pierre (2002): Habitus. In: Hillier, Jean and Emma Rooksby (eds.): Habitus: A Sense of Place. Aldershot: Ashgate, pp. 27-32.

Bourdieu, Pierre and Loïc Wacquant ([1992]1996): Reflexive Anthropologie. Frankfurt am Main: Suhrkamp.

Braten, Eldar (2005): Resurrecting "Java": A call for a "Java"nese anthropology. In: Antlöv, Hans and Jörgen Hellman (eds.): The Java that Never Was: Academic Theories and Political Practices. Münster: Lit Verlag, pp. 21-42.

Clarke, Adele E. (2005): Situational Analysis: Grounded Theory after the Postmodern Turn. Thousand Oaks Ca: Sage.

Davidson, Jamie S. and David Henley (eds.) (2007): The Revival of Tradition in Indonesian Politics: The Deployment of Adat from Colonialism to Indigenism. London: Routledge.

Durkheim, Emile (1984) (Original 1912): Die elementaren Formen des religiösen Lebens. Frankfurt: Suhrkamp.

Emmons, Robert A. and Raymond F. Paloutzian (2003): The Psychology of Religion. In: Annual Review of Psychology 54, pp. 377–402.

Fealy, Greg and Sally White (eds) (2008): Expressing Islam: Religious Life and Politics in Indonesia. Singapore: Iseas Publications.

Folkman, Susan and Judith T. Moskowitz (2004): Coping: Pitfalls and Promise. In: Annual Review of Psychology 55, pp. 745–74.

Fontana, David (2003): Psychology, Religion, and Spirituality. Malden, Mass.: Blackwell.

Frömming, Urte Undine (2006): Naturkatastrophen, kulturelle Deutung und Verarbeitung. Frankfurt: Campus.

Geertz, Clifford (1996): After the Fact: Two Countries, Four Decades, One Anthropologist. (Jerusalem-Harvard Lectures). Cambridge: Harvard University Press.

———— (1985): Local Knowledge: Further Essays in Interpretive Anthropology. New York: Basic Books Inc.

———— (1973): The Interpretation of Cultures: Selected Essays. New York: Basic Books, Inc.

———— (1960): The Religion of Java. Chicago: The University of Chicago Press.

Geertz, Hildred (1961): The Javanese Family: A Study of Kinship and Socialization. Glencoe: The Free Press.

Glaser, Barney (1978): Theoretical Sensitivity: Advances in the Methodology of Grounded Theory. Mill Valley, CA: Sociology Press.

Glaser, Barney G. and Anselm L. Strauss (1967): The Discovery of Grounded Theory: Strategies for Qualitative Research. Chicago: Aldine.

Greenfield, Patricia M., Keller, Heidi, Fuligni, Andrew and Ashley Maynard (2003): Cultural Pathways Through Universal Development. In: Annual Review of Psychology 54, pp. 461-490.

Grootaert, Christiaan, and Thierry van Bastelaer (eds.) (2004): The Role of Social Capital in Development: An Empirical Assessment. Cambridge: CambridgeUniversity Press.

Hadiyono-Prawitasari, Johana, Paramastri, Ira, Suhapti, Retno, Novianti, Lucia P., Widiastuti, Tiara R. and Nindyah Rengganis (2009): Social Artistry, Lokales Wissen und Konflikte nach einem Erdbeben. In: Zeitschrift für Psychodrama und Soziometrie 8, pp. 277–29.

Henley, David and Jamie S. Davidson (2008): In the Name of Adat: Regional Perspectives on Reform, Tradition, and Democracy in Indonesia. In: Modern Asian Studies 42, pp. 815-852.

Hill, Peter C., Pargament, Kenneth, Wood, Ralf W. Jr., McCullough, Michael E., Swyers, James P., Larson, David P. and Brian B. Zinnbauer (2000): Conceptualizing Religion and Spirituality: Points of Commonality, Points of Departure. In: Journal for the Theory of Social Behaviour 30, pp. 51–77.

Hobfoll, Stevan E. (2001): The Influence of Culture, Community, and the Nested Self in the Stress Process: Advancing Conservation of Resources Theory. In: Applied Psychology 50, 3, pp. 337-421.

——— (1998): Stress, Culture, and Community. New York: Plenum.

Hobfoll, Stevan E. and Petra Buchwald (2004): Die Theorie der Ressourcenerhaltung und das multiaxiale Copingmodell eine innovative Stresstheorie. In: Buchwald, Petra, Schwarzer, Christine and Stevan E. Hobfoll (eds.): Stress gemeinsam bewältigen. Ressourcenmanagement und multiaxiales Coping. Göttingen: Hogrefe, pp. 11-26.

Hobfoll, Stevan E., Johnson, Robert J., Ennis, Nicole and Anita P. Jackson (2003): Resource Loss, Resource Gain, and Emotional Outcomes Among Inner City Women. In: Journal of Personality and Social Psychology 84, 3, pp. 632–643.

Hörning, Karl H. and Julia Reuter (2004): Doing Culture: Kultur als Praxis. In: Hörning, Karl H. and Julia Reuter (eds.): Doing Culture. Neue Positionen zum Verhältnis von Kultur und sozialer Praxis. Bielefeld: transcript, pp. 9-15.

Kitayama, Shinobu and Ayse K. Uskul (2011): Culture, Mind, and the Brain: Current Evidence and Future Directions. In: Annual Review of Psychology 62, pp. 419–49.

Koentjaraningrat, R.M. (1985/89): Javanese Culture. Singapore, Oxford, New York: Oxford University Press.

—— (1960): The Javanese of South Central Java. In: Murdock, George Peter (ed.): Social Structure in Southeast Asia. Chicago: Quadrangle Books, pp. 88-115.

Kryspin-Watson, Jolanta, Arkedis, Jean and Wael Zakout (2006): Mainstreaming Hazard Risk Management into Rural Projects. In: Disaster Risk Management Working Papers Series 13. www-wds.worldbank.org (accessed 20.03.2011).

Lazarus, Richard S. and Susan Folkman (1984): Stress, Appraisal and Coping. New York: Springer.

LeVine, Robert A. (ed.) (2010): Psychological Anthropology: A Reader on Self in Culture. London: John Wiley and Sons.

Magnis-Suseno von, Franz (1981): Javanische Weisheit und Ethik: Studien zu einer östlichen Moral. München: Oldenbourg.

Marsella, Anthony J., Johnson, Jeanette L, Watson, Patricia and Jan Gryczynski (eds.) (2007): Ethnocultural Perspectives on Disaster and Trauma, Foundations, Issues, and Applications. Berlin: Springer.

Muhr, Thomas (2004): User's Manual for ATLAS.ti 5.0. 2nd Edition. Berlin. http://www.atlasti.com/de (accessed 16.03.2011).

Mulder, Niels (1994): Inside Indonesian Society: Cultural Change in Java. Bangkok: Editions Duang Kamol.

Mulder, Niels (1990): Individuum und Gesellschaft in Java. Bielefelder Studien zur Entwicklungssoziologie. Saarbrücken: Breitenbach.

—— (1978): Mysticism and Everyday Life in Contemporary Java: Cultural Persistence and Change. Yogyakarta, Indonesia: Gadjah Mada University Press.

—— (1975): Mysticism and Dayly Life in Contemporary Java. A Cultural Analysis of Javanse Worldview an Ethic as Embodied in Kebatinan and Everyday Experience. Diss. Ansterdam: University of Amsterdam.

Pargament, Kenneth. I. (1997): The Psychology of Religion and Coping: Theory, Research, Practice. New York: The Guilford Press.

Pargament, Kenneth. I., Koenig, Harold G., Perez, Lisa M. (2000): The Many Methods of Religious Coping: Development and Initial Validation of the RCOPE. In: Journal of Clinical Psychology 56, pp. 519–543.

Pemberton, John (1994): On the Subject of Java. Ithaca: Cornell University Press.

Prawitasari, Johana E. (2008): Religious Issue in Psychotherapy. In: Anima 23, 4, pp. 325-332.

Putnam, Robert. D. (2000): Bowling Alone: America's Declining Social Capital. New York: Simon & Schuster.

Reyes, Gilbert and Gerard A. Jacobs (eds.) (2006): Handbook of International Disaster Psychology. 4 Volumes. Westport, Connecticut: Praeger Publications.

Ricklefs, Merle C. (2008): Religion, Politics and Social Dynamics in Java: Historical and Contemporary Rhymes. In: Fealy, Greg and Sally White (eds.): Expressing Islam: Religious Life and Politics in Indonesia. Singapore: Iseas Publ., pp. 115-136.

———— (1979): Six Centuries of Islamization in Java. In: Levetzion, Nehemia (ed.): Conversion to Islam. New York: Holmes and Meir, pp. 100-128.

Sarason, Seymour B. (1974): The Psychological Sense of Community: Prospects for a Community Psychology. San Francisco: Jossey-Bass.

Schlehe, Judith (2010): Anthropology of Religion: Disasters and the Representations of Tradition and Modernity. In: Religion 40, 2, pp. 112–120.

———— (2008): Cultural Politics of Natural Disasters: Discourses on Volcanic Eruptions in Indonesia. In: Casimir, Michael J. (ed.): Culture and the Changing Environment: Uncertainty, Cognition, and Risk Management in Cross-cultural Perspective. Oxford, New York: Berghahn, pp. 275–299.

———— (2006a): Kultur, Universalität und Diversität. In: Wohlfart, Ernestine and Manfred Zaumseil (eds.): Transkulturelle Psychiatrie: Interkulturelle Psychotherapie – Interdisziplinäre Theorie und Praxis. Berlin, Heidelberg: Springer Verlag, pp. 51-57.

———— (2006b): Nach dem Erdbeben auf Java: Kulturelle Polarisierungen, soziale Solidarität und Abgrenzung. In: Internationales Asienforum 37, 3/4, pp. 213–237.

Schlehe, Judith and Boike Rehbein (2008): Einleitung: Religionskonzepte und Modernitätsprojekte. In: Schlehe, Judith and Boike Rehbein (eds.): Religion und die Modernität von Traditionen in Asien: Neukonfiguration von Götter-, Geister- und Menschenwelten. Berlin: Lit Verlag,. pp. 7-17.

Schuchardt, Erika (2002): Warum gerade ich ...? Leben lernen in Krisen – Leiden und Glaube. Fazit aus Lebensgeschichten eines Jahrhunderts. Göttingen: Vandenhoeck and Ruprecht.

Skinner, Ellen A., Edge, Kathleen, Altman, Jeffrey and Hayley Sherwood (2003): Searching for the Structure of Coping: A Review and Critique of Category Systems for Classifying Ways of Coping. In: Psychological Bulletin 129, 2, pp. 216–269.

Stange, Paul (1984): The Logic of Rasa in Java. In: Indonesia 38, pp. 113–134.

Star, Susan L. and James R. Griesemer (1998): Institutional Ecology, 'Translations' and Boundary Objects: Amateurs and Professionals in Berkeley's Museum of Vertebrate Zoology. In: Social Studies of Science 19, 3, pp. 1907-39.

Stewart, Charles (2004): Relocating Syncretism in Social Science Discourse. In: Leopold, Anita and Jebbe Jensen (eds.): Syncretism in Religion: A Reader. London: Equinox, pp. 264–285.

———— (1995): Relocating Syncretism in Social Science Discourse. In: Aijmer, Göran (ed.): Syncretism and the Commerce of Symbols. Göteburg: IASSA, pp-13-37.

Strübing, Jörg (2004): Grounded Theory: Zur sozialtheoretischen und epistemologischen Fundierung des Verfahrens der empirisch begründeten Theoriebildung. Wiesbaden: VS Verlag.

United Nations (2007): International Strategy for Risk Reduction. http://www.unisdr.org.

———— (2005): Hyogo Framework for Action 2005-2015: Building the resilience of nations and communities to disasters, 2005. www.unisdr.org/eng/hfa/hfa.htm (accessed 29.2.2008).

Wisner, Ben, Blaikie, Piers M and Terry Cannon (2004): At Risk. Second Edition. Natural Hazards, Peoples' Vulnerability, and Disasters. London: Routledge.

Woodward, Mark R. (1989): Islam in Java: Normative Piety and Mysticism in the Sultanate of Yogyakarta. Tucson: The University of Arizona Press.

World Bank, United Nations (2010): Natural hazards, unnatural disasters: the economics of effective prevention. http://www.gfdrr.org/gfdrr/nhud-home (accessed 20.03.2011).

Young, Allan (1995): The Harmony of Illusions: Inventing Post-Traumatic Stress Disorder. Princeton NJ: Princeton University Press.

Zaumseil, Manfred (2007): Qualitative Sozialforschung in klinischer Kulturpsychologie. In: Psychotherapie und Sozialwissenschaft - Zeitschrift für qualitative Forschung und klinische Praxis 9, 2, pp. 99-116.

———— (2006): Beiträge der Psychologie zum Verständnis des Zusammenhangs von Kultur und psychischer Gesundheit bzw. Krankheit. In: Wohlfart, Ernestine and Manfred Zaumseil (eds.): Transkulturelle Psychiatrie - Interkulturelle Psychotherapie. Interdisziplinäre Theorie und Praxis. Berlin, Heidelberg: Springer Verlag, pp. 3-50.

III. Politics of space:
negotiating reconstruction

The Attabad landslide and the politics of disaster in Gojal, Gilgit-Baltistan

Martin Sökefeld

1. Introduction

On 4th of January, 2010, a gigantic mass of rocks came down the slope above Attabad, a village in the high mountain area of Gilgit-Baltistan, northern Pakistan. The large-scale landslide filled the narrow valley of the Hunza-River, burying part of the village and the neighbouring hamlet of Sarat. It did not come unanticipated. Already years before widening cracks had appeared on the slope. Attabad had been evacuated but a number of families returned. The landslide claimed nineteen lives. While this was disastrous enough, a second disaster developed in consequence of the first. The debris created a huge barrier of more than hundred metres height and one kilometre width which completely blocked the flow of the Hunza-River and also buried the Karakorum Highway (KKH), the only road link to the area. Consequentially, the whole area upstream, the *tahsil* (subdistrict) of Gojal, was cut off from access to Pakistan. In the subsequent weeks a lake developed behind the barrier which continued to grow till August 2010. Until then it had reached a length of almost thirty kilometres. The lake inundated one village completely and four others partly. Large sections of the KKH came under water so that also communication between the villages was severely disrupted.

A growing body of literature of the anthropology of disasters has pointed out that "natural disasters" are in fact not simply "natural". Taking mostly a political ecology perspective which emphasises the close connection, interdependency and, practically, mutual constitution of "nature" and "society", it has been argued that disasters occur when events that are characterised as being "natural" (i.e. not man-made) impact upon vulnerable human, social spaces. The concept of vulnerability, in a nutshell, provides the link between the "social" and the "natural". "… Vulnerability is the conceptual nexus that links the relationship that people have with their environment to social forces and institutions and the cultural values that sustain or contest them", writes Anthony Oliver-Smith (2007: 10). It is always a particular complex of social, political, and cultural configurations that makes people in particular places vulnerable to specific events. Many of the people who are affected by the Attabad landslide hold a slightly different view of the relationship between "natural" and "man-made" disasters. For them, the disaster was natural in the beginning but subsequently turned into a man-made disaster. According to their perspective it became a man-made disaster because insufficient steps were taken by the regional and national authorities to prevent more damage. In local discourse about the Attabad disaster, references to "government" abound. In local perspective, the disaster has a very close connection with the political sphere. This chapter explores interconnections between "disaster" and "politics" in the case of the Attabad landslide. It focuses more on the second part of the disaster, the inundation of the villages, than on the first, the burying of Attabad village.

Prominent examples of anthropological studies of disaster have pointed out that in many cases the impact of a cataclysmic event initially erases social structure. Immediately after the impact social differences like class and hierarchy (probably to a lesser extent perhaps also gender) often give way to a liminal phase of *communitas* and solidarity (Oliver-Smith 1999; Schlehe 2006). If we take the convenient categorization of post-disaster action and experience into the three phases of "rescue", "relief" and "reconstruction", it is the rescue phase which is frequently characterised by solidarity and a lack of discrimination. Social structure is restored after a certain time, often with the onset of relief operations. While social differentiation re-emerges with relief it seems that political action related with disaster is more often connected with reconstruction efforts. In the story of the Yungay earthquake and landslide Anthony Oliver-Smith narrates that political mobilisation of the victims started in opposition to government's resettlement plans (Oliver-Smith 1986: 203ff.). Both Oliver-Smith and Schlehe write that interventions of agents from outside the affected community play a significant role in the re-emergence of social order and in the inception of political mobilisation.

In the case of the Attabad landslide and lake, however, things were different. The inundation of villages was not a sudden, unreckoned event that took the affected villagers by surprise. To the contrary, the disaster approached slowly in the shape of the daily rising waters of the newly formed lake. Many people told me that from day one after the landslide they expected their houses to be inundated. They had time to cope with the anticipated flooding. Although solidarity with the victims played an important role in the affected villages, there was apparently no phase of undifferentiated *communitas*. Instead, political action started almost immediately after the impact of the event.

I use a quite conventional understanding of "politics" in this chapter. While in much anthropological writing the concept of "politics" has been extended to refer to all kind of power relations, including, for instance, power relations within the "private" realm of the family, I limit my discussion to the sphere of public action in relation with government. This does not mean that I regard other realms of action, like those within the family, as non-political, but simply that there is sufficient "conventional politics" to be discussed in relation to the Attabad disaster.

My analysis of the disaster is based on two short research trips of altogether five weeks duration to Gilgit-Baltistan and Gojal in November 2010 and February 2011.[1] In the affected area I stayed in the village Gulmit which is the administrative centre of the sub-district Gojal. Reports on the Internet blog *Pamir Times* are a very important source, too. Further, my understanding is informed by my general acquaintance with Gilgit-Baltistan which started from my doctoral fieldwork in Gilgit in the early 1990s.

This article proceeds as follows: First I will introduce the area and the people affected by the landslide. Then I will narrate with more detail the unfolding of the disaster, the ways, in which people were affected and attempted to cope with it, including relief efforts by government(s) and non-governmental organisation. After that I will detail and analyse political action that emerged with the disaster but was "rooted" in pre-disaster political dynamics.

2. Gojal and Gilgit-Baltistan

Gojal is a sparsely populated arid high mountain area spread over 8,500 km². Villages are situated at an altitude between 2,300 and 3,200 meters. The population of Gojal is around 20,000 people.[2] Settlements are found in the main valley which is formed by the Hunza-River and in the side valleys of Shimshal and Chupursan. The main valley was connected by the KKH which runs alongside the Hunza River and crosses into the Chinese Province of Xinjiang over the Khunjerab Pass at an altitude of 4,690 m. The KKH is the only road link between Pakistan and China. The valley of Shimshal has a road link with the KKH and the main area of Gojal only since 2003. Economy in the area is largely agrarian. Cultivation depends on irrigation which is fed by melt water from glaciers.

While it would be wrong to insinuate that the area had been completely isolated before the opening of the KKH in 1978, it is certainly true that the Highway had a very important impact on society in terms of links and communica-

1   Without the help of many friends from the area such a short period of research would have produced very meagre results. I am particularly indebted to Fazal Amin Beg, Adil Shah and Zulfiqar Ali. The responsibility for any flaw of the article is, however, entirely mine. I am also indepted to the Swiss National Science Foundation for funding the trips.
2   Some documents give a population figure of more than 25,000, but this probably includes many people who have migrated from Gojal and now live in different cities of Pakistan.

tion. The Highway brought a significant transformation of economy in Gojal.[3] While before the opening of the KKH economy was largely nonmonetary and subsistence-oriented, transport facilities enabled the cultivation of cash crops. Before the disaster, the cultivation of potatoes was the most important source of income in Gojal. Other significant road-related sectors of economy were tourism (including trekking tourism) and small-scale trade with China. The Highway also facilitated migration to down-country Pakistan for the purpose of work and education. The KKH formed the backbone of economy in Gojal, yet the road link was always precarious. Because of the hazardous high-mountain environment landslides and rockfalls that block the KKH consecutively for days or even weeks are quite common.

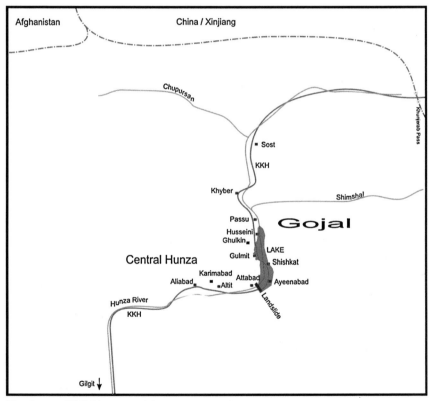

Sketchmap of Gojal (not to scale).

---

3 On social changes brought by the KKH see Kreutzmann 1991.

Politically, Gojal is part of Gilgit-Baltistan. In consequence of the Kashmir dispute, the region which until 2009 was called "Northern Areas of Pakistan" is under the control of Pakistan yet does not form a constitutional part of this country. After 1947, Gilgit-Baltistan was considered a "disputed area" the inhabitants of which have no right to participate in Pakistan's formal political processes. Most importantly, they do not have the right to cast their votes in elections for Pakistan's National Assembly (Sökefeld 1997a, 2005).

Until the 1970s, parts of Gilgit-Baltistan were administered by more or less autonomous local rulers. Gojal was part of the small kingdom of Hunza under the rule of the Mir of Hunza. The kingdom was divided into three parts: Shinaki, Central Hunza and Gojal. Hunza society was characterized by a strict hierarchy in which the lower strata were required to pay heavy agricultural taxes and to provide forced labour to the king. Central Hunza was the most privileged part of the state as taxation was considerably higher in Shinaki and especially in Gojal, yet rigid stratification prevailed also in Central Hunza.[4]

The Mir's exploitative regime was abolished only in 1974 by the Pakistani Prime Minister Zulfiqar Ali Bhutto. Bhutto initiated a first step in a series of political reforms of Gilgit-Baltistan. In 2009, the up until now last of these reforms brought a kind of limited self-rule to the area. Since then, Gilgit-Baltistan has a Legislative Assembly and a Government under a Chief Minister who is elected by the Legislative Assembly. At present, the Pakistan People's Party (PPP) has the majority in the Assembly and like the Government of Pakistan the Government of Gilgit-Baltistan belongs to this party. Political competences of both Assembly and Government of Gilgit-Baltistan, however, are very restricted. Beside the Chief Minister there is a governor who is appointed by the federal government of Pakistan. The more important administrative and political issues are under the authority of the Government of Pakistan through its Minister of Kashmir Affairs and Gilgit-Baltistan.

The relationship between Gilgit-Baltistan and Pakistan is still at issue. Most of the politically aware inhabitants of the area demand the full integration of Gilgit-Baltistan into the Pakistani State as its fifth province, beside the existing provinces of Punjab, Khyber-Pakhtunkhwa, Sindh and Baluchistan. With the reform of 2009, the political setup of Gilgit-Baltistan has become to some extent province-like, but still it is not a province. A vocal minority, however, rejects Pakistan completely and demands the area's full independence (Sökefeld 1999).

Gojal is a *tahsil* (sub-district) of Hunza-Nager District. The landslide occurred at the border between Central Hunza and Gojal. Attabad is the last village of Central Hunza and the area affected by the lake falls entirely within Gojal. The principal village of Gojal is Gulmit (population ca. 2,500), the *tahsil* head-

---

4   On historical social structure in Hunza and Gojal see Kreutzmann 1989: 166ff; 1996: 282ff.

quarters, which before the disaster had a bazaar of more than 130 shops. The village houses local administration and facilities like a post office, a bank and a hospital, which, however, is not permanently staffed. For many if not most services the people of Gojal depend on access to Aliabad, the principal place of Hunza, or Gilgit, the capital of Gilgit-Baltistan. Gojal is linked with both places by the KKH.

Another important village of Gojal is Sost where a dry port for China trade is situated. As long as the Khunjerab Pass is open, i.e. from May to December every year, goods are loaded from Chinese trailers onto Pakistani trucks and, to a lesser extent, vice versa at Sost. Large-scale trade with China is dominated by traders from down-country Pakistan, most importantly Punjabis and Pashtun.

Gojal is inhabited by people belonging to two ethno-linguistic groups, Wakhis and Burusho.[5] Wakhis form the majority and Burusho live in a few villages only. Historically, Wakhis constituted a subordinate segment of society and were subject to heavy taxation by the Mir of Hunza who himself was a Burusho. Both Wakhi and Burusho belong to the Ismailiyya, the Islamic community which is headed by the Aga Khan. Ismailis are organised through a hierarchy of Ismaili councils which provide religious and social services.

Religious affiliation is highly significant for the economic and social development of the area because it implies the very strong commitment of the various institutions of the *Aga Khan Development Network* (AKDN). For many individuals in Gojal and Hunza, the commitment to AKDN has a strong connotation of moral obligation towards the Aga Khan.

The most important of these institutions is the *Aga Khan Rural Support Programme* (AKRSP) which started to work in Gojal in 1983 by initiating the voluntary self-organisation of the local population into *Village Organisations* and *Women's Organisations* (VO/WOs).[6] VO/WOs are groups who collect savings and give loans to their members for business, educational or health purposes. The VO/WOs of several villages together form *Local Support Organisations* (LSOs) which undertake larger projects with the help of AKRSP.

Different kinds of formal voluntary organisation have become very common in Gojal. For example, there is in almost every village a community school which is organised and funded by the parents. These schools are generally believed to be of higher standards than government schools. While the language of instruction in government schools is Urdu, community schools are mostly English medium schools. Compared to other parts of Pakistan, the level of education in Gojal and Hunza for both females and males is extraordinarily high. In the younger generation, the literacy rate reaches hundred percent. The value of edu-

---

5 Beside there is a very small minority of Domaki-speaking people, comprising a few households only.

6 On the work of AKRSP in Gilgit-Baltistan see Wood, Malik and Sagheer 2006.

cation was very successfully inculcated by AKDN and most parents shun no efforts to provide their sons and daughters with a good education. Many young Gojalis study at universities in Gilgit, Rawalpindi, Lahore, or Karachi.

3. After the landslide: the unfolding of disaster

The impact of the landslide of 4th of January 2010 was immediately felt in many parts of Gojal. A large cloud of dust emerged from Attabad and reached even Gulmit at a distance of around 15 km. When people rushed to the site and saw the magnitude of the landslide, they were immediately worried. They realised that they were cut off from Pakistan and anticipated the formation of the lake which would threaten their villages. Fazal Abbas, a villager from Ayeenabad which is only a few kilometres from the site of the slide, told me that he and his co-villagers immediately started to prepare for the disaster. They moved their possessions to the upper parts of the village, dismantled the houses in the lower part in order to save precious construction materials like wooden beams and door frames and even cut the trees. About twenty days after the landslide the lake reached Ayeenabad and started to submerge houses. At the same time the section of the KKH between Ayeenabad and the barrier was flooded. It was no longer possible to reach the blockade by road from the Gojal side. That time, water level was increasing by more than 0.5 meters per day.

Because the road was blocked, about two hundred people from Gojal were stranded at Aliabad in Hunza. Gojal could only be reached by helicopters. Services were provided by the Pakistani Army and the *National Disaster Management Authority*, NDMA. Yet due to difficult weather conditions and limited resources helicopter sorties were often suspended for several consecutive days. The service was not enough to provide the people of the area with all necessities. Anticipating a prolonged blockade of the KKH, Gojalis had immediately rushed to the shops and stores to acquire provisions for their households. After a few days, shops were virtually empty.

On January 10, Qamar Zaman Qaira, the then governor of Gilgit-Baltistan rushed to Hunza and delivered speeches in Altit and Gulmit, promising that the Pakistani government would take all efforts for the relief and rehabilitation of the affected people. In the last days of the month, the *Frontier Works Organisation* (FWO), an engineering body of the Pakistani Army, started to work on the blockade, attempting to cut a spillway through the debris. A provisional access road was constructed to reach the lake from the KKH across the blockade. At this time, the spillway point was still more than eighty meters above the water level and the lake had already reached a length of eleven kilometers.

In mid-February the NDMA brought some motor boats in order to transport people across the lake. Yet the boats turned out to be old, leaky and quite unsafe. In March, traders eager to resume business with China launched larger wooden boats on the lake which also took passengers. By these boats Gojal was reconnected with Pakistan, but travel became troublesome, time-consuming and expensive.

Five villages were directly endangered by the rising water level: Ayeenabad, Shishkat, Gulmit, Ghulkin and Husseini. In late May the water started to spill over the debris. Ayeenabad was completely submerged now, as was the greater part of Shishkat. In Gulmit the low lying parts of Goze and the bazaar area along the KKH were inundated. The main parts of Ghulkin and Husseini are situated higher above the floor of the valley. Here only a few houses close to the erstwhile river were affected. Although the level of the lake had reached the spilling point it continued to rise due to increased inflow of melt-water from the glaciers during summer until end of August. The lake had a length of approximately twenty-eight kilometers then. In addition to destroying buildings, the lake also claimed fields, gardens and tree plantations. Especially many apricot and other fruit trees which play a significant role in local diet died because they do not endure wet conditions.

With reduced inflow from the glaciers in autumn, the water level receded to some extent. In Gulmit, some fields and buildings re-emerged, now covered by a thick layer of sand and silt.

Houses destroyed by landslide Attabad and Sarat	141
Houses destroyed by inundation Ayeenabad	32
Shishkat	130
Gulmit	61
Ghulkin	7
Hussaini	10
Further losses shops and businesses	130
school buildings	7
trees	ca. 80,000

Losses in Hunza and Gojal due to Attabad landslide and lake formation.
Source: Early Recovery Plan and Framework for Disaster-Affected Areas of Hunza-Gojal, AKRSP 2010.

Locally, two categories of disaster affected people are distinguished. *Directly affected* are all those who suffered loss of property (houses, land, shops) due to the landslide or inundation. Taking the loss of houses as indicator, around 380 families are directly affected. Because these people have lost their homes and had to move elsewhere they are also called IDPs (internally displaced persons), a designation which has become very popular in Pakistan after the displacement of thousands of people in consequence of the war between army and Taliban in Swat in 2009. Besides the directly affected IDPs there are those who are *indirectly affected* and this category comprises the whole remaining population of Gojal. They are affected by the serious consequences of disturbed transport and communication.

In winter, this travel routine was disturbed again. Because of extreme cold the lake froze between Shishkat and Husseini. From the spillway, boats could only go up to Shiskhat. In January, the Chinese goods were carried or dragged over the ice from Husseini to Shishkat where they were loaded onto the boats. Due to strong winds and high waves boat services were completely suspended for several weeks from mid-January and Gojal was cut off again. Several boats were damaged or even destroyed by ice. In February, the ice broke and boat transport was resumed.

Travelling by boat across the lake (Photo: Martin Sökefeld).

4. Economic consequences of disconnection

Because in many respects life in Gojal depends on the KKH, the obstruction of transport and communication had dramatic effects. I will focus here especially on economy/agriculture and education.

Gojal is no more a self-sufficient subsistence economy. Although most families still produce part of their food themselves, economy and life depend largely on monetary income. Income opportunities, however, were badly affected by the lake. This is most obvious in the case of tourism. In the past, many Gojalis were engaged in tourism as guides and tour operators or as restaurant or hotel owners. Tourists came from Western countries and Japan. International tourism was already badly hit by terrorism in Pakistan and the country's devastating security situation. After the Attabad landslide, tourism in Gojal was reduced to zero. On one hand, tourists simply did not travel any more to the area; on the other, tourism infrastructure has also been seriously damaged. In Gulmit, all hotels except one are closed. Several hotels have been destroyed by water.

Even more significant than tourism in terms of loss of income is the effect of the lake on agriculture. Over the last decades, the cultivation of potatoes as cash crop has become the most important source of income for Gojalis. Potatoes were bought by traders from Punjab or Khyber-Pakhtunkhwa who transported the crop to markets of down-country Pakistan. For Gojalis the profit from the sale of potatoes was sufficient to enable certain prosperity and also to meet the expenses for educating one's children.

After the landslide and the blockade of the KKH the farmers anticipated on one hand great difficulties in marketing their crops and on the other problems to import sufficient food into the area. When the agricultural season started in March, many farmers therefore decided not to plant potatoes for sale but wheat for home consumption. In Gojal, as in other parts of South Asia, potatoes are not regarded as staple food but as vegetable which is eaten in addition to staples like flat bread made from wheat, or rice. In early summer 2010 a food crisis seemed imminent because of the difficulty of food imports. It was, however, averted by food relief. In order to explain the critical situation, the coordinator of the World Food Programme, who is from Gojal himself, told me: "There was such a scarcity of food that the people started to eat their potatoes."

In spite of the much reduced supply of potatoes, traders offered only dramatically reduced rates in autumn. While one *būri* of potatoes (100 kg) had fetched more than 2,000 PKR[7] the year before, farmers were initially offered only 700 PKR in 2010. The traders argued that they were not in a position to pay more because they had to bear the much increased transport rates. The farmers were not in a bargaining position. At that time many of them had spent most of their savings and were desperate in need of cash; they were forced to sell their

---

7   Ca. 17 Euro. At that time one Euro equalled 115 PKR.

produce at almost any rate. Some farmers who were able to wait a little longer until selling their crops got around 1,200 PKR per *būri*. Because of both reduced crops and low rates, the income derived from potatoes decreased dramatically.

In Gulmit also the bazaar economy was badly hit. The central bazaar was situated along the KKH. Around 130 shops were inundated. Only few of the shops could be reopened at other places of the village. Most of the shopkeepers lost their income as did the shop owners who are normally paid a monthly rent by the shopkeepers.

Thus while on one hand most people lost much of their monetary income, local bazaar rates of most goods increased by at least thirty percent. This had dramatic effects on voluntary organisations. Almost nobody was in a position to put further savings into VO/WOs and those who had taken out loans were often unable to pay their instalments. Before the landslide the payment moral had been very good and loans had rarely been defaulted.

Many parents stopped paying their children's fees for the community schools. The headmaster of Al Amyn Model School in Gulmit told me that fees continued to be paid for only thirty percent of the pupils. While before there had been a close relationship between school and parents, most defaulting parents now felt highly embarrassed and avoided any contact. As a consequence, the school was unable to pay the teachers' regular salaries. After negotiations, the teachers agreed to work on reduced pay; otherwise the school would have to be closed.

The disaster also contributed to a severe energy crisis. Electricity for Gojal is generated by a hydro power station in Khyber village, further up the valley. In winter, one of the two turbines working at the power station broke down. Because of the lake it could not be transported to a workshop in Gilgit for repair. In consequence, shortage of electricity increased dramatically. While before electricity had been available for roughly half the day, supply was now reduced to six hours every three days.

As was emphasised before, most parents are willing to invest much money in the education of their children. Educating one's children well is among the most important social values in Gojal. Many students are sent to educational institutions in Hunza, Gilgit, or down-country Pakistan where they do not only have to pay for tuition fees but also for boarding in student hostels. Both students and parents fear that they have to discontinue courses sooner or later if families are unable to meet expenses. For most people in Gojal, this is the most disastrous consequence of the landslide. They say that while a destroyed house can be rebuilt after some years a lost education cannot be recovered. People fear that the disaster will have lasting effects on their children's future.

Beside all the negative economic consequences of the landslide it has to be mentioned that the disaster also created some economic opportunities. Most importantly, the loading of goods at Husseini village from trucks onto the boats provides income to local men. When I visited the place in February 2011, around 120 men were involved in this work. While most of the porters at this end of the lake are from the surrounding Wakhi villages, almost none of the porters at the spillway are from Gojal. Most porters earn between 1,000 to 2,000 PKR per day. Given that the recommended minimum wage in Pakistan is presently 7,000 PKR per month this is quite a good income. When the lake was frozen and goods had to be carried and dragged over the ice to Shishkat, opportunities were even better. At that time even white-collar employees like teachers took to portering. Men were paid 1,000 PKR per tour from Husseini to Shishkat and some men were able to do five tours per day.

Another economic opportunity is the boat traffic. Up until now, however, only two of the boats on the lake are owned and operated by locals. For tourism, the lake itself might once become an opportunity. This, however, would presuppose the reconstruction of infrastructure (hotels and restaurants) and much more convenient access to the area.

5. Coping with rising waters

This section deals with various efforts to cope and manage with the landslide and the lake. Some coping strategies by traders (establishing a transport system across the lake) and households (planting wheat instead of potatoes) have already been referred to in the last chapter. Concerning households, the relocation of housing of IDPs has to be added. When the water approached houses in Ayeenabad, Shishkat and Gulmit families moved their households. This was a collective affair in which the directly affected families were assisted by Ismaili volunteers. These volunteers were organised by the Ismaili Council which through local *Jamaat Khanas* (Ismaili religious and community centres) announced time and place of action. The volunteers helped the affected families to pack their belongings and to either store them at a safe place or shift them to their new, temporary place of accommodation. Further, they cut trees and dismantled houses before they were submerged in order to save construction materials like beams and door and window frames. For the concerned families this assistance was indispensable. Many affected family members felt paralyzed and were completely unable to take part in the dismantling of their own houses. Volunteers also cared especially for women and children and kept them away from the sites of disaster in order to prevent further trauma. While IDPs from Gulmit could be accommodated in the houses of relatives within the village, IDPs from Ayeenabad and Shishkat shifted to rented accommodation in Aliabad, Central Hunza.

Besides coping efforts at the household level, there were a lot of disaster managing activities undertaken by institutions like government, NGOs, and international organisations. Even in the case of a spatially limited disaster like the Attabad landslide, which, compared with other catastrophes like the Indian

Ocean tsunami of 2004 or the Japanese earthquake cum tsunami of 2011, affects a relatively small number of people, efforts towards disaster management create a very complex space of action in which many agents are involved. To a great extent, complexity derives from the fact that none of the institutions involved actually constitutes a clearly bounded, unified actor. Taking "government" as an example, there is the Government of Pakistan and the Government of Gilgit-Baltistan, there are governmental institutions like the *National Disaster Management Authority* (NDMA) and the army, and there are individual politicians like ministers and members of the Legislative Assembly. This complexity is often eclipsed in local discourse which mostly refers simply to "the Government", implying not only a homogeneous institution but also a difference or even an antagonism between government and the local, affected population.

Broadly, three realms of such disaster management activities can be distinguished: First, the efforts to cut a spillway through the debris of the landslide in order to release the water; second, efforts to provide affected households with relief goods, especially to prevent severe food crisis; and third, miscellaneous measures like a business revitalisation and a cash-for-work programme.

Efforts to cut a spillway through the debris started in late January 2010 when this task was given to the Frontier Works Organisation (FWO) of the Pakistani Army. Given the gigantic size of the blockade most local observers doubted that FWO was capable of completing this assignment. Still, while visiting the sites of disaster, politicians announced several times that the debris would be cleared within a few weeks. Most locals had much more confidence in the abilities of Chinese companies and engineers and demanded from the beginning that "the Chinese" should be engaged.[8] Yet this did not happen. Over the months, it became clear that FWO indeed was incapable to reach the set target. FWO succeeded in cutting a spillway through the debris, but this spillway was neither deep nor wide enough to prevent the further increase of the lake even after the spilling point had been reached. In the course of the year, the goal of action was changed. Now the idea is to lower the level of the lake by thirty meters so that most stretches of the KKH would come out of the water. In December, the Pakistani *National Highway Authority* and the *China Road and Bridge Corporation* signed a 275 million USD agreement for the reconstruction of the KKH. The project is scheduled for completion in two years, but given the hugeness of construction work involved this, too, seems hardly feasible.[9]

---

8 Chinese construction agencies have a lot of experience in the region. They built large sections of the KKH and are currently engaged in widening and reconstructing the Highway.

9 According to the agreement, seven kilometres of the KKH will be rehabilitated and seventeen kilometres have to be constructed anew. The project involves the construction of two tunnels of a combined length of 5.7 kilometers and seven high-level bridges (Chinese Embassy 2010).

In December 2010 and January 2011, FWO completely blocked the outflow of water from the lake in order to deepen the spillway. When the outflow was opened again, water remained largely at the same level. The procedure was repeated in February/ March 2011, without much effect.

Relief efforts concerned especially the provision of food items to the affected population. I am concerned here with relief within Gojal and not with the villagers of Attabad who were shifted to camps in Altit village in Central Hunza. In spring, initial relief was provided on a smaller scale by organisations like the Pakistan Red Crescent Society and FOCUS Humanitarian Assistance, the disaster management organisation of AKDN, to the directly affected families. It is said that also the provincial government of Punjab which is headed by the Pakistan Muslim League, the main contender of the PPP, sent a few trucks with relief goods to the area but I was unable to find out, who ultimately received these goods.

Later, relief was extended to the indirectly affected people in Gojal, that is, to the whole population of the *tahsil*. End of July 2010 the Chinese government offered relief to Gojal and during September around 3,000 tons of food items including wheat flour, rice, sugar, milk powder and cooking oil were transported to Sost dry port. It is said that these supplies were enough to feed Gojal for six months. Apparently, the Chinese government had intended to send relief to the flood victims in down-country Pakistan, but as the KKH was blocked and transport was difficult the Government of Pakistan asked China to give the relief to Gojal instead. The Chinese relief was distributed through a newly formed *Relief Committee* which was established by the government of Gilgit-Baltistan. Many people allege that the committee consisted of PPP supporters only and that it favoured party allies.

The Chinese relief consignment also contained fuel (petrol and diesel) and coal. Both were meant to be given free of charge to the households. Yet only the coal was distributed; it was badly needed in winter for heating purposes. The fuel, however, was claimed by the Government of Gilgit-Baltistan. Officials told that the fuel would be sold to meet expenses of the disaster but local people generally doubted this intention. They alleged, rather, that the fuel was given clandestinely to the government's cronies and that any profit derived from its sale went into the pockets of the officials. In any case, it remained unclear how and by whom the fuel was utilized.

Already in early summer 2010 the World Food Programme (WFP) had begun to plan a relief operation for Gojal but this operation was delayed by the floods that struck Pakistan in late July and August. Because in the meanwhile the Chinese Government had started to dispatch food relief to Gojal, the WFP reduced its own package for the area in order to prevent oversupply. Nevertheless, the WFP sent food items sufficient to feed Gojal for two months. This relief was mainly financed by USAID. The distribution of the WFP relief was assigned to FOCUS. As food was to be distributed according to household size, FOCUS relied on LSOs for the preparation of lists of beneficiaries. Ismaili Volunteers also helped in the distribution of relief.

Beside the works at the spillway and the different relief operations a number of smaller measures were undertaken to alleviate the consequences of the disaster. Most of them were either initiated by or implemented through AKRSP. Before the disaster, the organisation did not have an office in Gojal, but in June 2010 an Emergency Field Office of AKRSP was established in Gulmit. The most important projects implemented through this office were a business revitalisation programme and a cash-for-work project.[10] The business revitalisation programme was devised by the NDMA, funded by USAID and implemented by AKRSP with the assistance of MASO, the LSO of lower Gojal. The programme gave financial assistance to those entrepreneurs who had lost their businesses to the lake. Depending on the magnitude of their losses they received cheques worth one or two lakh PKR[11] which were intended to help them in restarting their enterprises. In October, 132 business owners, most of whom had lost a shop in the drowned bazaar areas of Gulmit and Shishkat, received cheques. Some of them were able to reopen their businesses in existing shop buildings which line Gulmit's polo ground or in newly built wooden cabins. While the shopkeepers were happy about the assistance, the owners of the shops buildings which had gone under water complained that they were excluded from the scheme. They argued that the shopkeepers had actually suffered little damage because they could easily shift their merchandise before the bazaar was flooded, but that the shop owners were the real victims, for the shop constructions could not be saved. Still they did not receive any compensation for their losses.

The cash-for-work programme was funded by the German *Diakonie Katastrophenhilfe* through the *National Rural Support Programme* (NRSP) and was implemented by AKRSP, again with the assistance of MASO. IDPs and (non-IDP) "ultra poor" families were the beneficiaries. One person per beneficiary household was employed for a period of thirty days at a wage of 300 PKR per day in different construction schemes. While IDPs constructed mainly cattle sheds, other beneficiaries were employed in the repair of link roads and canals. For IDPs who had shifted to Central Hunza also cash-for-training programmes were initiated in which mostly women participated. Altogether more than 400 individuals took part in this project. IDPs in Gojal were also provided with one health and hygiene kit per household.

Another measure that needs to be mentioned is the Government's subsidy for schools. Initially, the Chief Minister of Gilgit-Baltistan announced that the Government would pay the fees for all pupils and students from Gojal at educational

10 For both projects and other activities of the Emergency Field Office see AKRSP 2011.
11 One lakh equals 100,000.

institutions in Gilgit-Baltistan. This promise did neither include boarding costs nor expenses of students outside of Gilgit-Baltistan. Further, much less was paid than originally announced. While the Al Amyn Model School in Gulmit, taken as an example, suffered a loss of about twenty lakh PKR it received only a subsidy of ten lakh. Also the provincial government of Punjab promised to waive the fees of Gojali students at the educational institutions of Punjab but this did not materialise. Students were forced by colleges and universities either to pay or to quit.

From the perspective of the affected people, many things were promised but much less delivered. Large scale action like the relief operations by China and the WFP totally lacked coordination. In Gojal, announcements by the Governments of Pakistan or Gilgit-Baltistan were considered with mistrust and reservation. Almost all government action was connected to rumours and reports about corruption. In Gulmit and the neighbouring villages the perception of relief was mixed, at best. People told me repeatedly: "We do not want relief. We want the water to go away!"

6. Political action in the context of disaster

Public political action and mobilisation in response to the disaster started almost immediately after the landslide of January 4, 2010. On January 6, the *Progressive Youth Front* held a demonstration in a bazaar in Hunza against "government inaction". On January 14, the *Rābita Committee Mutasirīn-e Gojal* (Coordination Committee of the Affected People of Gojal) was established in Gilgit. On January 27 a "Save Gojal Ralley" was held in front of the Gilgit-Baltistan Legislative Assembly in which both the *Rābita Committee* and the *Balawaristan National Front* participated.

On January 10, Qamar Zaman Qaira, the PPP Federal Minister of Information who was at the same time Interim Governor of Gilgit-Baltistan, came from Islamabad to visit Gulmit and Altit in Hunza where he distributed cheques to the IDPs of Attabad. On January 12, Syed Mehdi Shah, PPP Chief Minister of Gilgit-Baltistan, visited IDPs in Hunza and distributed cheques, too.

These two sets of events open an arena of public political action and mobilisation around the Attabad disaster. On one hand, members of the Government visited sites of disaster or IDP camps announcing or at times publicly distributing government aid; on the other hand non-governmental actors organised public events to highlight the plight of the disaster victims and to blame the Government for not taking sufficient interest in and action against the disaster.

As mentioned in the last section, local discourse about the disaster constructs an opposition between the local, affected population and the government. In this context, "government" cannot be understood in strictly legal terms. In local discourse, "government" does not refer to the executive authority only but included also the members of legislative assembly. The term "government" (*hukumat*) rather refers to "the body of institutionalised politics up there", in Gilgit or in Islamabad.[12] Especially two members of the Gilgit-Baltistan Legislative Assembly (GBLA) played an important role in this discourse: Wazir Baig and Mutabiat Shah. Both of them belong to the PPP. Wazir Baig is the member of the GBLA for Hunza and Gojal. He received a large number of votes from Gojal and most of the people in the area are of the opinion that without their determined support he would not have won the seat. Wazir Baig became Speaker of the GBLA. From September 2010 to January 2011 he was also Acting Governor of Gilgit-Baltistan because the previous Governor had died. Wazir Baig is from Central Hunza. Mutabiat Shah who is from Gulmit is a member of GBLA, too. He has not been elected but was appointed as "technocrat member" of the GBLA.[13] In local discourse, both men are considered as persons who bear a special responsibility for the area, Wazir Baig because he was elected by Gojalis and Mutabiat Shah because he hails from the area. And both were harshly criticized for not taking sufficient interest in the disaster and not raising their voices for the affected people. Although at least the technocrat Mutabiat Shah is a "local" in the strict meaning of the term, his way of (in)action after the disaster rather confirmed the discursive construction of an opposition or even dichotomy between government and the locals: Having entered the sphere of government, he neglected his local obligations.

In spite of charges that the government failed to act appropriately, there was a repertory of action by government actors to deal with the calamity. Publicly, members of government showed their concern by visiting the sites of disaster. The normal course of action went like this: A politician (minister/ MLA/ governor, etc.) arrived at the site (spillway, IDP camp in Altit, or an affected village in Gojal), in most cases by helicopter and in company of media persons, he delivered a speech to the people, emphasised what action government had already taken, made announcements about future government action and distributed some material benefit (cheques, relief) to some of the victims before leaving the scene again.

While members of government ostensibly showed their activities and their sympathy with the victims through these on-site visits, most people commented that these were "a show only", intended to mask the government's actual neglect and inactivity. Already an early press conference held in Islamabad on January 16, 2010, accused the government of holding expensive photo sessions instead of taking effective action. Especially in the first months of the disaster, helicopter flights for politicians' site visits were heavily criticized because they reduced the sorties available for the transport of affected persons. As no boat service had

---

12 For the "vertical" conceptualisation of politics see Ferguson and Gupta 2002.
13 Two seats in the GBLA are reserved for nominated professionals or "technocrats".

been established at that time, air transport was needed to move people and goods between Gojal and Hunza.

The "show" did not always consist in distributing cheques. An author writing for the news blog *Pamir Times* told me that shortly before a visit of Qamar Zaman Qaira, the then Governor of Gilgit-Baltistan, to the blockade in spring 2010 the number of excavators working at the spillway had been doubled by the FWO – only to be reduced again after the Governor had left.[14]

Events of protest were not less frequent than the visits of members of government. Besides the three events referred to at the beginning of this section I would like to mention a few other prominent examples. Most of these events were organised by the *Rābita Committee*. On 4th of April 2010, a protest demonstration was held in Gulmit under the title of *Yom-e bedāri wa tahfuz-e huqūq-e Gojal*[15] in order to mark the completion of three months after the landslide. A member of the *Rābita Committee* told me that until this date there had been protests in Gilgit and in other places, but never in Gojal: "Wazir Baig and his fellows were saying that the people in Gojal were quiet, that protests were taking place only in Gilgit. Therefore we wanted to show that this was not an issue of Gilgit only but of the whole *qōm*[16]." Around 400 people participated in the protest which took place in front of the *tahsildar's* office. On the same date followers of the PPP held another public meeting in Gulmit in order to commemorate the death anniversary of party founder Zulfiqar Ali Bhutto.[17] My interlocutor from the *Rābita Committee* suspected that the PPP organised this meeting only after the demonstration had been announced in order to divide public attention. At the demonstration demands were raised for the release of the water and for the posting of medical doctors at Gulmit, among other things. The protesters shouted slogans against the government and a speaker threatened that the youth of Gojal would turn towards China if governments of Pakistan and Gilgit-Baltistan would not fulfil the local demands. A week later it transpired that FIRs (First Information Reports)[18] had been lodged against eleven of the participants of the demonstration. They were charged of blocking the KKH (which was blocked by water anyway) and disrupting social order.

---

14 Stories about Qaira's visits to the spillway have become a kind of folk genre in Gojal. At another visit to the blockade in March, Qaira announced that the government would open a spillway and release the water within three weeks. When an elderly man among the audience openly expressed his doubts in this schedule, Qaira barked at him: "Are you an engineer?" "No", replied the man, "but I know the area and I know that this is impossible."

15 "Day of vigilance and protection of the rights of Gojal."

16 *Qōm* is an ambiguous concept in Urdu which may refer to descent groups, ethnic or linguistic groups or the political nation.

17 After being ousted by a military coup under General Zia Ul Haq, Bhutto was put under trial and executed on 4th of April, 1979.

18 In Pakistan, an FIR is a document prepared by the police on some offence. Through issuing an FIR the process of criminal prosecution is set in motion.

Being disillusioned about the progress of the FWO's efforts to release the water activists planned a demonstration with hatchets and shovels at the site of the spillway. On June 17, 2010, several hundreds of protestors came from both sides toward the blockade. Police tried to stop movement from Central Hunza but failed. Also the spillway was cordoned off by the police but a number of people managed to break through and started to dig with their shovels, symbolically attempting to widen the spillway and to increase the outflow of water. Digging continued the next day but it was stopped after negotiations with NDMA and the Army which promised to intensify and accelerate works. Government imposed Section 144 of the Pakistani Criminal Procedure Code in Hunza which bans assembly of more than five persons in public. The Home Secretary of Gilgit-Baltistan, Asif Bilal Lodhi, termed the protestors as "agents of the enemies of Pakistan" and threatened them with detention and criminal cases.[19]

A final example of protest which I would like to mention was the first anniversary of the landslide which was announced as "Black Day". On 4th of January 2011, protest meetings were held in Gulmit, in Central Hunza and in Gilgit. Section 144 was imposed on Hunza again but protest meetings could not be prevented. In many places, a *qarardār* (resolution) was proclaimed and distributed in which the demands of the *Rābita Committee* were listed. In Gulmit, a demonstration took place in front of the *tahsil* office, speeches against the Government were delivered and tyres were burnt on the ice of the frozen lake. FIRs were issued against all speakers at the protest in Gulmit.

Protest action was not limited to the territory of Gilgit-Baltistan. Since the 1960s many Gojalis have migrated in search of work and education to the cities of Pakistan, especially to Karachi, Lahore, Rawalpindi and Islamabad. A strong network of kinship and village ties connects migrants in these places with Gojalis in Gilgit-Baltistan. In many cities, Gojalis formed local associations, especially student associations, which also organised public meetings concerning the disaster. On February 23, 2010, a "token hunger strike" was initiated in Karachi and on 28th of the same month two hundred students from Gojal staged a demonstration in front of the Lahore Press Club, demanding relief and rehabilitation for the disaster victims. At least three press conferences were organised in Islamabad. Young people from Gojal also called on persons of public interest in Pakistan in order to enlist their support. In Lahore, for instance, Gojali students met with Asma Jahangir of the Human Rights Commission of Pakistan. Further, activities were not limited to Pakistan. Also in New York migrants from Gojal organised meetings in support of the disaster victims.

---

19 *Pamir Times*, June 18, 2010. Available online: http://pamirtimes.net/2010/06/18/those-widening-the-spillway-are-agents-of-the-enemies-of-pakistan-home-secretary-gb/ (accessed April 18, 2011).

7. The role of the media

Public appearances by government officials and protest events are performances which aimed at gaining public attention either for the government's commitment or for the protestors' message of government's negligence. Accordingly, media attention was considered important and a short paragraph on media in the context of the Attabad disaster is in order. A number of protest events were explicitly held as press conferences and government officials on site-visits were usually accompanied by journalists. Yet the political economy of public attention in Pakistan was not very favourable for the concerns of the Attabad disaster. Generally, Gilgit-Baltistan does not receive much attention in the Pakistani public sphere. Most Pakistanis have at best very dim ideas about the high-mountain area. While national TV networks did not report on the disaster for almost three and a half months, some English language newspapers like the daily *Dawn* which had a local correspondent in the area reported from the beginning.

The most important news source concerning the disaster is the Internet-blog *Pamir Times*. *Pamir Times* had been established by two young activists from Gojal, Zulfiqar Ali and Noor Mohammad, in late 2006. The blog is organised as a community, non-profit enterprise which publishes all kinds of news items related to Gilgit-Baltistan in general and Hunza/Gojal in particular. Around thirty persons based in different places are registered as "community journalists" with *Pamir Times* who supply texts, photos and sometimes short videos. With on average more than 2,000 visitors per day *Pamir Times* has become the most important online news portal on Gilgit-Baltistan. Most of its readers are based outside of Gojal. Within the region, there is no Internet café and very few inhabitants possess web compatible mobile phones or mobile internet access. Yet *Pamir Times* very effectively links Gojali migrants across Pakistan – and the world – with their area of origin.

The outreach of *Pamir Times* goes beyond its direct readership as it is used as source by other media. Especially many photographs of the disaster were appropriated by other media, often without giving proper credit, and frequently the activists of *Pamir Times* were interviewed by newspapers or TV channels. Pakistani national TV channels like *Geo* started to report about the disaster only after it had become a more direct concern for Pakistan. In March 2010 an intense debate started about a possible breach of the blockade and a sudden outburst of the lake which would inundate hundreds of villages downstream the rivers Hunza, Gilgit and Indus. Such a breach would not have been without historical precedent: In the mid 19th century a landslide dammed the Hunza River at roughly the same place. That time the pressure of the dammed water broke through the blockade and a huge wave rushed down the valleys. A Sikh army which camped at Attock on the banks of the Indus was washed away.[20] From March on most TV channels and news networks sent their teams to Hunza and Gojal to report on the disaster and the feared danger. Yet media attention faded again when the Attabad disaster was eclipsed by the larger catastrophe of the floods that hit Pakistan from end of July 2010 onwards. Protestors hoped to revive public and media attention through the activities planned for the "Black Day" of the disaster anniversary on January 4, 2011. But on this day public interest was captured by the murder of Punjab Governor Salman Taseer in Islamabad.

While press and electronic media are significant for wider public attention of the disaster, news were conveyed to the more directly concerned public, i.e. most importantly people in and from Gojal, by means of text messages sent via mobile phones. Text messages informed about current developments and also called for action like participation in a protest. An exemplary text message reads like this:

> Dist. Amnstn H/N has started illegal shifting of the remaining qty of 3500 begs of chinese relief atta 2 downstrm despite ban on shift of any kind of relief goods from Gojal. We strongly condemn this animity against Gojali public. Pls raise ur voice. Pls fwd all.[21]

8. Agents of protest and opposition

While it may seem natural that the people affected by the disaster criticize the government and protest against insufficient support, it is still important to have a closer look at those who protested. The first protest mentioned at the beginning of this section was organised by the *Progressive Youth Front* (PYF). The PYF is headed by Baba Jan, a political activist from Nasirabad, Hunza, who in the political context of Gilgit-Baltistan is considered a leftist and nationalist. The PYF is not an organisation of people from Attabad or from Gojal. In Gilgit-Baltistan, "nationalist" is a designation for activists and groups who challenge Pakistan's control over the area. I mention this not to put the PYFs genuine concern for the disaster into question but rather to point out that from the beginning the politics of the Attabad disaster was framed within the coordinates of larger political contention of Gilgit-Baltistan. Even more outspokenly "nationalist", that is,

---

20 In order to prevent such a disaster, more than 20,000 people living along the rivers were evacuated for several weeks in 2010 until it became clear that the blockade was much too massive to break.

21 This message circulated on March 5, 2011 and was written in typical sms-language. A "translation" of the text would read like this: "District administration Hunza-Nager has started illegal shifting of the remaining quantity of 3,500 bags of Chinese relief flour [*atta*] to downstream despite ban on shift of any kind of relief goods from Gojal. We strongly condemn this enmity against Gojali public. Please raise your voice. Please forward to all."

openly demanding the independence of the Gilgit-Baltistan from Pakistan, is the *Balawaristan National Front* (BNF) that together with the *Rābita Committee* organised protest in Gilgit on January 27.[22]

In contrast to the BNF and PYF the *Rābita Committee* is a loose network of activists formed after the disaster by people from Gojal. While the BNF or other nationalist groups quite often stage protests in Gilgit, a joint event of BNF and activists from Gojal is much less likely. This is not to say that Gojalis are generally content about the political system of Gilgit-Baltistan. But their goal for political change is rather the opposite of what nationalists envision: Instead of independence, people from Gojal (and Hunza in general) mostly demand the complete integration of Gilgit-Baltistan as fifth province into the Pakistani state. Gojalis, including the activists of the *Rābita Committee*, are generally representatives of what the nationalists despise as *wafadārī* (loyalism). Although the Committee sometimes cooperated with the nationalists, its activists did not share their ideology. A young activist of the *Rābita Committee* told me: "In the freedom struggle of 1947 our elders decided for Pakistan. Although Pakistan keeps us in a colonialised status, we honour our elders' decision and opt for Pakistan."[23]

The *Rābita Committee* became the most important agent of protest in the context of the Attabad disaster. The common political denominator for the majority of its activists is that already before the disaster they were opponents of the current PPP-government. In the shape of the *Rābita Committee* a decade-old political antagonism came to the surface again. Many people in Gojal hold the PPP in high esteem because its founder Zulfiqar Ali Bhutto abolished the oppressive regime of the Mir of Hunza.[24] The Mir effectively controlled the movement of his subjects and people from Gojal were generally not allowed to leave the state. During the 1960s, however, the Mir's grip weakened to some extent and some Wakhis from Gojal who opposed his rule managed to escape to Karachi. In the city they aligned themselves with the PPP which was established in 1967.

However, not all inhabitants of Gojal were opponents of the Mir's rule. Among his supporters were Burushos settled in Gojal, some of whom were relatives of the Mir, and also some Wakhis who enjoyed privileges, especially in terms of taxation. The Mir's family wields considerable political influence even today and the old distinction of the Mir's opponents versus his supporters is still relevant. The Mir's opponents are generally followers of the PPP. Ghazanfar Ali

---

22 The BNF which has split into two factions is dominated by people from Punial and Ghizer districts. On nationalism in Gilgit-Baltistan and the BNF see Sökefeld 1999, 2005.

23 On the freedom struggle of 1947 in Gilgit see Sökefeld 1997b.

24 Bhutto was the first President and Prime Minister of Pakistan who took considerable interest in Gilgit-Baltistan. It can be assumed that he intended to turn the region into a province of Pakistan. On Bhutto and Gilgit-Baltistan see Sökefeld 1997a: 290ff.

Khan, the son of the last ruling Mir Mohammad Jamal Khan, was member of the Northern Areas Legislative Council (the predecessor of the GBLA) and also held the now abolished office of the Deputy Chief Executive of the Northern Areas. He sided with the Pakistan Muslim League and, after General Musharraf had seized power in Pakistan, with the PML-Q, that is the faction of the Muslim League which supported the military ruler.

After the PPP had won the elections of the GBLA and took government, the Mir's supporters became "the opposition". The opposition now dominates the *Rābita Committee*. Although also many supporters of the PPP find it now difficult to enthusiastically defend the Government's performance of disaster management[25], they mostly shun from aligning themselves with the *Rābita Committee*. Conversely, the *Rābita Committee* also tried to exclude PPP supporters, at least in its formative phase. Still, the Committee is not restricted to allies of the Mir. A few years ago, another political contestant surfaced in Gojal in the shape of the *Mutahida Qomi Movement* (United National Movement, MQM). The MQM came into being as the party of *Mohajirs* in Karachi and Hyderabad, i.e. as *Mohajir Qomi Movement* (Mohajir National Movement) but was renamed in 1997. Since then the party tries to position itself as a liberal, secular party for the whole of Pakistan and not simply as a parochial representative of Mohajir interest in Sindh. People from Hunza and Gojal who lived in Karachi established the party in these areas. Now, also followers of the MQM are part of the *Rābita Committee*. Also the first convenor of the *Rābita Committee* was a supporter of the MQM.

The *Rābita Committee* does not only criticize the PPP-government for taking insufficient action in coping with the disaster but also for corruption and for subverting local institutions. It is alleged, for instance, that the LSOs in Gojal are largely dominated by followers of the PPP. Charges of corruption refer most importantly favouritism in the distribution of relief in general and the "fraud" concerning the fuel gifted by China in particular.

However, not only PPP and government are criticized by the *Rābita Committee*. Although much less outspoken, critique is also extended to Ismaili institutions like AKDN and the Ismaili Councils. Here, the reproach is that at least initially these institutions largely kept silent in the public debate about the disaster. When in the first months after the landslide various options to prevent the flooding of Gojal were discussed and it was obvious that the government was not willing, perhaps for strategic reasons, to enlist the help of international organisations and companies, many Gojalis expected AKDN – and AKRSP in particular – to speak up for them. But AKDN kept conspicuously silent.

---

25 Defence of the government is rather lukewarm. For example, a staunch PPP supporter told me in an interview: "The government is assisting the [affected] people. But the process is very slow."

AKDN's reluctance to raise a voice conforms to the organisation's general policy not to act openly politically and not to confront government, in other words, to "work *with* government" (emphasis added) and not against it (Najam 2006). This can be seen as a general political maxim of Ismailis in Pakistan. The spiritual head of the Ismailiyya, the Aga Khan, demands that his followers are always loyal to their respective state and government.[26] An important rationale for this political caution is that in the country's religious and political context Ismailis occupy a precarious and vulnerable position. Ismailis are considered as being "heterodox" by many representatives of majority Islam and it has repeatedly been demanded to formally exclude them from the fold of Islam.[27] Society in Gilgit town is characterised by strict and often violent polarisation between Shias and Sunnis which radiates across the whole of Gilgit-Baltistan. So far, Ismailis have been successful in largely keeping aloof of sectarianism and this is another reason for keeping a low profile. Still, in a situation of extreme emergency when entire villages were drowning, many Ismailis in Gojal had expected AKRSP to offer more outspoken advocacy and support. In this context, the establishment of the emergency field office in Gulmit can be seen as an assertion that AKRSP "does not leave the disaster victims alone", as an employee of the organisation told me.

While a lot of relief work has been undertaken by AKRSP and other organisations, by China and also the government of Pakistan, the demands and expectations of the *Rābita Committee* and the affected people of Gojal are far from being fulfilled. While relief is generally appreciated, the principal demand is the draining of the water. This is also the first point in the long list of demands of the resolution published on the "Black Day" of 4 January 2011: "It has to be made sure that the water of Attabad Lake is immediately released."

Below the threshold of political activism there is the widespread feeling among the affected population of Gojal that the government does not really take interest in their calamity. *Koi pūchnewālā nahīn hai!* is a frequently heard expression of this sentiment: "Nobody cares for us!" Many people in Gojal assume by now that the government is not really interested in the draining of the lake and the recovery of lost land and houses but only in lowering of the water level to the extent that most of the KKH comes out again so that trade with China, which is the main economic interest in the area, can be resumed on its previous scale.

---

26 This is also an important reason for the reluctance of people in Hunza and Gojal to support nationalist organisations in Gilgit-Baltistan. It does not mean, however, that all Ismailis comply with this maxim. The main leaders of the BNF, for instance, are Ismailis, too, hailing, however, from other areas of Gilgit-Baltistan.

27 This happened to the Ahmadis in 1974 who since then suffer severe discrimination and persecution in the country.

Some activists of the *Rābita Committee* make a connection between protest and relief operations. For them, relief is intended also to silence the protest and to divert the people's attention from their real demand, the draining of the lake. "Indeed", one activist told me in November 2010, "there was no public protest in Gojal after the Chinese relief had arrived."

In any case, to publicly stage protest was a new key in the politics of Gojal. A member of the *Rābita Committee* told me: "It is part of our culture that we accept whatever the government says. But when it turned out that all government announcements and promises were fake we felt compelled to take to the road and to express our protest."

A younger activist added at another occasion:

> *In the beginning we had the problem that we did not know how to protest. On TV we see protests [in Pakistan] where people burn tyres and effigies and shout 'death to so and so!' But we never did such things ourselves. Our approach was always intellectual. We organised a meeting, invited some politician and conveyed him our demands. But this issue was so huge and our approach did not work. We had to learn to shout slogans and threats, to make much noise, etc. Unless you do this, nobody will listen and nothing will happen.*

But not everybody in Gojal appreciated this new mode of public politics. Especially elderly people were very proud of the peacefulness of the region and felt ashamed for youngsters who assembled on the road making noise and shouting slogans against the government.

9. Politics of disaster and the political dynamics of Gilgit-Baltistan

According to Jenness, Smith and Stepan-Norris (2006: ix), natural disasters offer a "particularly compelling empirical window" for the examination of social processes. What, then do the politics of the Attabad disaster tell about political processes in Gilgit-Baltistan and Pakistan? First of all, the politics of disaster take place within a context of previous and ongoing contestations. Although the damming of the Hunza River and the development of the lake was a severe disruption of routines and brought a radically new situation, public political action continued from previous constellations. The disaster was immediately drawn into the political game of pitting opposition against government. Instead of bringing new actors to the fore, the politics of disaster rather open a new arena for an ongoing struggle between old contestants. The dominant divide of politics in Gojal and Hunza, the opposition between supporters of the PPP and the partisans of the Mir dominates the politics of disaster, too. Beyond the political context of Hunza and Gojal also the contestants in the wider political framework of Gilgit-Baltistan find a new arena in the disaster: Nationalist groups entered disaster politics, trying to find new and quite unlikely allies. Still, the disaster brought also political change to Gojal, most importantly a change in the style of "doing politics". The self-representation of Gojal as an abode of peacefulness and loyalty was disturbed by young men shouting slogans against the government on a stretch of the KKH in Gulmit which had still escaped inundation. It is not yet clear to what extent such agitation will become part of the regular repertory of political action in Gojal, but in any case a precedent has been set. In spring 2011 there was also a public protest meeting by elderly people in Gulmit.[28] A friend wrote me that this protest was staged by erstwhile supporters of the PPP who had become disillusioned about government.

Further, the politics of disaster reveal the highly ambivalent conception of government which generally pervades politics in Pakistan. The government is known to lack resources and to be weak – it is generally accepted that the army and the bureaucracy are much more powerful institutions in Pakistan than government – but still it is burdened with high and mostly unaccomplishable expectations. While on the one hand public confidence in government is very limited, it is on the other hand held responsible for almost everything. Representatives of government do little to reduce expectations. To the contrary, they continue to make all kinds of promises. Public appearances of government representatives mostly take place in a mode of *elān*, of making announcements. Thus, members of government announced that the water of Attabad Lake would be released within two weeks, that the level of the lake would soon be lowered by thirty meters, that IDPs would be resettled, or that funds for compensation would soon be released, to mention just a few examples of still unfulfilled promises.

Practically, government did not matter too much in Gojal before the landslide happened. In fact, over the last three decades AKDN had established a kind of parallel administration which cared for almost all local needs that were neglected by the authorities: education, health, rural and community development, finance, and even, to some extent, infrastructure. The people of Gojal lived in a sort of state of benign neglect by the government of Pakistan, largely untouched by many issues which dominate politics in the country. Yet the landslide brought a disaster that was too large to be handled by AKDN. People who despairingly saw the water rise to devour their fields and houses expected that government prevented the unfolding of disaster but were utterly disappointed. Many promises were not delivered and relief efforts were pervaded by another constant of politics in Pakistan: corruption.

The contestation following Attabad also shows that politics in Gojal is not territorially confined. In consequence of migration a translocal network of activists now reinforced by electronic communication developed through which political action could be extended to the major cities of Pakistan. This enables a

---

28 http://pamirtimes.net/2011/04/23/elders-of-gojal-protest-against-the-government-for-not-fulfilling-promises/ (accessed May 9, 2011).

kind of public political representation in Pakistan which, in consequence of the internationally disputed status, the area lacks in the country's formal political setup.

In the perception of many people in Gojal and in Gilgit-Baltistan in general, the status quo of being dependent on, yet unrepresented in Pakistan considerably contributes to the region's vulnerability. While the flood-affected constituencies in Pakistan can expect their representative in the National Assembly to press issues of relief and reconstruction because he or she wants to be re-elected, Gilgit-Baltistan has no representatives to put forward demands at the federal level.[29] The reform package which was passed in September 2009 by the government of Pakistan under the bold name of "Gilgit-Baltistan (Empowerment and Self-Governance) Order" brought no relief in this respect. According to the perspectives of my interlocutors which also included PPP supporters, this package did not do more than change the name of "Northern Areas" to "Gilgit-Baltistan" and turning the Northern Areas Legislative Council into the Gilgit-Baltistan Legislative Assembly, without, however adding any legislative powers.

10. Conclusion

Politics is a significant dimension of the Attabad disaster. In local perspective, it is largely failed politics that ultimately turned a "natural" event into human calamity, and therefore the disaster has to be approached politically, not only by technical or managerial means. Yet politics of disaster do not constitute a new, detached arena of political contention. To the contrary, it almost seamlessly blends into general competition for power in Gojal and Gilgit-Baltistan.

The question of change and continuity is a recurrent theme in the sociology and anthropology of disasters. Being events of apparently utterly disruptive nature, many researchers expected that disasters generate rapid and pervasive social change. Yet many research outcomes show that after disasters continuity largely prevails over change (Henry 2011): "Disasters do not generate change in and of themselves, but rather intensify or accelerate pre-existing patters" (Committee on Disaster Research 2006: 166, quoted after Henry 2011: 224). The development after the Attabad landslide confirms this assessment. But continuity can also lead to escalation: On August 11, 2011, Gilgit-Baltistan's Chief Minister Syed Mehdi Shah was scheduled to visit Hunza yet a group of angry IDPs blocked the road in Aliabad, demanding the release of funds for compensation. They confronted the Police that wanted to clear the road. A shuffle ensued in the

---

29 However, one Member of National Assembly, Marvi Memon of PML-Q, has become a strong advocate of Gilgit-Baltistan in the Pakistani Parliament. After the disaster, Memon visited Hunza and Gojal several times and also met activists from Gojal in Islamabad. At the Pakistani national level she has become the most committed critic of the government's disaster management.

course of which a policeman shot into the group, killing two persons and injuring three more. In response, over the next days crowds of angry protestors led by nationalist activists attacked government offices in Hunza and demonstrations took place in Gojal, Central Hunza, Gilgit, Islamabad and Karachi. The government, in turn, arrested a number of activists. Thus, the antagonism between "the people" and "the government" became relevant and visible to an unprecedented extent.

At the time of writing the issue is still unresolved. As those who have lost their homes due to landslide or lake have not yet been resettled, a phase of reconstruction after the disaster has not yet been reached in any meaningful sense. Rather, the phase of relief which is characterised by dependency and uncertainty about what future will bring is prolonged. It has brought, in fact, only short-term relief and in the affected areas life continues in a kind of interim phase the end of which is not yet visible. The disaster is by no means over. The Attabad case shows, then, that not only reconstruction can be pervaded by political antagonism and contradictions of interests but that already before reconstruction takes place politics can be a significant dimension of disaster – perhaps especially when relief and reconstruction efforts are perceived as being delayed and utterly inadequate by the affected people.

11. Bibliography

AKRSP (2011): Six-Monthly Progress Report, June – December 2010. Emergency Field Office Gulmit: AKRSP.

———— (2010): Early Recovery Plan and Framework for Disaster-Affected Areas of Hunza-Gojal. Gilgit: AKRSP.

Chinese Embassy (2010): NHA, CRBC Sign Agreement for Reconstruction of Affected Portion of KKH. Press release, Embassy of the People's Republic of China in the Islamic Republic of Pakistan, December 21, 2010. http://pk.chineseembassy.org/eng/yingwenzhuanti/t779904.htm (accessed 12.04.2011).

Committee on Disaster Research (2006): Facing Hazards and Disasters: Understanding Human Dimensions. Washington DC: The National Academies Press.

Ferguson, James and Akhil Gupta (2002): Spatialising States: Toward an Ethnography of Neoliberal Governmentality. In: American Ethnologist 29, pp. 981-1002.

Henry, Jacques (2011): Continuity, Social Change, and Katrina. In: Disasters 35, pp. 220-242.

Jenness, Valerie, Smith, David A. and Judith Stepan-Norris (2006): Editors'
    Note: Thinking about "Natural" Disasters in Sociological Terms. In: Con-
    temporary Sociology 35, 3, pp. ix-x.
Kreutzmann, Hermann (1996): Ethnizität im Entwicklungsprozess. Die Wakhi
    in Hochasien. Berlin: Reimer.
———— (1991): The Karakorum Highway: The Impact of Road Construction on
    Mountain Societies. In: Modern Asian Studies 25, pp. 711-736.
———— (1989): Hunza. Ländliche Entwicklung im Karakorum. Berlin: Reimer.
Najam, Adil (2006): Working with Government: Close But Never too Close. In:
    Wood, Geoff, Malik, Abdul and Sumaira Sagheer (eds.): Valleys in transi-
    tion: Twenty Years of AKRSP's Experience in Northern Pakistan. Karachi:
    Oxford University Press, pp. 426-453.
Oliver-Smith, Anthony (2007): Theorizing Vulnerability in a Globalized World:
    A Political Ecological Perspective. In: Bankoff, Greg, Frerks, Georg and
    Dorothea Hillhorst (eds.): Mapping Vulnerability: Disasters, Development
    and People. London: Earthscan, pp. 10-24.
———— (1999): The Brotherhood of Pain: Theoretical and Applied Perspectives
    on Post-disaster Solidarity. In: Oliver-Smith, Anthony and Susanna M.
    Hoffman (eds.): The Angry Earth: Disaster in Anthropological Perspective.
    London: Routledge, pp. 156-172.
———— (1986): The Martyred City: Death and Rebirth in the Andes. Prospect
    Heights: Waveland Press.
Schlehe, Judith (2006): Nach dem Erdbeben auf Java. Kulturelle Polarisierung-
    en, soziale Solidarität und Abgrenzung. In: Internationales Asienforum 37,
    pp. 213-238.
Sökefeld, Martin (2005): From Colonialism to Postcolonial Colonialism:
    Changing Modes of Domination in the Northern Areas of Pakistan. In: Jour-
    nal of Asian Studies 64, pp. 939-974.
———— (1999): Balawaristan and Other ImagiNations: A Nationalist Discourse
    in the Northern Areas of Pakistan. In: van Beek, Martijn, Brix Bertelsen,
    Kristoffer and Poul Pedersen (eds.): Ladakh: Culture, History, and Devel-
    opment between Himalaya and Karakoram. Aarhus: Aarhus University
    Press, pp. 350-368.
———— (1997a): Ein Labyrinth von Identitäten in Nordpakistan: Zwischen
    Landbesitz, Religion und Kaschmir-Konflikt. Köln: Köppe.
———— (1997b): Jang azadi: Perspectives on a Major Theme in Northern Areas'
    History. In: Stellrecht, Irmtraud (ed.): The Past in the Present: Horizons of
    Remembering in the Pakistan Himalaya. Köln: Köppe, pp. 61-82.
Wood, Geoff, Malik, Abdul and Sumaira Sagheer (2006): Valleys in Transition:
    Twenty Years of AKRSP's Experience in Northern Pakistan. Karachi: Ox-
    ford University Press.

Internet Source
pamirtimes.net

Glossary of Acronyms

AKDN	Aga Khan Development Network
AKRSP	Aga Khan Rural Support Programme
FIR	First Information Report
FWO	Frontier Works Organisation
GBLA	Gilgit-Baltistan Legislative Assembly
IDPs	Internally Displaced Persons
KKH	Karakorum Highway
LSO	Local Support Organisation
MASO	Mountain Support Organisation
MQM	Mohajir Qomi Movement, later renamed as Muttahida Qomi Movement
NDMA	National Disaster Management Authority
PKR	Pakistani Rupee
PML-N	Pakistan Muslim League-Nawaz
PML-Q	Pakistan Muslim League-Qaid
PPP	Pakistan Peoples's Party
VO/WOs	Village Organisations/Women Organisations
WFP	World Food Programme

Representations and practices of "home" in the context of the 2005 earthquake and reconstruction process in Pakistan and Azad Kashmir

Pascale Schild

1. Introduction

Natural disasters challenge society by confronting people with enormous devastation. In October 2005 such a disaster occurred when a major earthquake hit parts of Pakistan and nearby Azad Kashmir. Largely as a result of building collapse almost 80,000 people died and over 3 million people were left homeless. The widespread collapse of domestic dwellings revealed houses and homes as major sites of devastation, reconstruction and recovery. Dealing with the destruction of houses and the loss of homes entails a difficult and complex process encompassing and affecting social, economical as well as political domains of society.

In the process of reconstruction not only households participate, whose homes were destroyed, but also state authorities and international organisations which impact on reconstruction by targeting the house and home in policies. With reference to several case studies from my fieldwork in Muzaffarabad (the capital of Azad Kashmir) I examine how local people represent and practice home in the context of the earthquake. This general context ranges from traumatic experiences of destruction to the difficult reconstruction process of houses and interactions with the state housing policy. I argue that the destruction and reconstruction of houses after the disaster highlights multiple ways in which "home" (*ghar*) is re-constructed not only physically but also socially and politically. My examination of "home" is guided by the following questions: How do people represent their homes after the experience of (complete) destruction and the reconstruction of their houses? What ideologies, values and practices are linked to these categories "home" and "house" in general? How does the housing policy target houses and homes and how do these political objectifications figure in people's representations of home? How are homes practiced after the earthquake in the everyday family life of households? And, finally, how are representations and practices of home related to one another and shape this basic category of social life?

Regarding the post-disaster "home", it is often assumed that state and international relief and reconstruction programmes caused the separation of households and the decline of the "joint family system". Contrary to this oversimplification, my case studies reveal that representations of homes and the practices of everyday family life are far more multifaceted, although they are affected by the state's housing policy, but rather in fragmentary than consistent ways. Local representations of home are highly ambiguous and frequently contradictory because they are strategically articulated in particular situations and specific contexts with multiple references such as conflicting household ideologies and common practices as well as disaster-related experiences of disruption, reconstruction and the state's housing policy.

2. The 2005 earthquake and the vulnerability of homes

The current literature on the anthropology of disaster puts forward a political ecological approach to studying and analysing disasters in specific social contexts (Oliver-Smith 1999). Natural hazards, such as earthquakes, floods, landslides etc. are neither exclusively natural nor social events, but complex incidences which occur at the interface of "nature" and "society" (Oliver-Smith 2002: 28). Thus, natural hazards do not inevitably constitute disasters. It is rather the social, cultural, political and economic context and the "historically produced pattern of vulnerability" (Oliver-Smith 1999: 29; Oliver-Smith and Hoffman 2002: 3), within this context, which turns the natural hazard into a disaster for a society.

As natural disasters are difficult to predict, anthropologists have been predominantly concerned with studying the post-disaster situation and the ways in which local people respond to a disaster and recover from it. Both, vulnerability (the scale to which people are, or have the potential to be, affected by a natural hazard) and social recovery and, thus, the ways people (can) deal with the consequences of a disaster, are interlinked and conditioned by social organisation, economic production, political power relations and cultural values (Oliver-Smith and Hoffman 2002: 8; Oliver Smith 1999: 29). Rather than single events, disasters are multidimensional and processual phenomena which encompass the social, political and economic context before, during and after the disaster's occurrence (Oliver-Smith 1999; Oliver-Smith 2002).

Because of the complex intersectionality of disasters and societies, the anthropological literature often refers to disasters as "empirical windows" on society. Disasters challenge society and thereby expose how social structures, political relations and cultural values function in ways more evident than under "normal" conditions.

[...] disasters disclose fundamental features of society and culture, laying bare crucial relationships and core values in the intensity of impact and the stress of recovery and reconstruction. (Oliver-Smith 2002: 26)

As Oliver-Smith and Hoffman (2002: 7-12) argue, the anthropological study
of disasters provides deeper insights into human sociability in general.
The earthquake in the Pakistan Province of Khyber Pakhtunkhwa (PKP) and
the, so-called, State of Azad Jammu and Kashmir (AJK)[1] occurred on October 8,
2005, with a magnitude of 7.6 on the Richter scale. The disaster caused the
death of almost 80,000 people and left over three million people homeless
(EERI 2006). People's general vulnerability to the earthquake were clearly exposed through the devastation in terms of fatalities (largely as a result of building collapse) and the scale of the damage to key infrastructure such as roads,
water and power supply, government buildings, schools, hospitals and people's
homes (EERI 2006). Oliver-Smith (2002: 36) argues to conceptualise this vulnerability as "social, political and economic power relations [...] inscribed
through material practices (construction urban planning, transportation) in the
modified and built environments."
The fact that almost 400,000 houses were destroyed or damaged (to various
degrees) and, as a consequence, many people died and were left homeless revealed that people's homes were extremely vulnerable to the earthquake disaster
(ADB and WB 2005: Annex7). This vulnerability is clearly inscribed in the poor
construction and hazardous sites[2] of houses in the city of Muzaffarabad. As the
political ecological approach puts forward, disaster studies must be concerned
with the examination of power relations and, thus, with the question of how vulnerability of groups as well as their power to determine the shape of reconstruction differ in a society (see Oliver-Smith 1999; Oliver-Smith and Hoffman
2002). Although I consider it important to analyse how the vulnerability of
homes and the allocation of resources for housing reconstruction are unequally
distributed among social groups, it is beyond the scope of this paper to examine
these differences in detail. I focus here on the general vulnerability of homes

---

1   Formally AJK is an independent state with its own government, parliament, and judiciary
    (only delegating defence, diplomacy, and currency control to Pakistan). But in fact, AJK is
    almost entirely dominated by Pakistan economically and politically. The Government of
    Pakistan allocates AJK's annual budget and appoints the most influential positions within
    the state's bureaucracy. Because of Pakistan's ideological stance on Kashmir as an inte-
    gral part of the nation of Pakistan, opposition politicians critical of Pakistan's territorial
    claim over AJK and interference in the state's affairs are put under pressure by the mili-
    tary and intelligence agencies present in the area (ct. Rose 1992).
2   The north of AJK and the earthquake affected Hazara District in the PKP are mountainous
    areas which are prone to landslides. In the 2005 disaster many houses and sometimes en-
    tire villages were destroyed by landslides caused by the earthquake.

which indicates some basic aspects relevant to my examination of "home" in the context of the earthquake and the reconstruction process.

First, the widespread destruction of houses reveals that homes and its residents, households, are among the most severely affected groups by the disaster. These households are confronted most with the consequences of destruction, the loss of human lives and property, and are mainly involved in dealing with theses consequences and recovering from them. As a site of destruction, homes also play a crucial role in immediate disaster response, initial recovery and long-term reconstruction. Second, the destruction of houses reveals the strong association between notions of "home" and its locality in a "house". This close relation of home and house is of crucial importance for people in Pakistan and Azad Kashmir. The destruction of houses, thus, provides an important framework of re-thinking, re-presenting and re-practising home. Third, a political effect of the destruction of houses was to make the "house" and "home" the objects of the state's reconstruction policy. The policy-related objectification of house and home affect people's notions and representations of home.

2.1 Households in the context of disaster

I conducted fieldwork in Muzaffarabad, the capital of Azad Kashmir, more or less continuously from October 2009 until May 2011. Muzaffarabad, a city with almost 100,000 inhabitants, was near the epicentre of the earthquake and, thus, one of the areas worst affected by the disaster. My fieldwork in Muzaffarabad commenced in 2009, four years after the earthquake. Although this clearly locates it in the context of long-term reconstruction, rather than in that of emergency rescue and relief, I encountered quite often narratives about the immediate disaster's aftermath which point to important meanings of home in relation to the traumatic earthquake experience of the destruction of houses and the loss of homes.

Many narratives of the earthquake tell about panic-stricken people who, first of all, rushed home in search of shelter and family members. The home (or what was left over from it) was *the* place to go in this situation of severe crisis. The home was maintained in the crises although it was "destroyed" insofar as the home lost its materiality and physical location: the house. In the days, weeks and months after the earthquake it was within these (maintained and modified) homes people managed the living in temporary shelters, such as tents and simple barracks, the cooking, sleeping and emotional care for traumatised household members. Later, homes and households engaged in the expensive and time-consuming reconstruction of their houses, which in many cases is still ongoing.

The intuitive orientation towards home in the immediate aftermath of the earthquake illustrates a "sense of belonging" to home as a place where people seek protection and emotional care. It is the home which turns out to be the most important source of belonging whereas other social affiliations and (more politicised) religious and ethnic identities seem to be rather negligible in this particular situation of extreme distress.[3]

In Muzaffarabad (as well as in Azad Kashmir and Pakistan in general) home (*ghar*) is closely associated with the material object, the house. Compared to the English term "household", the Urdu term "home" (*ghar*) accentuates the location, the domestic dwelling which is ideally a "house" (*makan*) of a massive construction (stone, bricks, cement blocks or mud). This physical connotation reflects a particular ideal of the concept of "home" (*ghar*), namely the notion, that the "home" is located in a "house" which belongs (by legal title) to its members. The ownership of a home is expressed by the phrase *apna ghar* ("home of one's own"). Nevertheless, many homes exist without a house of their own (in a house for rent or a tent) and, though, are referred to as *ghar*.

The local Urdu[4] spoken in Muzaffarabad does not differentiate between "home" and "household". The term *ghar* denotes both "household" and "home".[5] People refer to *mera ghar* ("my home") as the place where they live, eat, sleep and work together with other people. The home/household is actively maintained by the performance of domestic work, *ghar ka kam* (literally translated as "homework"), and by financial contributions. The (financial) maintenance of the home is expressed by the phrase *ghar ka intezam calana* ("to run the home"). When people explicitly denote the household members the term *gharvale* ("people of the home") is used which can mean the nuclear, extended or the joint family (comprising of a married father and his married sons or several married brothers) living together in a common home.

Anthropology has been concerned with homes and, predominantly, households for many decades. Instead of producing a universal definition of "household" anthropology attempts to analyse households with reference to local perspectives and practices in particular situations and specific social, economic and political contexts. Thus, definitions of household, which emphasise rather household ideology and composition, household activities or locations of household, must be applied in sensitive ways to local contexts.

---

3   The (temporary) decrease in relevance of elaborated identities (in contrast to a less politicised sense of belonging to home) probably points to a prominent finding of disaster studies about the emergency and relief phase, which is often characterised by spontaneous solidarity and a temporary decrease of social differences (Henry 2006: 14).

4   The local language spoken in Muzaffarabad is called Pahari (or Hindko as the same language is called in the Hazara District of the PKP) which is assumed to be a Panjabi dialect and, thus, close to the Urdu language. Urdu as the official language of Pakistan is spoken by most of the people in Muzaffarabad. The Urdu term *ghar* is also used in Pahari language, though, it is pronounced little differently without the "h" and with a long "a", such as *gaar*.

5   Although in official Urdu a separate term for household exists (that is *gharana*), I almost never encountered people using this term in everyday speech.

Regarding the ideology and composition of households it is widely assumed that households often, comprise "families", either nuclear, extended or multiple families (ct. Hammel and Laslett 1974). Families are ideological groups which are tied together by kinship and marriage. Nevertheless, the household needs to be analysed separately from "family" since families may live in two or more households, or one household may comprise more than one family (as well as non-kin members). Goody (1972 cited in Sanjek 2002: 286), for instance, defines household activities in terms of either dwelling (sleeping), reproduction (eating) or production and economic contributions. The examination of households' locality mainly deals with the physical space, the house or domestic dwelling, which affects and is affected by households' everyday life and history (Sanjek 2002: 285-287).

I agree with Sanjek (2002) that anthropology must direct its efforts to analysing households by means of emic concepts. But in addition, anthropology must also admit and approach the problem that one conclusive emic definition does not exist. In every context, even within the household, people's views differ on what the household is. Women and men, young and elderly, marginalised and more dominant people perceive, interpret and represent the household in different ways and, thereby, challenge and dispute each others' notions of home and the practices linked with it (ct. Carr 2005). Household is not a distinct unit but a social, ideological and political space of constant negotiation and contest between different interpretations, interests and practices and, thus, a space of ambiguity and contradiction in terms of separation and solidarity, conflict and mutual support.

The household's composition, ideology, activities and location are relevant when examining the general characteristics of households in a particular context. Thus, "household" could be defined as the flexible and dynamic arrangement of people (composition) who are related by notions of "kinship" (family ideology) and cooperation (activities) for a common "home" in everyday life (locality). This preliminary definition of household provides a starting point for further elaboration of the concept which accounts for local variations and discrepancies in imagining, representing and practicing home.

2.2 Locality of belonging: dialectic of house and home

Home features a sense of belonging to a social group (household) and a locality (house). Because of this close association of home/household with house, the earthquake revealed that home is not only a place of intimate care and protection but can become a site of brutal destruction and danger[6]. Although the home basically continued to exist and was maintained by people during emergency, relief

---

6 After the earthquake many people feared (the destruction of) houses and, thus, preferred to
   live in tents, even though their houses were not destroyed.

and reconstruction, the destruction of the house (and the death of household members under its rubble) entailed a sense of disruption and *loss of home*, frequently expressed in terms of nostalgia. According to Simpson (2005: 220) nostalgia is a means of dealing with grief and the traumatic experience of disruption caused by disasters.

Lovell's account for studying and analysing how notions of belonging and locality are mutually constructed points to the dialectical nature-human relationship reflected in these constructions. She deals with locality as "natural" landscape, both in its unmediated form and its culturally mediated constructs such as architecture, which provide the material for articulations of belonging. Although she is not concerned with domestic dwellings as such, her account suggests analysing home as the sense of belonging to a certain locality which transforms social groups and their environment "in a process of mutual definition" (Lovell 1998: 12). The house is thus not a mere symbol of home but a participant in the relationship between people and the (natural and built) environment.

In a similar way, Wilford argues to take the materiality of the (built) environment seriously and, thus, to resist taking it as a mere representation of socially constructed meaning. According to his theoretical stance, labelled by him as "new materialist", meaning and materiality are dialectically related and bring each other into existence (Wilford 2008: 648; 659). Analysing the destruction of houses in New Orleans by the Hurricane Katrina, he puts strong emphasis on the house as a "mediator object" (Wilford 2008: 650) between society and nature and assumes that "perhaps nothing is more fundamental to the meaning of nature than the human abode" (Wilford 2008: 650). Ontologically, the house structures the meaning of nature for society. The building of a house attempts to control nature by constructing "an inside in opposition to an existing outside" (Kaika cit. in Wilford 2008: 651). In addition to nature/human the materiality of the house also shapes the ordering binaries of public/private, shared/intimate, self/other and mine/not mine. The house, thus, not only participates in the relationship of the household with the environment but also in the relationship between households as well as the household and the larger society (Wilford 2008: 651; 655). The materiality of the house is crucial in producing and maintaining these ordering concepts. In a natural disaster this materiality is transformed. The destructed house challenges society by disrupting the sense of order. The materiality of the destructed house thus becomes itself meaningful for the home. Through the rebuilding process of houses materiality participates in the re-constitution of the *meaning* of home itself. The destruction of houses reveals the dialectic of house and household, materiality and sociality. "Natural disasters disrupt the relationship between humans and things and thereby offer [...] a glimpse of both sides of the dialectic: the differentiation of humans and things alongside their integration" (Wilford 2008: 659).

Taking the dialectic of materiality and sociality into account, the rebuilding of a house can not simply entail the reconstitution of the home as it was prior to the earthquake. The collapse and rebuilding of domestic dwellings, inevitably, raise questions concerning the constitutive relation of houses and homes. In many cases, people do not simply reconstruct the house they lost in the earthquake but make modifications which, in turn, lead to modifications of the home itself. Reconstruction requires of people to rethink their previous houses *and* homes. This dialectic process can be problematic and create conflict among members of the household. But apart from presenting solely a threat to the reconstitution of home, reconstruction can provide the opportunity for people to modify their homes and the practices linked with it. In any case, people's perspectives and strategies related to the loss and reconstruction of houses highlight multiple ways the "home" is re-presented, re-imagined and re-practiced after the earthquake. The experience of destruction, immediate disaster response and long-term reconstruction, thus, induce the re-examination of this basic category of everyday life.

2.3 Reconstruction and the "enlarged state"

The household's characteristics of composition, ideology, activities and location, as examined above, are missing a relational political dimension. Households are always positioned within the structures of the larger society. They are socially, economically and politically related to other households, social and political groups as well as state institutions and international organisations. The context of reconstruction in Pakistan and Azad Kashmir clearly reveals that "the state" expanded (among other realms) into the realm of "home". Thus, the ways people imagine, represent and practice home are affected by political representations and practices of state policies.

In the disaster literature, reconstruction in general is characterised as a highly political and conflictive process which is not limited to the physical reconstruction of damaged infrastructure and simply returning to the 'normality' as it was prior to the disaster (see Oliver-Smith and Hoffman 2002; Henry 2006). The post-disaster situation presents, to some extent, a "new" situation for local societies. After a disaster international, national, state and non-state organisations enter the affected areas, first, to provide rescue and relief services and, later, reconstruction assistance.[7] These "external" actors and their ideological agendas

---

7   With reference to the activities of state, non-state and international organisations the post-disaster context is conventionally categorised into the three phases of "rescue", "relief" and "reconstruction". In the short-term emergency phase activities are directed towards the rescue and immediate survival of people. Whereas in the relief phase, usually lasting several months, tent camps are established and people are provided temporary shelter, food, health care etc. Gradually, the relief distribution is stopped and people leave the camps

interfere in certain ways in a particular local context and, thus, affect its social, cultural and political configurations. Compared to short-term rescue and relief activities, long-term reconstruction projects are much more problematic and contested. Hence, "reconstruction" is by no means a self-explanatory term but subject to different interpretations related to the differing needs, interests and values of the various international, state, non-state and local actors. In the process of reconstruction state institutions, international organisations and NGOs participate alongside local political groups, families, households and individuals, trying to achieve their particular and sometimes conflicting reconstruction aims.

The new presence of "external" actors can reverse or strengthen existing relations of power. Former marginal actors can gain more influence due to new opportunities evolving from the access to external resources but they can also lose ground, due to the decrease of opportunities, and find themselves in a more marginalised position than before the disaster (Hilhorst 2003: 46).

As examined by disaster studies, the post-disaster situation often creates political opportunities for a state to expand its influence in a local context by positioning itself as the dominant actor of reconstruction (Simpson 2005: 220; Henry 2006: 13). In this way, a huge centralised bureaucracy of reconstruction develops, often in cooperation with the international system of relief which allocates the required funds. This "enlarged state" (Simpson 2005: 230) affects the existing state-society relations (Hilhorst 2003: 44-46; Oliver-Smith and Hoffman 2002: 10). Often pursuing a top-down approach and positioning people as passive recipients of reconstruction aid, state policies and administrative procedures confront people in various spheres of life, socially, culturally, politically and economically. They are targeted as beneficiaries or excluded from programs, thus, placed in competition with other potential beneficiaries. They are subjected to programs' terms and conditions, and confronted with specific dominant interpretations of reconstruction and related concepts of indigence and vulnerability (Henry 2006: 14-15).

The reconstruction context offers an opportunity to analyse the effects of "the state" on the everyday lives of people. Regarding the state housing policy after the 2005 earthquake, the question must be asked of how people's representations are influenced by the state policy's objectification of house and home. The households' re-examination of home in the process of dealing with the destruction and reconstruction their houses must be contextualized within the state's policy of housing reconstruction.

---

and return home. The phase of relief gives way to the long-term reconstruction phase, which in most cases lasts for years.

2.3.1 The housing policy
After the earthquake the government of Pakistan established a huge centralised
bureaucracy to administer the reconstruction process in the PKP as well as in
Azad Kashmir. The establishment of the so-called "Earthquake Reconstruction
and Rehabilitation Authority" (ERRA) clearly demonstrates that people are confronted with an "enlarged state" after the earthquake.[8] People's encounters with
"the state" were in particular numerous in the context of housing reconstruction.
ERRA set up a housing compensation program for a, so-called, owner-driven
reconstruction of domestic dwellings. An inspection team comprising of army
staff, and officials from local authorities, assessed the damage to houses classified as either slightly, partially or fully damaged. Ownership of the house had to
be proven by the means of the land record. ERRA issued the compensation documents which entitled their holders to an amount of compensation according to
the scale of damage to the house. Compensation was issued in four instalments
with, at most, 1,750,000 rupees (approximately 2,000 USD) for a fully damaged
house. The policy contained several restrictions. First, it held that a *roof* is entitled to compensation and, second, that a person is entitled to compensation only
once, even though she/ he may own more than one house[9]. The amount of
money was very low given the actual cost of building a house according to
earthquake-resistant type of construction. The value of the compensation amount
was further eroded because of rising inflation and the late payment of compensation, which was issued in instalments, paid over a period of two years.

Five years after the earthquake, many households have still not rebuilt their
houses, mostly due to financial difficulties[10]. During the period of fieldwork,
people stated that they have to spend at least 4 lakh rupees for the construction
of a small house (two rooms, one kitchen, one bathroom) of the confined ma-

---

8   Besides, by establishing ERRA to administer reconstruction in Pakistan as well as in AJK,
    Pakistan enlarged its influence in AJK. All the international funds were collected by the
    Pakistani government and managed by ERRA. Thereby Pakistan strengthened its political
    and financial grip over AJK and the reconstruction process respectively.
9   A person who owned more than one house was allowed to make ownership over, for in-
    stance, to a (married) son or a daughter who, by this act, became entitled to compensation.
10 The city dweller not only faced the financial difficulties of reconstruction, but also politi-
    cal restrictions on the reconstruction of houses. ERRA prepared a comprehensive Master
    Plan for the reconstruction and mitigation of the city's disaster vulnerability which in-
    cluded the acquisition of private land for the widening of streets and the construction of
    parks. In order to avoid the situation where newly reconstructed homes obstructed city
    projects, the construction of permanent buildings was banned for a year and a half. ERRA,
    instead, promoted the construction of so-called temporary shelters (SGI-sheet construc-
    tions).

sonry[11] construction type. In many cases, people lost, to a considerable degree, much bigger and more complex houses than a simple two room dwelling. Thus, in comparison with the value of the property destroyed, compensation amounted to no more than a small reconstruction subsidy. In fact, an ERRA official legitimised the program's low financial assistance by explaining to me that the amount was only a "reconstruction subsidy" which people, by mistake, perceived as full "compensation."

Once the implications of the housing policy became evident, reconstruction was clearly exposed as conflictive and contested. In the city of Muzaffarabad political pressure developed and was manifested in the form of protests held in 2007. Most of the opposition against the ERRA was focused on the inadequacy of the compensation. In conjunction with this, the policy was also criticised because the roof and the legal ownership thereof were perceived to be objective criteria for determining eligibility for compensation. The policy targeted roofs rather than the people who live under that roof, and according to the policy's critics, thereby was completely failing to cater for the social realities. In fact, the "one-roof-one-compensation-logic" assumed that the physical "house" equated exactly with the social "home". It was taken for granted that *one* house represents *one* home. Against this logic, the critics pointed to the fact that in a house different homes exist. One roof can be the roof for more than one household. Thus, compensating a house does not mean compensating a household. To illustrate the argument, those interviewed often gave the hypothetical example of two married brothers who live with their wives and children in a house they inherited from their father. The two married brothers lived with their families in different parts of the house and with separate kitchens. The two families lived in separate households, although they lived in the same dwelling. According to ERRA's policy only one son was eligible for compensation. But what about the other son and his family who also lost their home? Thus, the policy was blamed for discriminating against more than one (nuclear family) household living together in one house. Accordingly, people demanded the married man (or married couple), who represents the nuclear family, to be compensated rather than the house.

ERRA never officially abandoned its housing policy. Nevertheless, the policy was in effect relaxed later on in so far as the authority went on to compensate married men (of the same house) who claimed ownership of a separate "house" which could have been only one room of a house, a simple construction for storage, or a building for livestock. This practice reveals that not only the policy's equation of "house" and "home" but the category of the "house" itself

---

11 Confined masonry denotes an earthquake-resistant construction-type of reinforced concrete beams and columns which are added after the construction of the cement block walls.

is highly problematic. What is a "house"? In the context of the local living arrangements this question is not easy to answer as is shown by the example of the two brothers who lived in separate households in the house constructed by their father. The question inevitably arises whether the house is, in fact, two houses. Families often live in compounds of houses comprising a separate living room, bedroom, kitchen, bathroom and storage constructions. Are these, according to the policy, separate houses or not? If the bedroom of a married couple collapsed, but not the bedrooms of the parents, the brother etc., does the authority categorise this single construction as a house or a separate roof respectively?

I once talked to an ERRA official about such difficulties in determining a house. He appeared to be very convinced that there is no doubt about the house. He completely ignored my question of what a house is, possibly, because it struck him as absurd to question such a clearly identifiable object as a "house". He adhered unswervingly to the "one-roof-one-compensation-policy" and the house/roof as the legitimate criterion for compensation. The cases which didn't adhere to this logic were denounced by him as deviant.

The approach taken by the ERRA does not allow for the "house" as an ambiguous and elusive category. This, as a consequence, created space for interpretations, strategies and corruption. According to those interviewed, discrimination and bribery were rife in the process of getting claims for compensation recognised. Thus, another conflictive dimension of reconstruction is evidenced. People were placed in competition for reconstruction aid whereas differences of access to social and economic capital were revealed. Without connections to the relevant officers in charge of the compensations and without financial resources people were more likely to be rejected as beneficiaries of the housing program. Thus, also mistrust and suspicion was created among families and neighbourhoods. Even today, people blame others for practices of bribery (*rishwat*) and favouritism (*sifarish*) in the context of aid distribution for reconstruction.

3. Representations of home

In the context of the earthquake home is re-examined on different levels and by different actors concerned with the collapse and reconstruction of domestic dwellings. It is not only those who lost their homes who are involved in the reconstruction of houses, but also state authorities (and, to some extent, international organisations and NGOs). People are targeted as beneficiaries of technical and financial assistance by means of categories such as "house", "home" and "household". A policy which deals with houses objectifies the "house" in a certain way and treats people according to its definition. Thus, local notions and practices of home must be contextualised within the political context of state housing programs.

In this section I examine different local representations of home in the context of the earthquake, the destruction and reconstruction of houses. I analyse, in particular, how these representations coexist, merge or confront the policy-related concepts and political practices linked to "house" and "home".

3.1 Local representations of home: ambiguities and contradictions

In relation to the earthquake and its impacts on households it is widely assumed that the joint family system in the earthquake affected areas declined because of the nuclear family bias of the relief activities and reconstruction measures. Because reconstruction money and relief goods were distributed to married men, it is argued that formerly joint households comprising of a father and his sons (or several brothers) separated into single, nuclear family households.

The "one-roof-one-compensation" policy's practice differed from the official representation. In fact, compensation was more often issued to each married men in the house, rather than being awarded purely on a per roof basis. Thus, the policy was also perceived to be of a nuclear family bias.

Further, it is argued that the official policy discriminated against joint family households. Whereas nuclear families with a small number of household members (one couple with children) were compensated for their separate house, joint families, who lived under one roof, were given the same amount of compensation despite the larger number of household members (several married couples with children and, probably, the grandparents). This argument illustrates another way of criticising the housing policy for its focus on the problematic category "house" which fails to take account of social realities. The social reality at stake, in this criticism, is the joint family system which was neglected. Whereas in the policy's criticism, mentioned earlier, it is argued contrary and in favour of the nuclear family. The critics reject the equation of *one* house with *one* home, pointing to the existence of several homes in one house. The reality at stake in this criticism is the separated household of the nuclear family and not the joint family system. The two approaches reflect contradictory representations of households in the context of the housing policy. I will further examine this contradiction with reference to several case studies which reveal this contradiction as well.

During my research I heard the assumption about the decline of the joint family system expressed frequently, both in the media and in conversations with well-educated people in offices and universities. I decided to look into the issue of separation on the more empirical level of concrete households, common perspectives and daily practices. I came across several cases of local representations which seemed to corroborate the assumption about the decline of the joint family system. Married brothers who lived together in one house before the earthquake built separate houses for each brother, his wife and children after the earthquake. Although the members of these households rarely explicitly stated that they separated the household because of the nuclear family bias of relief distribution, overall they represented separation of the household as coinciding with the aftermath of the earthquake and the reconstruction of houses. I frequently heard statements such as: *zalzale ke ba'd ham alag ho gae hain* ("we separated after the earthquake") and *ham alag alag rahte hain zalzale ki vajah se* ("we live separately because of the earthquake"). People often perceived living separately as being undesirable, and at the same time hold nostalgic views about communal family life, when they were all living together: *pehle ham sab akhat☐e the* ("before, we were all together"). The earthquake is conceptually linked to the home by the means of correlated dual oppositions: before/after good/bad and together (*akhat☐e*)/ separate (*alag alag*).

The problem of this representation became obvious when I asked people to explain whether they separated *because* of the earthquake. I never got a plausible answer and instead often found people avoiding my questions or even feeling embarrassed by them.

3.1.1 Case study 1

The case of Akbar Sahib and his younger brother is illustrative in this regard. The two brothers migrated twenty years ago with their parents from Indian held Kashmir to Azad Kashmir. The father built a house in a refugee camp which he occupied with his two married sons. After the father's death the two brothers lived in the house until it was completely destroyed by the earthquake. As a further consequence of the earthquake, the refugee camp collapsed into the river and became uninhabitable. By the time I visited the family, the two brothers had almost completed the reconstruction of two separate houses in a new camp. As we were sitting on the veranda of the younger brother's house and talking about the construction work of the new house, I asked about the house the families occupied before the earthquake. Akbar Sahib explained that they all used to live together in one big house before the earthquake. But, he added with some regret, they separated after the earthquake when his brother and he reconstructed their own houses. When I asked about the reason for their separation Akbar Sahib referred briefly to the earthquake and the family's subsequent moving from one place to another until the government finally allocated them plots in the new refugee camp. I could not ask any follow-up questions because Akbar Sahib (who very obviously wasn't well disposed to discussing the issue any further) changed the topic of conversation instantly. His explanation was not very plausible. He could have reconstructed a house together with his brother. Thus, the explanation was rather an attempt to avoid the topic and his reaction demonstrated that the separation of households is considered a problematic and inconvenient issue.

Akbar Sahib's wife, Anser Bibi, presented me with a much more detailed picture of the household's history. Anser and I were sitting in her kitchen where she prepared the food. She told me about her life in general and, thereby, also touched the topic of separation. Her representation contradicted her husband's representation of separation as coinciding with reconstruction after the earthquake. She explained to me that they already lived separately before the earthquake. After the death of her parents-in-law, and almost five years before the earthquake, they arranged a second kitchen for her sister-in-law. They still lived in the same house built by Akhbar Sahib's father, but she and her sister-in-law started to prepare meals separately. Compared to her husband, she didn't seem to regret the separation at all. On the contrary she explained, with reference to her sister-in-law, that it was a good thing for a woman to live with her husband and children in a separate household.

Since Anser Bibi and her husband had no children, after the death of his parents he took a second wife who moved into the house. Because of frequent quarrelling between her and her co-wife, an additional third kitchen was arranged. The second wife complained for several months that she was mistreated by Anser Bibi and her close relatives who lived in the same camp. Out of consideration for his second wife, Akbar Sahib decided to leave the camp. He moved together with his two wives to a rented house in the city area, but had returned to the camp just prior to the earthquake.

Akbar Sahib's younger brother was also planning a future independent of his elder brother before the earthquake. His wife, Rubina Bibi, told me that they had bought land in the city area. Since the allocated land in the camp was given to the refugees for temporary usage and therefore they didn't have legal title to it, Rubina Bibi and her husband planned to leave these uncertainties of the camp and to settle on their own land in the city where they would be able to build a more permanent and secure home. Instead, after the earthquake they sold the land, bought a car and moved to the new camp where they recently built a new house next to Akbar Sahib. Rubina Bibi attributed the reasons for selling the land to the earthquake. She explained they had been terrified by the devastation the earthquake brought and deterred from their previous plans of constructing a permanent and expensive house which they, possibly, would lose again in another earthquake.

Today the two brothers find themselves living next to one another. Before the earthquake, this situation would have seemed highly improbable. What Akbar Sahib represents and regrets as the separation from his brother after the earthquake, might as well, with reference to the more detailed account of Anser Bibi and her sister-in-law about the family's complex history, be interpreted as a kind of reunion of the two brothers after the earthquake. The construction of houses next to each other allows the two families to see each other frequently. In the context of their current living arrangements, the two brothers have, to some extent, a closer relationship today than they would have had if the plans for their respective living arrangements had proceeded, uninterrupted by the earthquake.

3.1.2 Case study 2

In another case I was confronted with a quite similar contradiction concerning the representation of the household's history in the context of the earthquake and reconstruction. In my neighbourhood the Qasi family, comprising of four married brothers and their mother, lives on land inherited from the deceased father. Three of the brothers live in separate houses near the mother's living compound. The eldest brother, whose wife died in the earthquake, and who suffers from a severe eye disease, lives together with his mother, children and unmarried brother. When I met Nigat Bibi, one of the sisters-in-law, she told me that before the earthquake all brothers lived together with the parents in one big house (*bara ghar*). She accentuated proudly that they jointly prepared the food on one cooking stove (*ek hi cula*). When referring to the housing compensation, she stated that they were compensated only once because they all lived under one roof (*eh hi chat*). It was only after the earthquake that they separated and reconstructed separate houses. She explained the separation with reference to the death of her sister-in-law in the earthquake. It was she who managed the joint household and the related housework tasks. In the wake of the earthquake they were unable to maintain the household as it had been previously. *Bartan bahut ziyada hai!* ("the crockery is too much!"), she stated and, thereby, cited the logistics of cooking and washing up for such a large number of people as the main obstacles to maintaining the household in its pre-earthquake form.

I found Nigat Bibi's representation of the family's past challenged by other family members. Once, for instance, I was engaged in a conversation with Qasi Sahib, one of the married brothers (and Nigat Bibi's brother-in-law), and his cousin about the latter's challenging living conditions. Qasi Sahib told me that the cousin was very poor and didn't receive a prefabricated shelter provided to other earthquake-affected families because he couldn't afford the bribes involved. Qasi Sahib asked me to help his cousin by presenting the case to the authority's representative in charge of the shelters. I then started addressing the cousin directly about his case. In addition, he pointed to the housing compensation and stated that it was, like the distribution of the prefabricated shelters, highly unfair and discriminated against the poor. Among his wealthy relatives (the Qasi family) each of the four married brothers and their mother received separate housing compensation while he as a poor man didn't get anything at all. I was quite surprised because, by this time, I had thought the family had been compensated only once, as Nigat Bibi told me. I, then, turned to Qasi Sahib who confirmed his cousin's statement. Qasi Sahib specified that he and his brothers and mother had been living separately before the earthquake. They lived in separate houses with separate kitchens and, consequently, they were compensated separately.

3.2 Situations and contexts of representations

People represent "separate" and "joint" family households in particular situations and local contexts. The anthropological literature promotes the analysis of representations as situational and context-specific practices. Because the situations and contexts of representing the home differ, so do the representations of home. Thus, the represented home is revealed as ambiguous and contradictory. To understand these ambiguities and contradictions, the context and situation in which certain representations are articulated must be taken into consideration. Although all the representations of home, discussed here, refer to the earthquake they focus on different experiences, cultural values, social practices and state policies coexisting in society.

3.2.1 Household ideologies, practices and the experience of disruption

In case study 1 Akbar Sahib conceals the previous complexities of the household's situation and maintains a very simplistic representation of the household's together-past and separated-present whereby separation coincides with the earthquake. In his nostalgic view of the household's past Akbar Sahib doesn't refer to the household's actual organisation immediately before the earthquake but rather to the joint family household as the ideal of patrilocality. His wife's representation, on the contrary, is a very detailed account of the households' complex history. Her perspective on the home is more practically anchored and points to the pre-earthquake separation of everyday cooking, washing and cleaning between the two brothers' wives.

I learnt about this history in a situation when I did not question Anser Bibi directly about the earthquake. She told me her "life story" (as she called it) in a very informal conversation. The situation of conversation with Akbar Sahib was different. He was directly confronted with my question of why he does not live together with his brother (as he "should"). Although I did not intend to anticipate the patrilocal ideology and to put him in a suchlike situation he was forced by my questioning into explaining why he and his brother do not live in accordance with the joint family ideal. In this particular situation he maintained that the earthquake disrupted the joint household and hindered the brothers from living in accordance with the ideology of patrilocality. In so doing, he completely ignored the households' separation of kitchens which took place many years before the earthquake. The *one* house provided the basis on which he idealised the joint household of the past. It is the symbolic *one roof*, no matter how many *kitchens* it has, that represents in this situation the ideology of patrilocal residence: fraternal and paternal cohabiting, solidarity and cooperation in everyday life.

The difference between approaches to representing one's household, either orientated on ideology or the practice of housework, is to some extent related to gender differences and gendered practices linked to the home. The social division of work between the sexes attributes housework (such as cooking, cleaning and washing) to female household members. Thus, women, working at home, are probably more likely to representing home with reference to these daily practices. These practices are not only more relevant for most women's lives, than the ideology of patrilocality, from which they are mostly excluded by male representation, but are also very common, at least, in a city such as Muzaffarabad.

Nevertheless, an overall allocation of female perspectives in "practice" and male perspectives in "ideology" would be oversimplifying. Women's approach to home is not, inevitably, more "practical" than men's. The separate household represents as well as the joint household an ideal of how to organise daily family life. Like Anser Bibi, many women told me that they appreciate living in separate households comprising of the nuclear family[12]. Marginalised in relation to the dominant ideology of patrilocality the separate household represents a contrasting household ideal frequently expressed by women.

It would be wrong to state that the separate household represents a female ideal whereas the patrilocal joint household represents an exclusive male ideal. Case study 2 illustrates that women are also in certain situations likely to put forward nostalgic representations of a pre-earthquake joint household. Nigat Bibi's representation has a great deal in common with Akbar Sahib's ideological approach to the household's history in the context of the earthquake. She also refers to the glorious pre-earthquake past of the "one big house" when the (patrilinear) family was living together. To emphasise the solidarity among the female members she adds to the "one big house" the "one cooking stove", where the women had jointly prepared the food. Although women are often excluded from the representation of the patrilocal joint household it is also a female ideal of family life expressed by the means of the "one cooking stove".

The "one cooking stove" is revealed in Nigat Bibi's representation as a problematic and vulnerable space. Her account of the joint household is also practically anchored as far as she points to the fact that it was difficult to maintain a joint household because of the problems of performing and dividing household tasks effectively. Whereas, Akbar Sahib keeps the household members as actors completely out of the joint household, and instead explains its separation as determined by external circumstances under the catch-all title of "the earthquake" Nigat Bibi refers to internal dynamics of a household. The

---

12 The „nuclear family" in Muzaffarabad, expressed in Urdu by the use of the English term "family", is connoted slightly differently from the "nuclear family" in western societies. It clearly includes the aged parents of a woman's husband as part of the separated household (of the nuclear family).

"one cooking stove" not just figures as the symbol of the female "joint household" but represents a dynamic social space to be actively managed and maintained by real women as participants and actors of the household. The earthquake in Nigat Bibi's representation is revealed as a disruptive event for the household, manifested in the tragic death of her sister-in-law. Akbar Sahib's representation, is much more diffuse, but shares the overall experience of disruption regarding the household. Both, Akbar Sahib and Nigat Bibi express the loss of family's "togetherness" in nostalgic terms. Akbar Sahib's nostalgia about the joint household is not exclusively "ideological". He actively deals with the experienced loss of "togetherness" which equated with the collapse of the "one big house". As argued by Wilford (2008: 659), the transformation of materiality participates in the construction of meaning. Thus, the destruction of the house changes notions of home. The "one big house" does not only figure as a symbol of the ideal patrilocality but reflects a practical sense of "togetherness". Thus, the reconstruction of two separate houses changes the practical experience of being together and affects the everyday organisation of the two households.

3.2.2 Strategic representations of home in the context of the housing policy
In her representation Nigat Bibi denies that her husband and his married brothers received separate housing compensations. She claims, instead, that the family was treated as a joint household living under one roof and that was why they were compensated only once. She draws implicitly on the housing policy which equated the "house" with the "home" and confirms her version of the joint household with reference to the "one big house" and the alleged fact of only one compensation amount the joint family received. Qasi Sahib, her husband's brother, completely contradicted her statement and pointed to the fact that the four brothers and their mother were each compensated separately.

I became aware of these details by accident rather than design. They were additional information which came up in the conversation about Qasi Sahib's cousin who was discriminated against by the officials in charge of relief distribution and housing compensation. In this particular conversation Qasi Sahib tried to legitimise his own case vis-à-vis his cousin's. While his cousin did not receive the appropriate assistance from the authorities and was disadvantaged by corrupt practices, he was treated correctly. The brothers and their mother were compensated separately because they lived in separate houses and separate households before the earthquake.

The compensation policy figures in Qasi Sahib's representation more explicitly. He draws on the policy's underlying assumption that one house represents one household which is then, according to the policy, eligible for compensation. Without running the risk of being accused of "cheating", Qasi Sahib can neither represent the family as a joint household nor as separate households who lived in one house. To legitimise the fact that the brothers and their mother were compensated separately he must represent the households as separated regarding roof *and* kitchen.

Nigat Bibi's and Qasi Sahib's representation are both in line with the policy's one-roof-one-compensation logic, although they articulate by the means of this logic different versions of the pre-earthquake household. Qasi Sahib represents the households as "separate". He must strategically over-estimate the separation against the background of the compensation policy in order to legitimise separate compensation. Nigat Bibi, on the contrary, is not confronted that much with the compensation policy and practice of the state authorities. Our conversation is probably an opportunity for her to explain her people's "culture of the joint family" to a western woman about whose culture she assumes that families and even old parents live separate. Thus, she maintains, by means of the one-roof-one-compensation-logic, the joint family household as the way of living before the earthquake.

The housing policy implemented after the earthquake provide insights into the ways people's representations and practices of home are governed by the state's representations of home. The "enlarged state", as examined above, expanded into the "home", a domain which people clearly regard as "private" and, thus, separate from "the state". As the case of Nigat Bibi and Qasi Sahib illustrate, the state's housing policy is deeply inscribed in the representations of home. Although this inscription is rather implicit than explicit, it reflects the effect of a mechanism of government which Michel Foucault introduced as "governmentality" (see Foucault: 2006). Besides the reciprocal constitution of power and knowledge, governmentality points to the power/knowledge-mechanisms by which human conduct is governed by states, institutions, procedures, discourses etc. Foucault's interest is directed towards the question of how power/knowledge is materialised and manifested in the everyday lives of people. In other words, he attempts to conceptualise mechanisms of government which are neither limited to the state nor characterised entirely by coercion, direct control and regulation. According to his understanding, mechanisms of government cut across and expand into "private" domains of the family and household (Ferguson and Gupta 2002: 989; Lemke 2001: 191). As convincingly argued by Gupta, (1995: 377) "the state" is not a cohesive and unitary whole but constituted trough "a complex set of spatially intersecting representations and practices." As Gupta demonstrates with reference to ethnographical examples, the state is encountered by people as fragmentary, disaggregated and inconsistent. Similarly, the effects of "the state" on people's lives and the ways policies "govern" people are fragmentary, inconsistent and often unapparent to people themselves. Although Nigat Bibi and Qasi Sahib's representations (unconsciously) adhere to the state policy's notion of home, their representations contradict one another. This indicates not only that the effects of power are fragmentary and contradictory, but that "knowledge" or, more specifically, the mode of thinking "home" and "household" is restricted by the rationality of the housing policy and its attempt to objectify and standardise house and home.

3.2.3 Representations, practices and hidden strategies

Representations are always incomplete and politically restricted reflections of the practices and related (economic) strategies. According to the practice theory of Pierre Bourdieu (1976: 203-317), practical strategies are hidden by representations in order to "work" in daily life. Bourdieu states a difference between what people say about their practice and what they do in daily life. The practice deals in a strategic way with the manifold ambivalences, contradictions and constraints of everyday life. People develop and apply strategies with the help of their "practical sense" for daily situations and circumstances. Representations, in turn, are local theories of the practice, but, detached from the daily situations and circumstances of that practice. Representations, therefore, differ from the practice they claim to represent. This gap between theory/representation and practice/strategy is of crucial importance for social life. The representation hides the practice, but, this is why and how strategies developed by people (with the help of their "practical sense") are effectively applied in daily life.

The representations discussed pursue two main objectives which contradict one another. The representations of separate households tend to hide the economic strategy developed by actors in dealing with the compensation policy whereas the representations of joint households conceal the strategies applied in order to legitimise living arrangements which deviate from the dominant ideology of the patrilocal joint household.

The strategic representation of separate households conceals that it is, in fact, a strategy to claim separate compensation. This strategy of representation draws on a practice especially common in the city, but strategically overestimates households' division vis-à-vis the compensation policy. While the authorities were surveying the damage to houses and transferring money to the respective beneficiaries on "a one house, one home basis", it was economically advantageous to be of a separate home. This strategy was also applied in the protests against the housing policy. The critics pointed to the existence of separate households living under one roof and, thereby, aimed to shift ERRA's political practice towards compensating married men, and, thus, separate households of nuclear families, rather than roofs.

There is another criticism of the policy's equation of house with home which has a different objective. The detractors in the protests criticised the policy's neglect of separate households. Whereas this other criticism points to the discrimination of joint families by the policy and, thereby, tries to corroborate the assumption about the decline of the joint family system. The same rationale is present in the local representations which maintain that separation coincided with the earthquake and/or the reconstruction of houses. These representations are either strategic insofar as they legitimise living arrangements which are no longer in accordance with the dominant ideology of the joint household. This strategy completely ignores the agency of households and their members. People themselves developed the strategic claim of separate households in order to be compensated more than once. The compensation (as well as the earthquake and reconstruction in general) was a means by which actors organised and reorganised their homes. The post-disaster situation allowed for opportunities to separate instead of maintaining the former joint household. But, separation of households was in no way an unavoidable consequence of the earthquake or the reconstruction policy. A household could have strategically represented itself as two separate households (to claim separate compensation) but then reconstructed a common house together.

The assumption about the decline of the joint family system in the earthquake-affected areas is an oversimplification. This perspective, far from describing a social fact, alludes to a general ongoing debate in society about the social question of how families do and should live together. In fact this debate around the "home" and "household" has been ongoing over several decades, and is not simply a product of the post-disaster situation. Modernity and the dominant ideology of the nuclear family, for instance, constitute a global trend to smaller household size (Sanjek 2002: 287) to which relief and reconstruction assistance more than likely contribute. But this trend also produces diversity and a context where different household and family values and practices of home coexist and sometimes contradict one another.

Anyhow, the earthquake and the consequences of reconstruction assistance must be contextualised in the wider social processes of change as well as in the daily practices and the concrete histories of households.

4. Practices of home

Representations always coexist with actual practices. Because there is a difference between what people say about their practices and what they do in daily life, these practices are not revealed by representations. People don't theorise their daily practices. Thus, the ambiguous and contradictory representations discussed don't say much about how households practically organised their everyday life before the earthquake and how they (re-)organised it afterwards. To what extent are the members of a household together (*akhat☐e*) and, to what extent, are they separate (*alag alag*)? In which realms do households cooperate or divide housework tasks, financial contributions and responsibilities relating to the maintenance of their home?

Up to now I have discussed strategies of representations in the context of the earthquake and reconstruction and how the separation of households is addressed theoretically by local people with reference to household ideologies and practices, earthquake experiences and reconstruction policies. I now turn to the practice of separation itself. Since local theories of separation represent the actual practices always incompletely one must ask what "joint" and "separate" mean in practice to households and their members in the organisation of everyday life. I discuss, first, a household's history as social process and examine separation not as exceptional, but as a constituent aspect of this process. Second, with reference to some empirical examples I argue that this process is not linear but complex, insofar as "joint" and "separate" arrangements frequently alter and intersect within and between households. The practices of home reveal that the boundaries of separated households opposed to joint household are not clearly confined but highly blurred. "Separate" (*alag alag*) and "joint" (*akhat☐e*) are ambiguous local terms referring to many different practices and social relations within and between households in everyday life.

4.1 The social process of home

I defined households as the flexible and dynamic arrangement of people related to each other by notions of "family", "kin" and cooperation for a common "home" in everyday life. These social relationships alter over time through births, marriages and deaths of household members and, thus, the circumstances and conditions of cooperation change. Households modify their living arrangements over time and adapt, for instance, to emerging conflicts, shifting responsibilities and new space requirements. Homes, thus, are subject to a social process which is referred to in the anthropological literature as the "developmental cycle" of households (Sanjek 2002: 286).

Marriages, for instance, alter the social composition in a household by introducing many new relationships between the recently married woman and her in-laws (father-in-law, mother-in-law, husband's brothers, husband's sisters, husband's brother's wife and children etc.). At the same time the relationship between the married man and the household alters due to newly acquired responsibilities vis-à-vis his wife and children. Alterations in the already complex social composition of household relationships are often problematic. Apart from frequent quarrels concerning the division and performance of housework, especially between the wife and her mother- and sisters-in-law, the household experiences after the marriage and birth of children tensions and conflicts regarding shifting responsibilities and loyalties. A married man is considered to be responsible for his wife and children and, to some extent, must shift his attention away from his parents, brothers and sisters. He lives together with his father in a joint household whereby they share the cost of running the house and kitchen like electricity, water, gas and food. But, apart from this, he is responsible for meeting all the additional expenses for his wife and children like clothes, education and pocket money while the father has the same duty vis-à-vis his wife and children up to their marriage.

It is a common pattern that the elder married son separates from the parents' household after the marriage of his younger brother. Thereby he gets a piece of land allotted by his father in order to construct his own house. In other cases the separation of brothers coincides often with the death of their parents, especially the father. As the moving in of members, the loss of household members affect the social relationships and power relations within a household and the disputes, tensions and conflicts existing in it. In the context of the death of the father his sons are exempt from the subordination to the father's authority. This situation creates an opportunity to rethink and renegotiate the household's arrangement and organisation of daily life.

The social consequences of incidents such as marriage and deaths do not lead automatically to the separation of households. There are many more circumstances, tensions and conflicts which affect the decisions and actions taken by household members in order to separate. But, in any case, households separate sooner or later. Separation is a constituent aspect of the social process of joint households.

The earthquake, reconstruction of houses and the consequences of housing compensation must be contextualised in this social process, the practice and the concrete histories of households. Reconstruction and separation do not coincide ad hoc but are shaped by the pre-earthquake household situation and the scale of separation already carried out by then.

The disruption of the social process of homes by the earthquake became especially visible in the destruction of the houses and living compounds. These are closely linked to the household's history. When people talk about the houses they had lost in the earthquake I found them often contrasting their actual living arrangements by claiming a huge dimension and beauty of their former houses. While these houses are very much idealised the actual living arrangements or reconstructed (parts of) houses after the earthquake are decently described as operational. People express by the means of nostalgia, grief at what they have lost in the disaster and the impossibility of reconstructing within a few years a house that has, in many cases, taken them (and their parents) decades of effort, money, time and commitment. The house and living compound participated in the household's past and is a constituent part of its specific history and future. The construction of a house is processual in the way the home and household are processual. House and home, as I argued with reference to Wilford (2008), dialectically reflect one another. Thus, separation is also materially inscribed, although, it can take many forms and be carried out gradually and either physically by constructing a separate house or practically by separating the cooking and adding a kitchen construction to the already existing living compound.

4.2 The intersection of "separate" and "joint" household practices
Separation as part of the social process of households is not strictly linear or circular. The term "developmental cycle" is somehow misleading because it suggests a simple cycle: households separate, grow to a joint household until they reach a certain size and separate again etc. I prefer the term process which includes the synchronic dimension. Households continuously oscillate between fission and fusion since "separate" and "joint" practices exist simultaneously within and between households.

The joint household features separate practices in terms of responsibilities and financial contributions. Each married man of a household is considered responsible by himself in providing, for instance, clothes to his wife and children. In the local language women express this dependency on particular men by the phrase *mangna* ("to demand"). A wife "demands" from her husband and not her father-in-law while an unmarried daughter "demands" from her father and not from her married brother. These ascribed responsibilities become blurred and ambiguous when, for instance, the father dies. The (married) brothers are in this situation obliged to provide the living for their unmarried sisters and younger brothers together. But, at the same time their main responsibilities towards their wives and children remain.

In a case from my neighbourhood, this ambiguity caused a conflict among the sisters and the wife of a man, which finally led to the separation of the household. The two married brothers were living in a joint household with their mother and sisters after the marriage of the younger brother. The recently married couple separated the cooking after several months and the young wife, Mariam Bibi, arranged the food for her husband and herself in a separate kitchen. She then became pregnant and joined the kitchen of the household of her in-laws again after the birth of her son. It was comfortable for her to have support in cooking, washing and cleaning since she was busy with her newborn. I visited the family frequently during this time and was quite astonished to find Mariam Bibi one day cooking on a gas stove at the corner of her dining room. Obviously, she had quickly arranged a kitchen. She told me that she had a dispute with her mother- and sisters-in-law. Since her husband was a short-tempered person the dispute turned into a fight between her husband and his elder brother. The latter took sides with his mother and sisters while her husband, in turn, was loyal to her. She had therefore separated her household once more and arranged a kitchen immediately. When I asked her about the cause for the argument with her in-laws she told me that her sisters-in-law had blamed her for having scorched the food. Mariam Bibi denied the accusation and acquainted me later on with her suspicions about the real cause of the argument. She believed that the sisters-in-law were in fact jealous and angry because she got a new outfit from her husband the same day while the sisters didn't. She condemned the jealousy of her sisters-in-law and justified her privilege by explaining to me that, after all, she was the wife.

The separation which took place has a history. It is, of course, neither the scorched food nor the new outfit exclusively which led to separation. Nevertheless, I believe that "home" (*ghar*) is negotiated in the every day by the means of such arguments about loyalty and responsibility which lead to practical modifications of the household's organisation as the separation of domestic work. Such modifications, as revealed in this case, are sometimes carried out quite frequently. The frequent switching between a "joint" and "separate" organisation of everyday life demonstrates that "separation" is not an absolute decision and action conducted by household members once and forever, but, a rather flexible means for frequent re-modification of the household's organisation.

Immediately after the conflict the two families avoided interaction for several weeks. Thus, Mariam Bibi, once more arranges the cooking and housework separately from her in-laws. This situation was quite difficult for her since she now has a little son to look after as well. The sisters-in-law told me that they even ignored Mariam Bibi's son for about a week after the conflict. But they gave up ignoring him and start to look after him again. Mariam Bibi's situation has therefore relaxed to the extent that she can, from time to time, give her son to her sisters-in-law to look after while she is doing her housework. The absolute separation carried out right after the quarrel already diluted somewhat after a couple of weeks and the separated households again cooperate in certain situations. Although the separation concerning the housework is relatively strict the two households are involved in joint activities. Once, I visited the family and found the sisters-in-law busy cooking food for guests. They told me that Mariam Bibi's husband invited an officer and his assistant from a local state authority to his house. The authority was conducting a survey in the neighbourhood in order to identify beneficiaries for prefabricated shelters. Her brother, the sister explained to me further, will try to get himself as well as the mother and elder brother registered as beneficiaries for the shelters.

The cooperation between the two "separated" households was very close in this situation of interaction with a local state authority. The brother from the one household invited the guests to his own house whereby the food for the guests was prepared in the kitchen of the other household. The two households agreed to be represented by the younger brother and, thus, closely cooperated in order to get registered for the receipt of three separate shelters. Thus, separation is neither an absolute and irreversible decision nor a restricted practice from which cooperation and joint activities are entirely excluded.

4.3 Home as resource of reconstruction and social recovery
Separate and joint households represent two extremes of an enormous spectrum of practices, dynamic social arrangements, and flexible organisation patterns of

households. This flexibility of households in re-arranging and modifying their daily life, both temporarily and permanently, turns out to be a very important resource in order to respond to the earthquake disaster and recover from it. With reference to another case study from my neighbourhood I briefly illustrate this flexibility apparent in the organisation of households in the context of the earthquake.

A joint family of parents, three unmarried daughters and a son as well as a married son with his wife and three children live together in a small rented house. The married son already lived separately with his wife and children in the same house before the earthquake while his parents and unmarried brothers and sisters lived in their village. During the earthquake the house in the village was destroyed and the family built a simple SGI-sheet construction. But, as the mother told me, their life in the mountainous village became very difficult in this type of house especially during winter. In addition to the difficult living conditions she told me about a conflict incident in the village in which her unmarried son was involved. The family decided to leave the village almost two years ago and moved into the small rental house of their married son's family in Muzaffarabad city. A return to the village is not a viable option for the mother. She told me instead that the household would try to buy some land in the city to build a house together someday.

The flexibility of "separate" and "joint" practices enables strategies for adapting and adjusting to new circumstances emerging from social crises like the earthquake. A household organisation can alter from a rather "joint" to a more "separate" arrangement of living, both temporarily and permanently. But, a family can as well dissolve their "separate" living situation, reunify in a "joint" household and plan a future together.

5. Conclusion

The earthquake severely affected households by destroying their houses and, thus, revealed homes as major sites of devastation, reconstruction and recovery from the disaster. Reflecting an important "sense of belonging" to a locality in the built environment, the home is re-examined and re-constituted in the process of housing reconstruction. As local representations of home indicate, this process is ambiguous and contradictory. The ways people represent their homes in the context of the earthquake allude to the society's general struggles over different household ideologies, values and conflicting practices of home. These social struggles are re-exposed in the disaster's aftermath and merge with disaster-related experiences of disruption and grief, reconstruction of houses and interactions with the state housing policy.

Although represented as binary opposites "separate" and "joint" households in practice highly intersect diachronically and synchronically. Households separate as part of a social process constituted by births, marriages and deaths of household members. Separate households participate in joint activities with other households as well as joint households divide responsibilities such as housework and economical contributions. The array of these practices of home as well as their local representations in particular situations and contexts are still to be compiled and analysed in more detail especially in reference to the disaster context and its effect on these practices and representations.

In this regard, I consider it especially important to examine further how people's vulnerabilities to the earthquake as well as their access to reconstruction resources differ according to their social, economical and political positions in the society. An analysis in this regard will provide deeper understanding of how people represent and practice home in relation to social, economic and political differences as well as how (disparately) people encounter and deal with the "enlarged state", evolved in Pakistan and Azad Kashmir after the earthquake.

With reference to the international and national states' disaster response, it is also crucial to ask of how people's vulnerabilities are reduced, reproduced or even exacerbated in the process of reconstruction. The financial difficulties of people in Muzaffarabad regarding the reconstruction of their houses indicate that the housing policy failed to reduce the vulnerability of those homes which lack necessary social, economic and political capital.

In addition to the structural vulnerability, my paper also points to the general questions of how people experience grief and cope with it in the disaster context. "Home" is an intimate place of affection, protection and mutual care where people, indeed, experience grief and the disruption of their lives most starkly but where they also cope with these traumatic experiences in the first instance. Thus, as my case studies evidenced, the experience of grief and the sense of loss of home are closely linked. Grief and disruption are frequently expressed by the means of nostalgia about the destruction and loss of the house as well as the separation of the family. By referring to the destroyed and lost house/home people demonstrate their grief. In the same way, by stating that family life profoundly changed after the earthquake people explain the sense of disruption experienced in the disaster. For coping with grief and disruption, the ambiguity of the category "home" as well as the flexibility of people's living arrangements are crucial. I believe that this ambiguity and flexibility provide important resources for households for adapting to changing circumstances, conflicts and crises such as a disaster. "Home" is re-shaped, strategically applied and politically exploited in the course of reconstruction, but, at the same time, enables ambiguous interpretations and flexible practices which, in turn, shape and contribute to the process of social recovery from disaster.

6. Bibliography

ADB and WB (2005): Pakistan 2005 Earthquake: Preliminary Damage and Needs Assessment. Islamabad: Asian Development Bank and World Bank. http://www.adb.org/Documents/Reports/pakistan-damage-needs-assessment.pdf (accessed 10.7.2011).

Bourdieu, Pierre (1976): Entwurf einer Theorie der Praxis: Auf der ethnologischen Grundlage der kabylischen Gesellschaft. Frankfurt am Main: Suhrkamp.

Carr, Edward R. (2005): Development and Household: Missing the Point? In: GeoJournal 62, 1-2, pp. 71-83.

EERI (2006): The Kashmir Earthquake of October 8, 2005: Impacts in Pakistan. Oakland, Earthquake Engineering Research Institute. http://www.eeri.org/lfe/pdf/kashmir_eeri_2nd_report.pdf (accessed 15.11.07).

Ferguson, James and Akhil Gupta (2002): Spatializing States: Toward an Ethnography of Neoliberal Governmentality. In: American Ethnologist 29, 4, pp. 981-1002.

Foucault, Michel (2006): Governmentality. In: Sharma, Aradhana and Akhil Gupta (eds.): The Anthropology of the State. A Reader. Oxford: Blackwell, pp. 131-143.

Gupta, Akhil (1995): Blurred Boundaries: The Discourse of Corruption, the Culture of Politics, and the Imagined State. In: American Ethnologist 22, 2, pp. 375-402.

Hammel Eugene A. and Peter Laslett (1974): Comparing Household Structure over Time and between Cultures. In: Comparative Studies in Society and History 16, 1, pp. 73-109.

Henry, Doug (2006): Anthropological Contributions to the Study of Disasters. In: McEntire, David (ed.): Disciplines, Disasters and Emergency Management: The Convergence and Divergence of Concepts, Issues and Trends from the Research Literature. Emmitsburg, Emergency Management Institute, Federal Emergency Management Agency FEMA, pp. 1-28. Electronic Book: http://training.fema.gov/emiweb/edu/ddemtextbook.asp (7-5-2011).

Hilhorst, Dorothea (2003): Responding to Disasters: Diversity of Bureaucrats, Technocrats and Local People. In: International Journal of Mass Emergencies and Disasters 21, 1, pp. 37-55.

Lemke, Thomas (2001): 'The Birth of Bio-Politics': Michel Foucault's lecture at the Collège de France on Neo-Liberal Governmentality. In: Economy and Society 30, 2, pp. 190-207.

Lovell, Nadia (1998): Introduction. In: Lovell, Nadia (ed.): Locality and Belonging. London: Routledge, pp. 1-24.

Oliver-Smith, Anthony (2002): Theorizing Disasters: Nature, Power, Culture. In: Hoffman, Susanna M. and Anthony Olivier-Smith (eds.): Catastrophe and Culture: The Anthropology of Disaster. Santa Fé: School of America Research Press, pp. 23-48.

————— (1999): "What is a Disaster?": Anthropological Perspectives on a Persistent Question. In: Oliver-Smith, Anthony and Susanna M. Hoffmann (eds.): The Angry Earth: Disaster in Anthropological Perspective. London: Routledge, pp. 19-34.

Oliver-Smith, Antony and Susanna M. Hoffman (2002): Why Anthropologists Should Study Disasters. In: Hoffman, Susanna M. and Anthony Olivier-Smith (eds.): Catastrophe and Culture: The Anthropology of Disaster. Santa Fé: School of America Research Press, pp. 3-22.

Rose, Leo E. (1992): The Politics of Azad Kashmir. In: Thomas, Ragu G. C. (ed.): Perspectives on Kashmir: The Roots of Conflict in South Asia. Boulder: Westview Press, pp. 235-253.

Sanjek, Roger (2002): Household. In: Barnard, Alan and Jonathan Spencer (eds.): Encyclopedia of Social and Cultural Anthropology. London: Routledge, pp. 285-288.

Simpson, Edward (2005): The "Gujarat" Earthquake and the Political Economy of Nostalgia. In: Contributions to Indian Sociology 39, 2, pp. 219-249.

Wilford, Justin (2008): Out of Rubble: Natural Disaster and the Materiality of the House. In: Environment and Planning D: Society and Space 26, 4, pp. 647-662.

The anthropology of a "disaster boom" economy in western India

Edward Simpson

1. Introduction

Ten years or so ago, an earthquake in western India claimed around fourteen thousand lives. Many people I already knew from previous research in the region (specifically, the district of Kutch in the state of Gujarat) died, more lost relatives, homes and possessions.[1] In this chapter, I chart some of the key moments in the life of the aftermath in order to highlight some of the contours of a 'disaster boom' economy from an anthropological perspective. Over the last decade, events in Gujarat have at times strongly resembled things that have happened after other disasters elsewhere. Immediately after the earthquake, traditional social distinctions collapsed; later there was mourning, nostalgia and a general reflection of regional identity; then came political protest and the reformation of social distinctions along the lines of caste, class and religion; but, perhaps most strikingly of all, the disaster gave birth to a consumer and then an industrial revolution.

Others of course have noted that catastrophe often leads to boom/revolution (Kendrick 1955; Seidensticker 1990 for instance). As a telling example, it is interesting to note that the anthropologists Barbara Bode (1989) and Anthony Oliver-Smith (1986) independently documented the changes, both subtle and stark, to follow an earthquake in Peru in 1970: the former writing with the subtitle 'destruction and creation', the latter with 'death and rebirth' to reflect the ethos of life after the disaster. In the conclusion of this chapter, I turn to examine the sociology of the destruction-boom sequence in the Gujarat case. Drawing on the political economy of John Stuart Mill (1848) and the neglected sociology of Samuel Henry Prince (1920), I suggest that, among other things, the moment of disaster is itself an accelerated moment of consumption, the veracity of which

---

1 I conducted doctoral research in the region in 1997 and 1998, and have spent periods from as short as two weeks to as long as six months in Bhuj every year since the earthquake of 2001. This research was supported by a Nuffield New Career Development Fellowship (NCF/00103/G) and an award from the UK's Economic and Social Research Council on its Non-Governmental Public Action Programme (RES-155-25-0065-A).

becomes part of a hyperbolic capitalism which reverberates and amplifies in the aftermath.

In what follows I review in general terms some of what happened to those who survived and the moral frameworks they developed to re-interpret life in the town that fell down. Then I turn to examine some of the implicit assumptions (neo-liberal and otherwise) that have driven state-led post-earthquake reconstruction, particularly in relation to the creation of a particular kind of consuming citizen, (sub)urban design and the rapid expansion of industry. The final part of this chapter, as I have indicated, turns to look at some aspects of the history of disaster capitalism. I remind readers that post-disaster booms are old and regular features of catastrophe, and that these cannot simply be attributed to the universal expansion of capital but must be understood as parts of the moment of sublime destruction and the collective emotional and sociological responses of the beleaguered.

2. Aftermath

In Gujarat, the dust eventually settled after the earthquake and the emergency response teams left for home (on the earthquake from the perspective of bureaucrats see Reddy 2001, Mishra 2006). Then people faced the task of rebuilding their houses, livelihoods and relationships in a landscape depleted of many familiar people and things. The initial report from the development banks suggested the cost of reconstruction was $2,274,000,000, although this figure seems rather arbitrary[2].From the outset, as it had done in the past, the state was expected to compensate, which it financed in part with a $390,000,000 loan from the Asian Development Bank repayable over 30 years.[3] The disaster quickly brought about regime change at the top of the state government. The coup was conducted in the name of 'efficiency' in relation to earthquake reconstruction policies. The moderate Keshubhai Patel was replaced with the darling of the Hindu right, Narendra Modi. The following year (2002), Gujarat was wracked by the worst communal violence for many years; many commentators reckoned Modi was somewhere between supine and complicit in the bloodshed.

On the ground, after the earthquake there was scramble for the rights to give and receive aid, which was intimately connected to the machinations of party politics in Gujarat as well as to various manifestations of modern Hinduism and politicised nostalgia (see Simpson 2005; Simpson and Corbridge 2006). In the countryside, villages were rapidly rebuilt by private agencies with financial as-

---

2   Gujarat Earthquake Recovery Program Assessment Report, A Joint Report by the World Bank and the Asian Development Bank To the Governments of Gujarat and India, March 14, 2001.

3   Asian Development Bank, Project Number: 35068, Loan Number: 1826, September 2008, India: Gujarat Earthquake Rehabilitation and Reconstruction Project.

sistance from the state (see Mahadevia 2001). Facilitated by the public-private partnership schemes, the reconstruction of rural Kutch fell to a wide range of interest groups ranging from toothpaste manufacturers to fundamentalist religious groups. In the process, a variety of new political and religious ideologies were inscribed on rural Kutch (see Simpson 2004). Organisations of the Hindu right in particular were granted relatively free access to Kutch but the ruling nationalist government. Such interventions were frequently associated with various types of religious and political proselytisation. At the time, these efforts seemed pernicious – exploitative of the vulnerability of the stunned and dispossessed – particularly when driven by the political and religious aims of the private organisation and their auditors rather than by the requirements of the villagers. For a while, religion and religious ideas were far more visible colonisers of the shocked population than capital. Signs littered the countryside advertising religious organisations and their development projects. Notable in this regards was in the village of Indraprashta in central Kutch which was fronted by a huge signboard stating 'purity is power'.

The foundation for rural intervention was laid by the first reconstruction package announced by the government. It allowed 'suitably qualified' organisations to reconstruct villages in new locations.[4] On the whole, villagers did not want to move and they demonstrated their anger. The distance new sites would put villagers from agricultural land, temples and graveyards were commonly cited reasons for not moving, as was a more general fear of the loss of the symbiosis of people and place. The revised package outlined a 'village adoption scheme' in the form of a public-private partnership; if such arrangements reflect neo-liberalism then this aspect of the reconstruction of Kutch was firmly such. The private partner was to reconstruct the village, with up to half of the basic cost being provided by the state. Where possible, the new village would be on the same or similar site.

The reconstruction of rural Kutch is a fascinating story, incorporating the extremes of aftermath-cynicism and humanitarian excellence. A vast number of Indian and overseas aid organisations, religious groups, political parties, state governments, social campaign movements, construction contractors and captains of industry stepped into this mêlée. In total, nearly 50,000 houses had to be built anew in over 600 locations.[5] Basic infrastructural and building regulations were specified by the government, but in practice these were often disregarded. The

---

4   Initial schemes were announced in 'Package 1' Revenue Department resolution No.XLS-162001-207-(4)-S.3 and General Administration Department resolution No. EST-102001-830- KH, dated 8.02.2001. 'Package 2' Announced under Revenue Department resolution No.XLS-162001-201-(4)-S.3 and General Administration Department resolution No. EST - 102001 - 830 - KH, dated 8.02.2001.

5   Figures derived from Appendix II, Coming Together (4th Edition), August 2002. Abhiyan, GSDMA and UNDP.

scheme opened the way for a whole range of interest groups to reconstruct villages, largely as they saw fit. The aftermath seemed to spell the end to village life as it had been known in Kutch. From the air, the extent of the reconfiguration of the countryside is still clearest. Many villages are now of two parts, the old section clustered around temples, water tanks and divided by roads and caste, and the new part some distance with houses laid out in neat grids; the passing years are however softening the distinction.

The adoption scheme polarised interest groups in many villages and, in some cases, villages divided into predominantly caste based sub-settlements. In other cases, wealthier sections of the population spurned offers of reconstruction – viewing offers as paltry and the style, size and quality of the houses as inadequate. In such cases, some have opted for a financial rather than reconstruction package. In yet other cases, some groups within villages opted for self-funded reconstruction programmes, unfettered by the budget restrictions and bothersome surveyors of the government.

The adoption scheme also produced competition between private interest groups for particular villages and between villages for the attention of particular sponsoring organisations. High-profile villages, such as those at the supposed epicentre, along the main highway, and of tourist interest were particularly sought after. Similarly, villages with certain social, caste and religious configurations were more appealing to some private agencies. Conversely, villagers quickly learned that if they supported an initial expression of interest from a private organisation, they were not bound to their commitment. If they could see that a neighbouring village was benefiting from a greater endowment than their own, they could approach that generous organisation and request them to come and reconstruct their village, and a kind of philanthropic gazumping game started. Both kinds of competition ultimately relied on the approval of the government and this fact has had clear consequences.

Organisations with cultural and political affinities to the ruling government were able to adopt high-profile villages, such as those at the epicentre and along the main highway into Kutch, forming something of a saffron strip on the road between Bhuj and Bhachau. In the initial flush of construction at least, these villages also readily received government services, such as an electricity supply. As we shall see later, these areas also became the locations of massive industrial investment; those organisations that sponsored the construction of the villages between Bhuj and Bhachau now also have their names and values intertwined with the flush of new wealth. In contrast, remote villages tended to be allocated to organisations with smaller budgets and less political ambition, but most obviously to organisations which had the least cultural and political affinities with the ruling government, Catholic organisations being notable in this regard.

Fundamentally, interventions were based on the restoration of what was lost, rather than on the basis of need or for the promotion of greater social equality.

Compensation levels were based on pre-existing property relations: landless labourers being restored with less than marginal farmers who were restored with less than small farmers and artisans who were in turn restored with less than those owning more than four hectares of land. Through the public-private partnership scheme many private organisations took on the functions resembling those of the state and in the process the state was re-imagined at the grassroots level. Layers of brokers and managers entered rural life to negotiate resources and to settle disputes; often such people were dressed in the robes of religion or the cloth of party politics. As I have said, organisations which had clear affinities with the ruling party were granted particular favour.

With a few years hindsight however these kinds of intervention seem quite innocent in scale and ambition when held up against broader movement of capital and industry into Kutch on the back of government subsidies in the name of post-earthquake reconstruction.

In Bhuj, the administrative centre of Kutch from where most of my data is drawn, the disruption in the early years (loosely 2001–2004), as seems to be the case in a great many disasters, is/was often referred to as the 'second earthquake' because the Government of Gujarat seemed to impose the weight of its understandable hesitancy on the beleaguered. It is not an exaggeration to say that surviving the earthquake became a full-time job for quite some years after the actual event. Many busied themselves with the details of the various compensation packages; some studied the new building codes; for others, there was time for little more than staying one step ahead of the bulldozers that had been drafted in to reconstruct the town.

The complexities and bureaucracy of urban reconstruction, the sheer size of the task, and clear divisions in popular opinion undoubtedly contributed to delays in policy design and implementation. There were some, for example, who thought Bhuj should be built anew elsewhere; others favoured rebuilding it the way it had been. For a long time, however, there appeared to be no plan and people became angry; then it was decided to rebuild Bhuj where it had been and to allow room for future expansion and the resettlement of the homeless. Then, as the months passed, concern shifted to building temporary shelters, the implications of having lost documents relating to property and finance and the lack of governmental coordination. Later, it was planning considerations, policies for rubble clearance, levels of compensation and baffling questions about the rights of tenants and apartment owners that preoccupied many. There were surveys and resurveys, and a mystifying system of damage classification was introduced which encouraged some of the propertied classes to do damage to their homes. Later still, there were gnawing questions about the location and design of permanent housing and the scale and scope of new infrastructure.

Meanwhile, as these important debates raged, large numbers of people found good reason why their homes, which had survived the earthquake, should not be destroyed to make way for new roads of the new town planning schemes or to conform to new safety regulations. It appeared to many people as if the Government was taking away property to make way for these new roads without compensation. This caused widespread anxiety and resulted in a number of suicides. To an extent, the tragedy of these suicides stood as a metaphor for the popular perceptions of maladministration and alienation at the time. Gangs of profiteering contractors, mostly from distant parts of northern India, descended on the town at night with noisy machinery and hoards of (supposedly) wild and unruly labourers from Madhya Pradesh to demolish what remained of the old town. The local citizenry, often quite sensibly feeling alienated, engaged in enthusiastic letter writing campaigns, hunger strikes and protest marches against delays and injustices. Some called for an independent Kutch or for Union Territory status, free from the shackles of step-motherly Gujarat (Kutch is an administrative district of the state of Gujarat). For many, the perceived neglect of ruined Kutch by those with the power in the east of the state was merely the extension of a well-established syndrome.

While many worried about money and shelter, for others there was an unfamiliar superabundance of what might be thought of as 'disaster boom' cash which fuelled the first waves of the consumer revolution. Resources had poured in from all over the world, generosity motivated in part by the presence of a large and influential Gujarati Diaspora in the UK and USA. Kutch had also been given a great material boost in the form of tents, sewing machines and other items intended to regenerate livelihoods. In the early days, the government gave emergence daily cash payments to all. There was a wide-range of state-led compensation schemes for lost livestock, body parts, property and life. With a view to kick-starting the local economy, the state also made available a large number of small enterprise grants; it is widely recognised that many applications to the scheme were fraudulent. And therefore, while it might not have been fully successful in its aims, the scheme contributed quite directly to the rejuvenation of the economy.

Although the property compensation payments were eventually staggered in line in certified construction, those who had regularised property but had previously scraped by on low wages suddenly had tens of thousands of Rupees from government compensation schemes. The money was, of course, intended for new housing, but agencies selling motorbikes and television sets prospered. For others, cash sums from insurance companies and the Government came at a very high price: the loss of relatives and/or body parts.

Many, encouraged by an emergent class of brokers, made false claims for compensation and for enterprise grants. Creative corruption was an open secret in which it often seemed as if the whole town was complicit; very few could point accusatory fingers at others from the security of innocence. Some corruption was simply so outrageous, especially in relation to contracts for rubble clearance, that it made people chuckle and this had a cathartic effect. In 2004, government audits revealed some of the false claims and many beneficiaries, having already disposed of the cash, faced the worry of further debt or imprisonment as well as the problems associated with a lack of funds for the construction of a permanent shelter. In 2010, the recriminations continue with quite high level bureaucrats facing prison for their part in various forms of organised corruption.

Gujarat is well known for its enterprise, and this largely withstood the impact of the earthquake well. Unless things were utterly destroyed, most newspapers and hotels only lost a few days business. Private transport operators, English speakers, and those with access to building materials were initially in high demand. There were no food shortages, although the bazaars of the old town were closed for a while, improvised markets for food stuffs and other commodities soon started to appear. Mobile phone operators hastily opened networks to 'improve communications'.

There was of course a consumer boom across western India before the earthquake, computers and the internet had just reached the provinces and many people were feeling richer, but the boom was greatly intensified by the concentrated and focussed investment the disaster brought. Other people I know in other parts of Gujarat had heard Bhuj was booming and at that there were jobs in construction and opportunities in property development, and they came in their droves in search of a fortune.

In 2001, the then Prime Minister Atal Bihari Vajpayee promised a 'New Kutch'. In 2004, Narendra Modi, the Chief Minister of Gujarat, told Kutchis to put the earthquake behind them because the rest of Gujarat had forgotten about it. In 2006, Modi started to liken Kutch to Singapore; according to him, New Kutch, the engine of Gujarat's double digit growth, was complete. Then, six years after the earthquake, Bhuj, and the neighbouring towns of Anjar and Bhachau were essentially construction sites of vast proportions. The story of post-earthquake reconstruction was clearly far from over despite the pronouncements of the politicians – but their words did reflect a tremendous investment in the region and the promotion of a feel-good economy. The sense of mourning remained pervasive and sometimes people felt that life was futile; on the whole however, life went on under clouds of dust, amid an endless and mentally draining cacophony of roars, thumps and tapping sounds. Elsewhere, thousands of families continued to live in shelters that were originally intended to be temporary (there were often quite counter-intuitive reasons for this – a common one is that families rented out there new property).

The aftershocks continued, but fruits of post-disaster policies, described in what follows, were beginning to ripen. Much of what can be seen in Kutch is new, if not exactly reminiscent of Singapore.

3. The new Bhuj

For at least the last hundred years Bhuj has been expanding, the population gradually settling outside the walled confines of the old town, tentatively at first and then in something of a rush. After the earthquake, the pace and scale of change has been greatly increased. The distinction between the old walled town and the post-1960s suburbs has been softened. New thoroughfares and round-abouts carry new names, particularly those of the various Hindu congregational orders which descended on the town to great effect with relief after the earth-quake and found a generally receptive audience among the townsfolk. On the back of a cash rich economy, boutiques, supermarkets and sweet shops with shimmering glass and metal facades sprang up, again reflecting the new possi-bilities for consumption.

The most striking thing however about the new Bhuj is the size of the place, as vast swathes of suburban land, the work of government relocation strategies as much as private developers, stretch many kilometres to the south. Now, it takes far more than an hour to walk from the southern edge of the old town to some of the more distant relocation sites. These suburbs are overlooked by the new Hill Garden (at present named after the Chief Minister Narendra Modi), a haven for flirtation and merriment on Sunday evenings. Acres of tarmac have been laid to form concentric four-lane roads surrounding the town most, eerily, unused by traffic but good for cricket and evening strolls.

New Bhuj has of course come at a cost and for a while the words 'town planning' (or simply 'TP') and 'cutting' (literally the process of cutting away the bits of buildings that stand in the path of new roads) were often heard in Bhuj. TP is popularly associated with road widening and the laying of new roads and has become a part of daily life, rather like the threat of malaria. TP is associated with modernisation and the imposition of order, and, to a much lesser extent, with the management of future disasters. For most people, TP is a given and non-negotiable set of parametres, widths and other measurements. Thus, popu-larly, there was no popular conception of good or bad town planning and, con-sequently, little protest against decisions that are clearly misguided or discrimi-natory.

People were of course wrong or solely speaking for dramatic effect when they said that town planning was a post-earthquake blight. The rulers of the for-mer princely state did it in their own way, just as the bureaucrats of the era im-mediately after independence did it in theirs. The municipality had indeed drawn up plans to improve a number of roads in the town before the earthquake. These plans were abandoned in the rubble. The key difference however between then and now is that TP has given the town a new character that has largely usurped the older symbols of its identity. Most of the key roads that snaked through Bhuj before the earthquake have been widened and, as I have mentioned, renamed. Three major new looping access roads were bulldozed through the old town.

The local administration championed the cause of road widening, suggesting that it affected everyone equally and no party, however influential, would be spared. The evidence rather suggests, despite the conspicuous cutting of the property of the wealthy in the main bazaar and along Hospital Road, that the potential loss and violence of TP really was like the threat of malaria and depended on how wealthy and influential you and your neighbours were as to how vulnerable you were to its dramatic effects.

If the scale of post-earthquake development plan for Bhuj was influenced by projected future population growth it remains difficult to imagine where so many new people could come from. Today, in 2010, now some ten years after the earthquake, many of the new housing colonies are half populated and some of the shopping centres bordering the ring roads have never opened. Perhaps the silent influence of military strategists, with an eye on the border with Pakistan, is behind the scale of new infrastructure; or, perhaps, the planners and consultants in the subcontracted pay of the Gujarat Urban Development Company simply got carried away by the aesthetics of segmented and concentric models of urban development – like those they had studied in school. Either way, the acres of deserted tarmac that encircle Bhuj carry weighty symbolic capital if not traffic.

In 2006, I asked a sample of thirty Bhuj residents whether their town was better before or after the earthquake. Half of those asked pointed to the miles of new road as irrefutable evidence that the town was much better than before; local people like them; they think new roads and infrastructure of all kinds are a very good thing – in this sense they are all environmentalists – environmentalists who want to create a new environment in which nature is tamed and concreted over; roads may have been done to them by the state and the contractors, but the population is enchanted by their produce.

That roads should have become such a positive totem is an interesting sociological fact in and of itself. In this case, roads are public spaces that have created new links as well as new divisions between people and places. They have also connected particular kinds of places such as the new airport to the best hotels, and the wealthy suburbs to the lakeside parks, for example. Roads have not similarly paved the way from the slums to the washing ponds in the northeast of the town, for example, despite the fact it is a very well used route. The new roads have only been laid in the directions that some people want to go; for other people – notably, those without the resources to own a vehicle or to live in a planned suburb – roads have become obstacles. Some roads were built regardless of public protest, and Muslim cemeteries, most notably, as well as other religious sites were bulldozed. In short, roads have created particular routes through the town, and in the process other kinds of geography have been marginalised.

The casual visitor who is accustomed to thinking about a urban space in terms of roads, could quite innocently leave at the end of their stay thinking that the citizenry are all reasonably well to do and enjoy riding motorbikes. This visitor will simply not have seen the most densely populated parts of Bhuj to the north, that are not only without roads but have been made all the more inaccessible, and thus invisible, by new roads.

That the development plan did little directly for the poor and marginal and little more for the minority communities is not of course particularly surprising, but the issue of roads also leads me to the overarching but implicit assumption contained within the development plan for Bhuj and this also affects those for whom the plan was hatched.

In the development plan, the majority of families are clearly intended to become suburban dwellers, dependent on their own motorised transport to reach their places of work. This is perhaps a price the planners had to pay in order to reduce the population density, especially in the old town. However, it seems particularly retrograde in terms of environmental impact and for the consumption of fossil fuels, at a time when in other parts of India people are looking for cleaner ways to move – or, more importantly, not to move within cities.

It is important to understand that many of those who have now settled in the suburbs are not wealthy people and have pitched up there because of government compensation payments, assistance from NGOs or a combination of both. As I have mentioned, travel to and from the suburbs is expensive and living there and working in the old city can carry an impossibly high cost. The fruit and vegetable sellers who make the long journey from the wholesale market to the suburbs also sensibly inflate their prices. Other goods and services simply are not available because of the low population density which again necessitates that people make expensive journeys. Therefore, some families have literally been impoverished by their move to these suburbs. This seems to have a greater negative effect on women who are forced to stay at home more (because the rickshaw is so expensive and it is not practical to walk) than on men who tend to be the main wage earners and can ride a motorbike if they own one.

The broad roads (at least the radial ones to the south) now justify themselves because they have become necessary to cope with the increased level of traffic that the suburbs have created. However, rather than attempting to reduce the flow of traffic through the town through pedestrianisation, traffic calming measures, public transport provision or clustering land use zones, the development plan exacerbates pre-existent problems, such as levels of traffic, through the act of suburbanisation, rather than attempting to change anything significantly, let alone radically.[6]

---

6   My knowledge of town planning in Bhuj comes mostly from conversations with staff of the Environmental Planning Collaborative (EPC), a not-for-profit organisation based in

Planning of the suburban government relocation sites has been conducted with wilful neglect of local ideas about social organisation, especially caste. Perhaps this is too harsh, simply the view of an anthropologist who tends to see the significance of social matters above all else; or, perhaps, the planners felt that caste is anachronistic in modern India and that people would simply forget about it along with their own social history once they were given a house big enough for a nuclear family in the suburbs. Casteless planning in this context however seems to assume very improbable things. Of course, the caste of neighbours is not of vital importance for everyone, and such a concern is often analytically (but revealingly) inseparable from the desire to have pleasant and familiar neighbours. But, I assert with some confidence, for many more people in towns like Bhuj caste is a good measure of the person and therefore remains very important when deciding where to live if there is any choice in the matter. Evidence from elsewhere in Kutch suggests that the planner's casteless suburb (note I do not use the word 'secular' here because very few Muslims have opted for life in the southern suburbs which are distinctly Hindu in character) was not only a cause of considerable anxiety for many people, as they waited years to know who their neighbours would be, but it will ultimately prove to have been, in part, a waste of time for the following two reasons.

First, in the countryside most villages were rapidly constructed anew through public–private partnerships (Simpson 2004). Many of these villages were built for reasons of economy and simplicity in grid patterns. In some cases, cursory attention was given to caste, religious and kin sensitivities of the local population. However, on the whole, and perhaps rightly so, haste and need were placed over and above social dynamics and houses were often allocated on the basis of a lottery. Like others, I was initially dismayed by the unimaginative and barrack-like nature of the new villages; villagers were mostly unenthused too. Yet, in the countryside, the ensuing years have seen many thousands of intimate deals brokered and struck, as houses and neighbours have been exchanged to make villages that make more sense in the villagers' terms. New housing patterns have been imposed on the grid as people have taken control of their space and have chosen who they want to share it with. Houses have been extended and merged,

---

Ahmedabad, and with officials of the Bhuj Area Development Authority (BHADA). My comments on planning are not addressed at the work of these organisations, but at the broader frame in which they had to carry out their work. It is also important to note that the models of infrastructural development seen in Bhuj are not isolated from planning policy as practiced elsewhere in the country. The Jawaharlal Nehru National Urban Renewal Mission of the Ministry of Urban Employment and Poverty Alleviation and the Ministry of Urban Development, for example, is a national scheme designed to improve governance and prosperity in major urban areas. This mission, along with others, also places great emphasis on the development of infrastructure as an objective rather than simply a means to other ends.

and compound walls have been built around particular houses. More telling however, walls have also been constructed around groups of houses, sometimes across the roads of the grid, blocking the way, to enclose kin or caste groups and livestock. In many villages, the grid is now barely discernable, as property has been re-organised by what might be thought of as the pressures of a non-economic and social market.

Second, while the suburban 'relocation' sites were the most prominent and regulated of the reconstruction schemes, there were lesser schemes run more along indigenous lines; among them there are two quite clearly discernable trends. First, caste organisations, notably those with clearly-defined patterns of leadership and religious affiliation, constructed their own suburbs, often including places of worship, schools and primary health facilities. Secondly, in other settlements, such as those that started life as temporary shelters but were later made permanent, successful attempts have been made by non-governmental organisations such as Kutch-based Abhiyan to consult people as to who they want as neighbours in order to design and distribute land and property accordingly. The results of this consultation mostly indicate bias towards caste preference.

In sum, there seem to be two major assumptions about how the world should be embedded in the urban plan for Bhuj. First, not everybody needs to be planned for equally. Secondly, those who have been planned for the most, so to speak, will lead particular kinds of life which will involve suburban living, largely devoid of pre-earthquake social history.

4. An industrial revolution in the countryside

I now want to take you out of Bhuj, east through the village of Madhapur and the chaos and dirt of the service area for the many hundreds of trucks that ply between the opencast lignite mines in the west of Kutch to the power stations and factories in the east of Gujarat. Further out of the town, the air freshens as the mechanical workshops and tyre centres give way to private resettlement colonies, a village for orphaned children and handicraft parks. Further still, the road splits, one branch heads south-east towards Anjar and the commercial port centre of Kandla-Gandhidam, the other heads slightly north of east towards Bhachau. A few years back, the road to Bhachau passed through a vast landscape of acacia, broken by the occasional village and agricultural oasis. The road served hinterland villages and was a corridor for nomadic pastoralists.

After the earthquake, the Government of Gujarat's Industries and Mines Department developed the 'Incentive Scheme 2001 for Economic Development of Kutch District'.[7] The scheme was intended to make 'the economic environment

---

7 See Government Resolution No.INC-10200-903-I dated 9-11-2001 for the regulations for exemption from Central Excise; see Government of India, Ministry of Finance, Depart-

of Kutch district live' by encouraging new investment in the region. Industrial development was also promoted by ministers from the Government of Gujarat at a number of trade fairs and at the 'Resurgent Gujarat' event of 2002. The Incentive Scheme offered both excise and sales tax concessions for new industrial units commencing production in Kutch before the end of 2005 for a period of five years.[8]

In 2003, the road to Bhachau was still silent; the only sound was the wind in the acacia. In 2004, three years after the earthquake, a ceramics firm had started to construct a manufacturing plant at the Bhachau end of the highway. By the end of 2005, there were around 25 medium- to large-scale units dotted along the length of the road. The rate of growth along the highway has been phenomenal; in other parts of Kutch it has been even higher, notably around the government port complex at Kandla and the private Adani-owned port at Mundra (further encouraged by recent Special Economic Zones). The road to Bhachau now passes through a surreal landscape. It is still rural, but not confidently so, as it is lined with generous slabs of factory, belching chimneys, impeccably neat compound fences and splashes of urban-style colour of the post-modernist architecture of office blocks and reception halls. Today, the road is also unmistakably busier with traffic and there is discussion of 'four-laning'. Here, as elsewhere in Kutch, the Incentive Scheme seems to have been a rip-roaring success as there are now huge industrial complexes and infrastructure has been considerably improved. And, as those who designed the policy might have anticipated, financial benefits, as well as effluent, are leaking, if not exactly trickling down, into the countryside.

On paper, the scheme is excellent and generous; there is provision for pollution control and for compulsory future investment by beneficiaries. The documentation however reveals that the policy is not even implicitly geared towards the creation of local industry, but is an invitation for existing industries from elsewhere to open plants in Kutch with direct and indirect government assistance. This too is something of an open secret, and a former minister associated with industrial development told me that it was the best he could expect from the policy was for Kutch to host the industry and its people to provide ancillary support in the form of haulage and the like. The clouds of very black smoke emitted from factories, especially those in the far east of Kutch, suggest that pollution is rather poorly controlled.

Along the road to Bhachau, there is the occasional flurry of people scrambling for transport when shifts in some of the factories change, but it remains oddly quiet. Local people are employed in many of these factories on assembly

---

ment of Revenue, New Delhi, dated the 31st July, 2001, Notification No. 39/2001- Central Excise.

8 The deadline for new industry to commence production was originally 2003 but this was extended to 2004 and later to 2005.

lines and the like, but much of the manufacturing is heavily mechanised, with little need for manual labour. There were rumours, probably true, that such factories were established for short-term profiteering, making use of a cooperative government and the ready (and thus indirectly subsidised) supply of land, water and power. For a few years rumours circulated that some of these factories claim to produce goods in Kutch for the tax concessions when in fact they produce elsewhere. However, alongside the creation of Special Economic Zones around the ports of Kutch and the facilitation of industry and big business by the state government much of the district has been transformed into a huge industrial park.

Through these and other industrial development policies large parts of Kutch are being degraded, other parts are being covered with the concrete of industrial infrastructure that allows for the processing, manufacture and transport of various materials that come and go through the ports. The profits of this activity, like most of the goods produced, are consumed elsewhere.

The occasional voice of protest has been heard, some complained about the destruction of mangroves in the Gulf of Kutch, and a politician from Mundra protested that pollution had rendered agricultural lands unproductive. There has however been little or no public debate about rapid industrialisation and scant regard for glaring ecological and other environmental concerns, notably the ruthless exploitation of ground water and ensuing salination.[9] Few have questioned the implicit assumption that industrial development should be given precedence over other concerns, and no one has suggested that industrial development is undesirable as an end in its own right.

Many local people have become wealthy from the promise of industrial development. The already-wealthy have become wealthier through land speculation of all types. Land in Kutch has also become a very attractive investment for people from elsewhere. Farmers have sold their lands, especially near highways, given the wild premiums they can demand. The villagers are then able to turn their backs on laborious agriculture in favour of wage labour in the local factory or the provision of an ancillary service of some description.

Clause 8e of the Initiative Scheme demands: "As per the employment policy of the Government of Gujarat, the unit availing of the incentives, will have to recruit local persons for a minimum of 85% of the total posts and for a minimum of 60% of the managerial and supervisory posts." Members of Group 2001, a citizen's pressure group based in Anjar, conducted surveys of the new industry in 2003 and found that most factories employed high percentages of 'outside' (that is to say non-Kutchi) labour. Group 2001 also suggests that since the earth-

---

9   In Bhuj, in the 1990s, a group of concerned folk discussed responsible industrialisation and took legal action against a number of firms. Their efforts were however re-focused by the earthquake and their discussions and activism lapsed.

quake the population of Anjar has increased by one fifth, and most of the new residents are non-Kutchis. And, as might be reasonably expected, this has created new tensions, resentments and a hyperbolic increase in the fear of crime and violence.

The standard argument in defence of the abuse of clause 8e is that local labour is unskilled, and therefore it is necessary to import labour from elsewhere. There is undoubtedly an element of truth in this but it is also equally true that outside labour is cheaper and the government is not enforcing 8e. The general point here is that those affected by the earthquake benefit far less from this industrial development in the short-term than might be supposed and in the longer-term such rapid industrialisation may well have highly detrimental effects on the health of the District.

If complex systems of production need dirty areas for processing and production and clean areas for the consumption of the goods they produce, Kutch is moving rapidly from being a marginal clean area to an important dirty one on the national stage. The relatively free access the state has granted to minerals and land has enabled large industrial enterprises to remove wealth from Kutch in the name of post-earthquake development. The costs of this shift seem to be carried primarily by the state, the environment and, perhaps in the long run, by the local population.

5. Shock doctrine

In the years of reconstruction that followed, new roads were bull-dozed through the symbolic core of Bhuj, changing the shape and boundaries of the traditional neighbourhoods, resettlement programmes redistributed people to new settlement colonies, radically altering the distribution of castes through the town. Once the capital of a feudal and ritually elaborate kingdom, Bhuj was redrawn by planners as a post-colonial town in which transport corridors and shopping centres oriented the town rather than the palaces of its rulers and the location of its religious sites. The plan was not only for the town, its effect was to demand a new type of citizen: a mobile suburban consumer. Elsewhere in the region, the state reduced various industrial taxes and facilitated the transfer of land and other resources to the industries of India (see Sud 2007 for a general discussion of this), creating an enormous if uneven boom. In sum, for those, such as myself who have watched for long enough, the moment of catastrophe brought about not sudden sustained change as initial analysis (including my own 2004) would suggest, but gradual and fundamental changes in society.

How can we account for the sequence of catastrophe and revolution or boom? Recent literature, notably Naomi Klein (2007) on 'the shock doctrine', has suggested that interventions in the aftermath of natural disasters are strategic and generic extensions of the type of free market capitalism outlined by the Chicago School led by Milton Freedman in the 1950s. In Klein's view, American capitalism uses the public disorientation of collective shocks to control and profiteer.

At first glance, and despite the lack of American's, Klein's model of 'the shock doctrine' seems a productive way of beginning to understand what has happened in Kutch after the earthquake. The state and capital joined hands in an environmentally catastrophic land grab, laws were discretely changed overnight after the earthquake as the government essentially colonised a troublesome political zone, which not only nestles against the border with Pakistan but also had the audacity to demand separate political status in the aftermath of the earthquake. From the centres of power and government in the east of Gujarat, the state used the disaster to gain a firmer footing in the outlying west of the region, an area which has been culturally and linguistically distinct in the past.

There were technocratic, linguistic and bureaucratic aspects to the neo-colonialism of the state. The new hospital was mounted on springs which distracted many for some time as the awe of imported technology pacified the beleaguered. As I have discussed, the swift instrument of town planning cut through old societies and taught people how to read maps and to envisage life in suburban housing colonies. Signs of reconstruction started to appear in Hindi, the national but previously invisible language in the region; religious sects who used the Gujarati language for liturgy seemed to be particularly encouraged by the state to move into the affected region where the primary language was Kutchi. More generally, the state expanded dramatically in size and function, and even when public-private partnerships were formed the state remained the senior partner. Freedman favoured the free operation of markets, with little or no interference by the state; the effectiveness of the state in post-colonial India suggests the shock doctrine is not quite of the type Klein outlines, despite the similarities in outcome. In these neo-liberal times of disaster, the state did not retreat but instead served as a broker and facilitator of private interest.

Experts on the sociology of natural disaster often use the expressions '*crise relavitice*' and '*tabula rasa*' to refer to society in the aftermath. The former refers to the barebones of a society stripped of its cultural niceties, the latter to the unformed, featureless mind as found in the philosophy of John Locke. Klein too finds the metaphor appealing drawing on the experiments of psychologists who labored with the idea that if the personality of a patient could be erased then it could be built back without fault or disorder. However, to begin the story in aftermath is to forget the sublime awe of the disaster itself as well as the time of innocence before the disaster.

In what is perhaps the first modern sociological study of a disaster, Samuel Henry Prince described the consequences of "the greatest single explosion in the history of the world" (1920: 27). The disaster was caused by the collision of two ships, one laden with explosives, in the port of Halifax, Canada. He convincingly shows how the matrix of custom is shattered or disorganised rather than exposed by catastrophe (as later writers have argued), as mores are broken up and scattered left and right. The post-disaster moment is one of chaos – not featureless, nor a simpler form of everyday complexity. The language of Prince's 'disaster psychology' (1920: 44) is reminiscent of Klein's as he discusses the role of the 'stun' or 'shock reaction at Halifax' (1920: 36) in relation to the rapid reconstruction of the city. Prince however also stresses the creative emotional and psychological repercussions for society of disaster. These range from "primitive behaviours" such as hallucination, delusion and pugnacity to the gregarious instinct, and the generosity which refurbishes civil society and fuels the growth of new industry.

Some time earlier, the philosopher John Stuart Mill (1848) had also wondered why countries rapidly recover from states of devastation. For Mill the destruction is nothing other than the accelerated consumption of what had been previously produced which would have been consumed anyway. Therefore, in some senses, disaster is a moment of hyper-consumption which necessarily acts as powerful economic stimulus – complicated by extreme emotional conditions and the confused and resource-rich interventions of governments and other agencies.

In very different ways, Prince and Mill draw our attention to the fact that the shock of disaster is not primarily in the aftermath, but in the event itself which reverberates into the emotional and political relations of society and into the opportunities for capitalism amid the ruins. Much of the confrontational politics surrounding the disaster were conducted on regionalist lines. The worst affected region claimed a separate linguistic and cultural identity from the rest of Gujarat and the Government of Gujarat in turn used the opportunity of the earthquake to strengthen its presence in the region. In this sense, the earthquake had a dynamogenic effect, congealing emotion and trauma in the form of a nostalgic regional identity which in turn came into conflict with the actions of the state, and it was primarily this conflict that animated the post-disaster landscape, aided by the rapid consumption of Bhuj by the earthquake. Alongside this, as the spaces for new kinds of citizens were created, and the dead were converted *via* compensation payments into consumer items, the public economy passed from sorrow to joy.

6. Bibliography

Abhiyan, GSDMA and UNDP (2002): *Coming Together*, 4th Edition.
Bode, Barbara (1989): No Bells to Toll: Destruction and Creation in the Andes. New York: Athena.

Kendrick, Thomas D. (1955): The Lisbon Earthquake. Philadelphia: J.B. Lippincott Company.

Klein, Naomi (2007): The Shock Doctrine: The Rise of Disaster Capitalism. London: Allen Lane.

Mahadevia, Darshini (2001): Privatising Earthquake Rehabilitation. In: Economic and Political Weekly 36, 39, pp. 3670-3673.

Mill, John Stuart (1848): Principles of Political Economy with Some of their Applications to Social Philosophy. Boston: C.C. Little & J. Brown.

Mishra, Pramod Kumar (2006): The Kutch Earthquake 2001: Recollections, Lessons, and Insights. New Delhi: National Institute of Disaster Management.

Oliver-Smith, Anthony (1986): The Martyred City: Death and Rebirth in the Andes. Albuquerque: University of New Mexico Press.

Prince, Samuel Henry (1920): Catastrophe and Social Change: Based upon a Sociological Study of the Halifax Disaster. New York: Columbia University.

Reddy, L.R. (2001): The Pain and the Horror: Gujarat Earthquake. New Delhi: Aph Publishing.

Seidensticker, Edward (1990): Tokyo Rising: The City Since the Great Earthquake. New York: Alfred A. Knopf.

Simpson, Edward (2005): The "Gujarat" Earthquake and the Political Economy of Nostalgia. In: Contributions to Indian Sociology 39, 2, pp. 219-49.

―――― (2004): "Hindutva" as a Rural Planning Paradigm in Post-Earthquake Gujarat. In: Zavos, John, Wyatt, Andrew and Vernon Hewitt (eds.): The Politics of Cultural Mobilisation in India. New Delhi: Oxford University Press, pp. 136-65.

Simpson, Edward and Stuart Corbridge (2006): The Geography of Things that May Become Memories: The 2001 Earthquake in Kachchh-Gujarat and the Politics of Rehabilitation in the Pre-memorial era. In: Annals of the Association of American Geographers 96, 3, pp. 566-85.

Sud, Nikita (2007): From Land to the Tiller to Land Liberalisation: The Political Economy of Gujarat's Shifting Land Policy. In: Modern Asian Studies 41, 3, pp. 603-37.

World Bank and Asian Development Bank (2001):Gujarat Earthquake Recovery Program Assessment Report, A Joint Report by the World Bank and the Asian Development Bank To the Governments of Gujarat and India, March 14, 2001. http://www.preventionweb.net/files/2608_fullreport.pdf.

IV. Constructing local meanings

Lightning, thunderstorms, hail: imaginations, religious interpretations and social practice among the Quechua of the south Peruvian Andes

Axel Schäfer

1. Introduction

News spread quickly through the little Andean village of Choaquere. Lightning had struck again in the upper neighbourhood, where just a few weeks ago it had already killed a boy and some of his sheep. People explained his death through the fact that the boy forgot to hide his shining gold necklace and transistor radio, both items that are said to attract lightning.

This night however, although the house of the Farfán family was spared, lightning had split a tree behind it and killed their two milk cows, which had sought shelter underneath. Everybody expressed sympathy with the family, and speculations ran high about the reasons for this misfortune.

Although compared with other disasters the damage seems minor, the analysis of the family's resources shows how devastating the consequences were for them in both economic and social terms. Due to their poverty, the death of their only cows signified a substantial economic loss since the milk was an important part of their subsistence; sold as cheese it secured them a continuous income. Above all, as the cows had represented the capital of the family as 'a bank on legs', their death deprived them of an important source of cash.

In addition to weakening their subsistence, the lightning had also taken their means of production away from them. They were no longer able to use their cows for ploughing but were now forced to rent animals from their neighbours to carry out this task. The incident implied further difficulties since, because of ritual prohibitions, the family was not allowed to make use of the animals now they had been killed, let alone to touch them. They were also prohibited from approaching the spot where lightning had struck. The particular sulfuric and burning smell (*qoyo*) after a lightning strike can lead to illness, like toothache or split skin between the fingers. Pregnant women are also placed in danger if they see lightning, as it is believed to cause disabilities or peculiarities in newborn children, like additional fingers or harelips.

Lightning, together with hail and storms, are ubiquitous in the high mountain region of the central Andes. Up to 120 thunderstorms can be expected annually in any particular place in the Peruvian Highlands (Gade 1983). Especially during the rainy season from December to March, lightning can be seen nearly every day dancing on the highest peaks. Sometimes thunderstorms bring destruction to the farmsteads and villages of the valleys. But most often it is the unsheltered huts of the herders in the summit region (Puna) that are hit. Careful estimates indicate that every year in the Central Andes at least three hundred people are killed by lightning and about six hundred injured (ibid.: 773). Everywhere houses can be found that have been damaged or destroyed completely by a bolt of lightning. A single bolt of lightning can kill a whole cattle herd when the animals squeeze together out of fear or seek shelter. Moreover, lightning always occurs together with hailstorms (*runtu*), which can themselves cause deadly injuries and can destroy an entire harvest in a few minutes. Every family in Choaquere has been seriously affected by such events in one way or another. As a result, an approaching thunderstorm can cause incredible fear. Families huddle together in their fragile huts, praying on their knees or even burning offerings of copal and tobacco.

Because of such existential experiences, lightning occupies a central place in the Andean cosmology. Especially amidst the population of the Puna region, there is a rich knowledge with regard to it. As in some other cultures, lightning is regarded as an ambivalent powerful force. On the one hand, it is conceived as a malevolent spirit, related to the underworld and dark forces, which must be ritually appeased. On the other hand, its positive characteristics are that it can bring rain, fertility and wealth, or elect ritual specialists and healers. Or as Rösing (1990: 9) states for the Callaways in Bolivia lightning is the "cause of disease but also a sign of vocation" (translated by U. Luig).

The present article analyzes this dual concept, its roots and the resulting practices. As a starting point a short introduction to Andean cosmology is given, followed by a description of the role and complex character of lightning in pre-colonial mythology. These facts are supplemented by recent anthropological documentation, including unpublished data from my fieldwork in Cotabambas Province in Apurímac Department, Peru. This procedure allows historical and regional continuities and changes to be identified. The last part of the article analyses how these beliefs and material consequences are reflected in concrete ritual practices by returning to the example of the Farfán family, already introduced above.

2. Andean cosmology

Like their ancestors, the Inca, the rural population of the southern Andes believe that the world is constructed of three layers. *Kay pacha* (literally, 'this world, this time') means the here and now, the domain of humans, who move about the cultivated surface of the earth. Above *kay pacha* opens up the upper world or *hanan pacha* (Nuñez del Prado Béjar 1970), also called *gloria* in Cotabambas. This is the region of pure light, occupied by the sun, the stars and planets, and is dominated by the Christian god and his saints. Beneath the *kay pacha*, lies the *uku pacha*. This inner or lower world is dark; it is associated with the female Earth goddess, *Pacha Mama,* and the bones of the ancestors (Steele 2004). Due to its underground water, springs and lakes it is considered the source of fertility (Harris 2000).

In colonial times this Inca construct of the earth was enriched with Catholic conceptions of heaven and hell. The chronicler Garcilaso de la Vega (1983) translated *uku pacha* as *supaypa huasin*, or house of the devil or hell. But despite these Christian influences, popular belief preserves a more ambivalent understanding of underground forces, which are seen as dangerous, but also as able to produce positive effects. These ideas may result from the close relationship of rural people with the earth and its productive forces. They might also have to do with the popular Andean belief that the world has undergone several stages of destruction and recreation (Minelli 1999). In this process, the Inca and their gods were not completely eliminated, but driven underground (Casaverde Rojas 1970). From there, these 'unbaptized' ancestors and demonized gods of autochthonous mythology unfold their powers, being much more closely related to their present offspring, the indigenous people (*runa*), than the distant Christian god in heaven and his white saints (Harris 1983).

The reciprocal influences of the three worlds are only possible because they overlap in manifold ways in their spatial, temporal and spiritual transitions. Topographic elements like caves, springs and lakes are thought to be entrances to the inner world (Steele 2004), where the gods of the underworld are contacted through rituals performed by ritual specialists. In Cotabambas these specialists, are called *yanapunku*, 'dark door'. They prepare complex offerings on portable altars, normally oriented towards the sunset. In contrast the *altumisayuqkuna*, the hosts of the highest altars, are oriented towards the sunrise. Generally their altars are dedicated to holy mountains and lightning, both of which are believed to be mediators between sky and earth (Ossio 2002). These two types of local ritual specialist occupy the highest ranks. Below them follow the *pampamisayuqkuna*, the hosts of the lower altars. They serve the not so dangerous powers of the present world, the *kay pacha*.

The connections between these worlds are intensified at certain times. On the one hand, this can be observed daily. Rituals for the upper world generally start after sunrise and before noon. Those for the underworld often take place during the night or at least in the afternoon. Some rituals are best performed on certain days of the week. None should be performed on a Sunday, the day reserved for the Catholic mass. Moreover, the rituals are generally seen to be most effective at two times of the year, i.e. at carnival and in August. People think that the earth is open and hungry at these times (García Miranda 1989). This idea is probably related to the Andean climate, which is characterized by a two-season model (Mitchel 1991). August marks the end of the driest and coldest period of the year. The land seems like a frozen desert. Angry storms drive dust over the harvested fields, where emaciated cattle are rooting out the last straw. The first precipitation falls as hail and endangers the seeds. In September, rainfall increases gradually, reaching its maximum in January and February. The peak of the rainy season around carnival time, which marks the start of the harvest, is a rather critical period. Thunderstorms, floods and landslides menace the mountain people and their fields, the latter now nearly ripe for harvesting. As a result, ritual activity is closely linked to the seasons, with climate change and corresponding agricultural tasks.

3. Lightning in chronicles and anthropological documentations

3.1 Lightning as Yllapa: etymology and meaning
The Quechua of southern Peru have various names for lightning, but the general term is *Yllapa* or *Illapa* (Rösing 1990; Claros Arispe 1991). It designates not only lightning and thunder but also other kinds of flashes, like that from artillery. *Yllapa* is also the name of the god of thunder (Lira 2008).

Etymologically *Yllapa* consists of two parts, the root *ylla* and the suffix *-pa*. *Ylla* is often used for things celestial, like dawn *(illáriy,)*, morningstar *(auquilla)* or the pleïades *(larilla)*. Generally, it means to shine or to gleam. The suffix *-pa* expresses repetition and renovation. In this sense *Yllapa* means repeated sparkling (Tschudi 1853). In fact, in Cotabambas *ylla* denotes the first flash of an approaching thunderstorm. Only if repeated does it turn into *yllapa*.

But *ylla* is also the name of stone amulets representing cattle or desired goods. In nearly all rural households a miniature of its animals can be found. They are thought to embody and preserve the spark of life of a species, its *ánima* or vital force (Tschopik 1968; Isbell 1985; Arroyo Aguilar 1987; Flannery, Marcus and Reynolds 1989). Accordingly a person who owns an *ylla* (*yllayoc runa*) is rich and happy. *Ylla huacin*, a house with *ylla*, describes a household with legendary luck and plenty (Holguín 1989). Such *ylla*s are normally found on sacred mountains and were activated by lightning (Flores Ochoa 1976; Lira 2008). In fact, as a nominal suffix *-pa* indicates the origin, basis or beginning of something (Lira 2008), of property and belonging (Soto Ruíz 1993; Lira 2008). In this context *Yllapa* can be understood as the originator and master of such *yllas*. As such, he can be identified as the creator of luck, health, fertility and abundance, especially if related to livestock.

3.2 Yllapa: *the Inca thundergod in the chronicles of early colonial times*

Fortunately, the written records with regard to the god of lightning reach back into the early colonial period. Polo (1916: 281 my own translation) reports: "in his hand was rain, hail, thunder and all the other things that belong to the region where the clouds form." Explicitly, Polo depicts *Yllapa* as the originator of rain, bucketing water down from the Milky Way. Cobo (1964:160 my own translation) also gives a detailed description: "they thought of him as a giant figure, build from stars, holding in the left a stub, in the right a slingshot and dressed with shining clothes. It flashed when he turned to fire the slingshot, whose crack could be heard as thunder. He did this to let it rain on earth, dispersing the water of the Milky Way, the celestial river." In another variation, Vega (1983) describes *Yllapa* as a furious man who destroys the water jar of his sister the moon[1] with his slingshot, thereby creating rain. This seems to justify his high rank in the hierarchy of the gods. In the chronicles lightning normally appears as the third highest god of the Inca pantheon. He was venerated in the *coricancha*, the central temple of the Inca in Cuzco: "the room next to the one for the stars was dedicated to lightning, thunder and bolts of lightning. These three things were understood and united by the Inca under the name of *Yllapa*" (Vega 1983: 138 my own translation).

Polo (1916) and Guamán Poma de Ayala (1980) report that the Inca figure of lightning was represented as a father with two sons. In colonial times this 'tripartition' was compared to the Holy Trinity (Yaranga Valderama 1979), which was assumed to have already been known to the local people. Such divisions are reflected in iconography. Shrines to lightning are often the split stones (Véricourt 2000; Bouysse Cassagne 1988) frequently found on mountain peaks (Arriaga 1992). Such stones serve as guardians and occupy the territory of an adherent group. Frequently they are thought to be mineralized ancestors, illustrating the founding myth (Duviols 1976; Véricourt 2000).

Several sources underline these close relations with the ancestors. Lightning itself is seen as the mythical ancestor of the Lama herders of the central Peruvian highlands (Duviols 1976). The dead bodies of the Inca kings were designated with the same title (Guamán Poma de Ayala 1980, see also Illustration: 351). Herders in the southern Peruvian Andes believe ancestors to be the senders of lightning that kills (Flores Ochoa 1976). Lightning therefore stands in a very intimate relationship with humans. The real or fictive kinship underwrites the

---

1   In the Andes the moon is considered feminine, being closely related to moisture and the underworld.

mutual responsibilities. Ancestors are seen as the originators of agricultural growth and fertility (Arriaga 1992; Anónimo 1968).

3.3 Conceptions of lightning in recent records

The Inca distinction of different meanings for lightning is made clearer by a consideration of recent concepts. At present, indigenous people in the Cuzco area distinguish between male and female lightning: the feminine *rayo*, which falls vertically almost soundlessly and the male *relampago*, which falls diagonally accompanied by deep thunder (Urton 1981). Informants in Cotabambas report a similar classification. Left lightning (*lloque yllapa*) crosses the sky diagonally and is thought to come from the underworld. Right lightning (*paña yllapa*) falls horizontally and is related to the upper world.

In Cotabambas bolts of lightning, depending on their colour, are said to deposit gold or silver in the ground. Frequent thunderstorms are the attribute of the holiest and highest peaks. Consequently these are believed to be filled with precious metals, expression of their power. A similar idea can be found at Cerro Hermoso in Bolivia. At the entrance to the mine of Coque Checa small metal grains are worshipped that are believed to be the bullets of lightning and can be used as instruments for telling the future. Montesinos (1882: 117) reports that: "witches search for them where lightning strikes with mighty thunder", and Lehmann-Nitsche (1995) tells us that such balls are found in 'lightning tubes', resulting from strikes in sandy soil. These balls testify to the contact between celestial energy and the earth and are believed to bring about human and agrarian fertility.

Other traditions in Cotabambas show even more complex differentiations. The Catholic Virgin Santa Barbara is seen as the mother of three brothers of lightning. She stands for soft snow. The oldest son represents non-hazardous lightning with fertilizing rain. The middle son brings moderate rain that wets superficially but does not reach down to the roots. The youngest son represents deadly flashes of lightning and hail. Rösing (1990) has witnessed the rituals of Bolivian healers. They address many different forms of lightning, such as the lightning of glory, of the dark world or of the sacred places. In Cotabambas *gloria* is used for the stars and the upper world in general. Likewise Aguiló (1985: 110) associates it with the sun. Maybe these traditions show a more precise understanding of the Inca *intiyllapa* or lightning of the sun mentioned in the chronicles. This could be understood as a celestial, benign form of lightning that takes place in the clouds and does not hit the earth. It is differentiated from an obscure, underworldly or feminine form of lightning that brings death and destruction.

Some forms of lightning are purely destructive: some are benevolent as well. For example, lightning acts as a vocation for those individuals who survive its attack. To be confronted with a flash of lightning is understood as a form of initiation and a privileged relationship (Albornoz 1989; Murúa 1946). Such individuals will become wiser and receive additional capabilities, like the ability to read coca leaves. They will become *altumisayuq* or ritual specialists if they survive three successive hits. The third time, lightning leaves you its balls as an instrument and badge of your profession. Such an election can already occur in one's mother's womb, testifying to a kin-like relationship. On these grounds the 'children of lightning' can influence *Yllapa* and gain access to his powers. Thus *Yllapa* is closely related to Andean ritual specialists and healers. Véricourt (2000) relates that Bolivian healers experience hits of lightning as a sexual act and even compare it to rape. This underlines the intensity of the link and at the same time the fertile ideas.

Ideas about lightning as an embodiment and mediator of complementary oppositions are symbolically linked to animals in indigenous chronicles. Wild cats, like pumas and the mythical *qoa* or *ccoa*, are closely associated with lightning. From their long tails squirts the rain. Hail falls from their nostrils and lightning from their eyes (Bolin 1998; Anónimo 1968). One explanation might be that the towering cumulonimbus clouds from which rain and hail fall as a dark sheet recall a cat with its tail hanging. The *ccoa* is believed to be the cat of the mountain gods and to live in waterholes and springs. As hail he robs the farmers of their products. He also selects sorcerers and gives them their power (Mishkin 1940). Cobo (1964) says that he had spots in all colours. Such traditions are still alive in Haquira, Cotabambas. The eighty-year-old Theofila reported a meeting with the *ccoa*. As a child herding sheep she climbed a wild and distant mountain, where she got caught in a violent thunderstorm, watching the spectacle from under her manta. All of a sudden, high in the deep dark, she saw the *ccoa*:

> *He was like a skin. Head, tail and limbs were hanging, like a 'lamina' [metal plate or religious painting]. Coming deeper and deeper, he grew smaller. Then, suddenly, he fell to the earth and changed into a real animal. He was lead-coloured, with golden and silver spots. The whiskers were enormous, bigger than the face. Swiftly, he vanished in the high grass. They say they live in the springs, are like their gods. Right at the moment when he touched ground the storm reached its climax. Furious lightning flashed with mighty thunder and accompanied by heavy hail. I was very afraid, because the ccoa had scared me, and I covered myself completely. I began to sweat, chewed coca leaves and prayed to mother earth and the mountain gods. Thus I saved myself. When the thunderstorm ceased, everything was white from hail. Only the spot under the manta (cloak) was clean. There I found this stone (Interview with Theofila 2009, Haquira).*

Theofila was referring to a small, dark, ferrous stone with rusty spots. Its form reminds of a four-legged being. She kept it in a *chuspa*, a bag normally used for coca leaves. Like a magnet it is said to attract luck or goods, and above all it is used for curing.

In conclusion here, it is clear that lightning in the Andes is experienced as an ambivalent force. On the one hand it is related to dark powers and can be harmful and destructive. On the other hand it initiates and empowers Andean healers and ritualists. It is seen as a celestial and benign sender of rain and abundance. The modern concept of lightning apparently reflects centuries-old mythological patterns, which since colonial times have become blended with Catholic ideas and figures.

3.4 Santiago Yllapa: lightning in the Catholic understanding of the Andean population

A number of saints are said to command lightning (Claros Arispe 1991; Véricourt 2000; Rösing 1990; Cortés Ortiz and Pantoja Revelo 1995). The selection depends on local traditions and the kind and timing of the request, leading to an established overriding hierarchy (Schäfer 2007). Transregional prestige and special powers of control are owed to the Apostle St. James (Rösing 1990; Claros Arispe 1991; Véricourt 2000). Therefore, and because he plays a prominent role in the final case study, his connection with lightning will now be analysed in greater detail. In Spain and Latin America this saint is known as Santiago and appears as a Christian knight on horseback wielding a sword. He first became popular as the patron of the war against the Arabs and the defender of the pure catholic faith, being introduced to South America with the same function. The introduction of Santiago into the Andes and his identification with the Inca god of lightning is already mentioned in the chronicles of the colonial period. Castro (2004: 242f.), Choy (1958) and Lefranc (2006) have analyzed these sources in great detail. According to them his reception in the Andes is best explained by esoteric Iberian traditions and related to his functions in respect of missionary or political strategies.

Other approaches refer more to the creative processes of appropriation by the indigenous population and the intimate embeddedness of the saint in Andean cosmology (Gade 1983; Huhle 1994; Schäfer 2007). This makes it clear that in the rural context the ideological content of the figure has faded or been reinterpreted. The inherent violence of the saint has become widely detached from the official dogmatic and historical events in Iberia. Instead it has been used for the interpretation of natural, social and even cosmic processes. This can be observed by analysing the saint's identification with lightning.

They say near Cuzco: "Santiago is a Spanish saint who always arrives with the lightning" (Bolin 1998: 127). In the lowlands of Bolivia too lightning is identified with Santiago (Aguilo 1985). In ritual, lightning is represented by a painting of the saint (Véricourt 2000), while he is depicted as a weather god on the portable altars of the herders of Bolivia's *altiplano* region. Rain pours off his cloak, and he throws thunderbolts with his sling (Mendizábal Losack 2003). In Argentinian legends he shoots the golden bullets of his musket against the celestial snake, which brings hail (Franco 1944). Such bullets are venerated in Bombori, a Bolivian pilgrimage centre dedicated to Santiago. The saint himself displays a diabolical character, which becomes visible through his connection with the underworld and his patronage over the local mining (Véricourt 2000). Similarly in Cotabambas, it is said that Santiago shoots gold and silver into the mountains with his sling. Santiago "knows how to make the *warak'a* [slingshot] roar like thunder" (Bolin 1998: 127). He controls rain, hail, wind and hence agriculture. For this benevolence he expects due veneration, which is given him by the Aymara of the *altiplano*: "Santiago must always be respected and honoured [by sacrifices]. If this isn't done several times annually, he won't send rain, but much hail that kills the plants. He can punish us, destroying our houses, killing our animals or sending a hurricane that desolates the fields" (Girault 1988: 68 my own translation).

Among Andean healers and religious specialists too, Santiago has often substituted the old *Yllapa*. Yaranga (1979) gives a description of a ritual near Cuzco to banish mischief from people and their possessions. Here *Qhaqia* Santiago is evoked together with *Yllapa* and Santa Barbara as a tripartite figure. Yaranga explains this as a result of the introduction of the Catholic trinity. But, as mentioned above, the three names could also relate to different kinds of lightning originating at different levels of the world.

In fact the plurality of the old *Yllapa* has been transferred to Santiago in many regions. The two biblical apostles of the same name were bound together as brothers. St. James the Greater is understood as *Quraq* Santiago, being the older brother. St. James the Lesser is seen as the younger brother, therefore as *Sullk'a* Santiago. Sometimes there is even a tripartite differentiation. Between the two brothers stands a middle Santiago, who keeps the balance in the centre. The other two have contrasting qualities and domains. The older one, who is also called the right one, is benign and associated with the light, while the younger one or left Santiago is believed to be dangerous because he is related to the underworld. The latter holds his sword in his left hand (Gretenkord 1997). Instead of a horse he rides a mule, which symbolizes infertility, which in turn is associated with the devil (Interview with Andres Carbajal Mallqui 2010, Urcos).

Some recognize Santiago by glamorous silver decoration, the metal being related to money (*plata*, or silver, is the Spanish word for money) and the devil (Interview with Abel Castro Mendez 2009, Cuzco). Similar traditions can be found in Cotabambas. In Challhuahuacho, Cotabambas, the three brothers are venerated at a triple parted rock. The right Santiago (*paña* Santiago) and the middle one (*chaupin* Santiago) are thought to be inoffensive and benign. Their storms can be appeased by a mere offering of coca leaves (*coca kintu*). They are celebrated on the official feast day in July and stand for the Christian cult, accepted by the Spaniards.

In contrast the younger Santiago (*lloque Santiago*) or Left Lightning is seen as the godfather of the *Pacha Mama*, the earth. He therefore receives an offering on the day of the godfathers, the second Thursday before carnival. Normally at this time the rainy season reaches its climax with frequent thunderstorms. Every family whose members, houses or belongings have been touched by lightning must participate. It is said that he always comes from the left and causes death and destruction. But when he has killed and there is no witness around, successively one of his brothers will strike. They reunite the pieces and resuscitate the victim.

Such a lightning strike is understood as an intimate divine act. Therefore, its victim must be ignored by bystanders, lest victims should have to face the contact with 'heavenly energy' alone and without defending themselves. A person who was resuscitated in this way can become an *altumisayuq*, the highest rank of Andean ritualists and healers. Because of the connection with the left Santiago, he can now see "the bad things about the persons, the sicknesses, the dark things that can't be seen, the malicious" (Interview with Luca Condori Navas 2009, Haquira).

But if there are eyewitnesses around, the person who has been killed will not come to life again. Similarly western medicine will make the incident more dangerous. This belief explains a practice that seems shocking and inhuman to foreign onlookers, illustrated by the following account from an elderly informant of Chumbivilcas, Cuzco:

> We visited the feast of the Mary of Mount Carmel in Sillota [a small village in the Highland of Espinar, Cuzco]. The weather was nice and the sun was shining. But then, in an instant it changed. It got dark and lightning struck. Various dancers were hit. People yelled immediately: 'Do not look! Do not look!' since it is said that you have to turn your back as if not paying attention. But it was already too late: lightning had hit all of them, it was horrible! Some were screaming, crying for help, but the onlookers did not react because of their belief that one must not touch the injured. The effect has to vanish of its own accord. When my nephew was brought to the medical station by his Arequipan friends, he died. As townspeople they didn't know that one shouldn't look, were unable to stand it without interfering, so he died. (Interview with Ciro 2009, Quiñota)

In sum, it can be said that the Christian Saint Santiago has been equated with the Andean *Yllapa* in nearly every detail. He is venerated as a weather god who controls several climatic phenomena, like thunder, lightning, storm and hail, but also rain that creates fertility. That makes him significant for farmers. He is thus an ambivalent and divided figure. One brother is connected with the sun, the firmament and lightning that does not hit the ground.

But there is also another Santiago related to the underworld and its obscure powers. This form can kill and destroy, but it also enables contacts between man and the sacred sphere. Hence he is the mediator of healers and ritualists.

4. Coping strategies: the *qhaqiamisa* ritual in Cotabambas

Until now I have described the different religious concepts of lightning from the early times of the Incas to the present, underlining the unity of its conception despite the very diversity of its appearances. Coping with disasters of lightning, hail or thunderstorms is a very individual or family-oriented act of placating the gods which requires ritual instead of administrative action, in accordance with the religious conceptualization of its agent.

How the above concepts are symbolized in ritual practice will be analysed in the following text, taking the killing of the two cows of the Farfán family in the little highland village of Choaquere as an example. The rituals that followed were divided into three parts: purification, identification and appeasement. In the first place the ground touched by lightning was purified and the injured animals buried. Next, the powers responsible for the destruction were identified. Finally an altar with sacrifices was prepared to appease these powers and prevent them from killing again. This last measure also had a preventive aim because it is locally believed that one misfortune always attracts more misfortunes.[2]

This fatal dynamic can only be halted by a succession of rituals generating a 'change of luck', which must be guided by an *altumisayuq*, a local religious specialist of the highest rank. Weighed down with a young sheep, rum and coca leaves, the father of the Farfán family and his brother visited the old Alejo, a well-known *altumisayuq* of the region. Although his homestead is far away, he promised to help by accepting the gifts.

Early next morning he came to the 'site of the disaster'. Before approaching it directly he greeted it from the distance, offering a coca *kintu*, three perfect coca leaves. At the spot he recited prayers, sprinkled and drank rum honouring the powers of the place and chewed coca leaves. After these preparations he carefully worked out the trace of the lightning as evidence of who had sent it and diagnosed *Lloque* Santiago, the left lightning, as the offender.

In the meantime, from a distance one son of the family dug a deep hole, where the worst hit cow and the scorched earth was placed. The parts of the less burned cow could be used after ritual purification, so the animal was dissected on the spot. The scorched parts and the bones were left in the hole, which was then carefully closed and covered with stamped earth and stones to keep away dogs and wild animals, which could transfer impure particles near to humans.

---

2 In the present case lightning was not experienced as punishment. But in Challhuahuacho, Cotabambas (2009), a small statue of Santiago peregrino is venerated as Just Judge and said to kill thieves with storm and lightning. Rösing (2008: 72) reports that in Bolivia damage by lightning is explained by offerings having been omitted. She also documented rituals and prayers that invoke catholic saints to punish theft with lightning. Urton (1981: 139) suggests that near Cuzco killing by lightning is sometimes explained as an answer to moral transgressions.

Meanwhile Alejo and his assistant smoked some cigarettes and drank rum, a conventional act of purification observed at all burials and wakes. The bag with the meat that could still be used, was carried to a nearby creek and left in the water. After a day it can be prepared salt free and eaten without risk. Both men returned to the Farfán farmstead to eat a small meal, and the date for the sacrifice was agreed. Alejo explained that the following ingredients should be provided: port wine, corn beer, white and red pink blossoms (*tika clavel*), incense, silver and gold foil (*qolqe libro, qori libro*) and water collected before sunrise (*agua virgen*).

4.1 Identifying the originator of the lightning bolt

On the morning of the day indicated, Alejo arrived at the family's homestead and was offered a warm meal, coca and cigarettes, but he refused the food and smoke because they would compromise the effect of the ritual.

In the middle of the yard and in the direction of the sunrise, Alejo spread out an alpaca blanket to serve as an altar. Around it the adult men of the family assembled, having removed their hats. First Alejo had to identify who sent the lightning. From his raised right hand he let coca leaves fall on to the blanket. Then he requested two coins from the household head, which he threw three times on to the coca. Muttering to himself, he drew his finger from the coin along the invisible lines of the coca leaves, apparently interpreting the position of the leaves with regard to the coins. Finally he confirmed his assumption that *Lloque* or the Left Santiago has sent the lightning. Again he mixed the coca leaves and tossed them three times. This time he wanted to know the reason for Santiago's fury, but had no luck. He explained that the *Lloque* Santiago "is like a maniac that acts without discernible reason." Relations with him are difficult: "He is like a small child who doesn't understand properly." Therefore Alejo now prepared an offering for him, as well as for his brother, the *Paña* or Right Santiago, who is believed to have a moderating influence on his younger brother. In the case of fields desolated by a thunderstorm, people don't tidy up before noon because it is hoped that the older brother can see the damage and force the *Lloque* to compensate their owners.

4.2 The altar for the lightnings

The altar (*misa*) consists of a blanket woven from camelid wool which points towards the sun. The offering is prepared in the centre (*chaupin*). The altar has a head (*uma*), foot (*chaki*), a left (*lloque*) and right (*paña*) side. Quite often these directions are defined not from the point of view of the human participants but from that of the godly addressee. The four corners embody the axes of the four sectors (*suyus*) of the earth.[3] They thus provide a model of space for the ritual (Polia 1994).

Alejo piled up the coca leaves on the left side and covered them with his inverted *qocha* (literally 'mountain lake'), a drinking bowl with four spouts. In front of him he lay down a bag containing his ritual instruments and ingredients. On the right side he placed two *queros*, goblets filled with port wine and pink blossoms. The *queros*, the wooden cups, symbolize the act of ritual drinking, which establishes all social relations. They are always used in pairs, thus expressing community. However, the two are differentiated by form or size, so that the arrangement allows hierarchies to be illustrated and displayed. The Incas used them as instruments of power, while the Spaniards banned them as mediums of indigenous identity and alliances. The *qocha*, the largest drinking jar, belongs to this constellation, too, illustrating feminine aspects. *Qocha* is the name of the high mountain lakes that represent the beautiful daughters of the mountain gods, which unite with poor herders and bring them abundant animals.

At the bottom of his altar, Alejo placed a pair of coins and sea shells (*llampu*), the latter being Santiago's key attribute. The coins represent his payment and the wealth of the family, whereas the shells are the attributes and instruments of the autochthonous religious specialists. The two shells are different in shape, colour and origin. Their ritual name *llampu* means soft and tame (Holguín 1989), which explains their ritual function. The abrasion produced by a blade simulates the rain that satisfies the thirst of the gods and appeases their rage. Tabooed or impure places, persons or items are cleansed in this way.

At the top of the altar Alejo placed white, yellow and black corncobs, and in the centre a dried husk of corn (*p'anka*) and a piece of chest fat from an alpaca (*pecho wira*). Finally, in order to sacralise and purify the whole arrangement he burned incense on a flat stone so that the smoke enveloped the altar and the participants. Then he knelt down and appealed to the Christian God and *Pacha Mama* for permission (*licencia*) to conduct the sacrifice. Then he invited the central addressees. As with normal people, this is done with the aid of coca leaves. He held coca *kintus*, three perfect coca leaves, between the thumb and forefinger of his right hand and offered them to the front and the sides. Underpinned by libations of corn beer he summoned *Paña* Santiago to the right side of the altar and *Lloque* Santiago to the left side. Taking the two goblets, he poured port and corn beer in their directions and drank to them. Then he handed coca and drinks over to the other men. Later, other *apus*, holy powers, like mountains and mountain lakes, were invited too. Every member of the family arranged a coca *kintu*. Following Alejo's instructions, each leaf was dedicated to one *apu* by saying his name and looking and raising the coca leaves in his direction. The

---

3   This seems to be an ancient model. The Inca called their empire Tahuantinsuyu, 'realm of the four quarters', with its ritual centre in Cuzco, literally 'navel'.

leaves were pasted together with alpaca fat and three coca seeds (*coca ruru*). Then the *kintus* were collected in a husk of corn.

This husk formed the 'golden plate' (*qolqe platu*) for the preparation of *Paña* Santiago's meal. Carefully and slowly Alejo placed white pink blossoms, small pearls of shell (*chiuchi*) and small pieces of gold and silver foil into it. Subsequently he detached three kernels from every corn cob, starting with white, followed by yellow and black. Before adding them, he raised every colour to his mouth, kissing them and blowing on them. After every ingredient there was a pause (*samay*) in which the participants sprinkled wine and beer over the sacrifice and the four corners of the altar. Everyone drank and chewed coca. At the end, Alejo grated two seashells (*llampu*) with his knife and the coins over the 'meal'.

Then Alejo moulded a miniature bull out of alpaca fat to represent the cattle of the family that lightning shouldn't hit again. He used pierced white shell pearls[4] for the eyes (*chiuchu*), which are seen as the entrance path of black magic and therefore must be specially guarded. White pinks were glued to the loins to increase the animal's fertility. Saddlebags were made of three coca seeds and pieces of silver and gold foil. The finished animal was placed on the altar facing towards the top and grated shells and coins also placed over it. Gold and silver foil, ritually called the 'golden and silver book' (*qori libru, qolqe libru*), is bought only for ritual usage. It symbolizes the wealth and prosperity that the animals will bring to the farmstead. The joining of the two metals is found again with the coins. Only two *nuevo soles* coins are used, which combine gold and silver sections.

After a longer pause, the whole offering was repeated in the same way on the other side of the table, and the 'plate' for the *Lloque* Santiago was prepared in nearly the same manner. Instead of the white pink blossoms, red ones were added, and the bull placed beneath the husk was also decorated with red pinks. Its eyes were made from brown mustard seeds (*piñi*) instead of white shell pearls, their protection coming from their sting. This time the sacrifice was not sprinkled with port wine but with corn beer drunk from the *qocha*, the large drinking bowl.

Once the sacrifice had been completed, the whole family, including women and children, were gathered together. The arrangement at this stage is depicted in Figure 1. Everybody had to drink and sprinkle port wine from the two goblets over the four corners of the blanket and its centre, the sacrifice. When the men were alone again, the offering for the *Paña* Santiago was sent to *qori marka*, meaning Cuzco, and that for the *Lloque* to Potosí (*putusci*). Therefore the

---

4   As with Llampu, the sea shell material is meant to appease the thirst of dangerous powers. The pierced holes remind one of eyes, and because of their quantity and movement, the tiny pearls are said to confuse malevolent magic and lead it astray.

drinking bowl was filled with corn beer. Beginning with Alejo, all men made a libation by means of a coca leaf and recited the addressees, Santiago, the holy mountains and the lakes, who were asked to guide the offering safely to its destination. Luck and prosperity was wished for the family and its farmstead. After the request, everyone had to drink up the bowl in one go, turning it so that the lips wandered over all four spouts. Then the bowl was put upside down on to the pile of the coca leaves. Eventually Alejo examined the moist rim to see where and how many leaves had stuck to it. These had to be eaten by the drinker without the use of the hands and without losing a leaf. When all participants had had their turn, Alejo 'sealed' the sacrifice by tying up the husks and placing them on the edges of the altar. The bulls were also protected with cotton and tied up in husks.

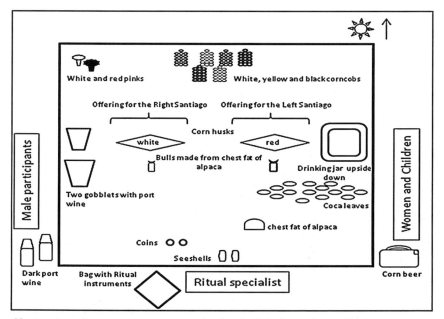

Qhaqia misa or *Rayusqa*: arrangement on the altar for lightning or the two Santiagos.

Without interpreting the information he had obtained from the coca leaves, Alejo suddenly took the bullwhip (*chicote*) which hung around his neck, the attribute of village superiors representing authority, order and also the punishment that prepares for remission.

With the exclamation "Is it possible?" (*'Áchu'*) Alejo hit himself and all the participants hard on the back. The explanation for this is "to support the *Pacha Mama* (mother earth) in her suffering" (Alejo 2009, Choaquere). No signs of pain were allowed. Afterwards, as in the Eucharist, the participants embraced and clapped each other on their sore backs.

.Eighty-year-old religious specialist Alejo relaxing after the end of the ritual.

Now a fire of dried dung was set in a ceramic jar. Alejo collected the sacrifice and left the circle, to burn it at some distance and thus avoid polluting the participants. At the same time the smoke of the sacrifice could ascend without disturbance to its addressees. Before and after the fire, incense was thrown into it. The corn kernels should pop at least three times. If not, additional kernels are added. It is considered a bad omen if they spring from the fire. The glow was stoked carefully until the offering has burned completely. Only then Alejo did return to the others. The remaining drinks were finished in a relaxed atmosphere. Eventually Alejo left the site of the ritual provided with food, coca leaves and crops by the grateful family. After his departure he wrapped the cooled down ashes of the sacrifice into a rag, to bury it at a 'silent' place on his way home.

4.3 Interpretation

The offering was explicitly prepared as a splendid meal for the gods of lightning, to appease their hunger and thirst. This explains the main ingredients, like coca leaves, fat and corn. Corn is a highly valued foodstuff. It does not grow in the rough climate at higher altitudes but is a product of the lower, more protected valleys. It is costly when exchanged. Therefore corn is associated with agricultural wealth, warmth and the sun. Already in Inca times it was a central element in sacrifice. The same goes for its liquid form, corn beer, which is produced from germ buds, which are chewed by the women to start fermentation. Thus, it represents fertility. In the ritual, the pale corn beer is contrasted with the dark, sweet port wine. The latter must be bought at high cost and is hold as the beverage of Mestizos, those associated with urban and Spanish culture.

Fat, especially chest fat, is seen as concentrated vital force. The Andean vampire does not suck blood but fat. Human fat is the only cure against the galloping consumption caused by contact with the 'gentiles', mythical ancestors or revenants. In ritual it is also essential because it facilitates the complete combustion of the sacrifice.

The coca is thought to be the plant of the gods. Its consumption helps against hunger, tiredness and altitude sickness. It is evergreen and permits several harvests each year. Hence, it embodies the vitality of the vegetal realm (Allen 1997). In ritual it often symbolizes the pastures of the cattle.

The corn husks in the centre of the altar form the equivalent of a plate on which the offering is presented. In many ways the sacrifice is related to pleas for protection, fertility and general well-being. This blessing is directed towards the members of the family, the farmstead and the primary basis of subsistence, the cattle. The effort to include all the relevant powers is obvious, because men's lives and activities develop at an interface between antithetical realms, the upper and under worlds, east and west, right and left. All powers have to be appeased by specific offerings. Therefore, on the altar a model of the universe, its different worlds, regions and powers like lightning and thunder is constructed. 'Natural phenomena', like the rainy and dry seasons, day and night, together with fundamental principles like dual, tripartite and quadripartite divisions are displayed in their order and complementary relations, but the central concern is with the cosmological powers. Lightning (symbolized by Santiago), mountain lakes (*qocha*) or mighty mountains (*apu*) establish and control contacts and transitions among different orders of being. The resulting forces can have particularly devastating effects. As the ritualist represents and classifies them on the altar, he not only portrays but also constitutes the required order. In the words of Lévi-Strauss (1968: 10), he "assign(s) every single creature, object or feature to a place within a class." To seal the complex balance, all human and divine participants have to gather in a common meal to codify their mutual responsibilities on the basis of a personal relationship.

The arrangement of the instruments and ingredients of the ritual follows a numerical order. The contrast between the colour or shape of its elements illustrates duality: the number two is therefore of great importance in the ceremony. This duality is played out in the offerings for the Santiagos using different colours of pink blossoms to stress their difference. White exemplifies purity. It is the colour of the Christian god and is used for the Right Santiago, related to the celestial lightning. The red for the Left Santiago symbolizes fire and rage, echoing the character of the addressee, as he is associated with lightning that kills. The same procedure applies to the bulls, which are made out of fat using white shells for the bull of the Right Santiago and brown and hot mustard seed for the bull of the Left Santiago. A similar logic is observed with the liquid offerings I have described above.

During the sacrifice, the duality becomes a trinity. Ritualists insist that one must always offer 'a pair and one' to secure luck. The three coca leaves of the *kintu*s are explicitly dedicated to the three parts of the world. Then they are united with three coca seeds, associated with the earth and its fertility. Green leaves and coal-black seeds are glued together with amber-coloured fat, symbol of the vital force that stands for the world of the living. Such a reference to three spheres is expressed by the corn too. White kernels are always used for offerings to the upper world. Yellow maize serves for this world, while black maize is associated with the obscure powers of the underworld. The centre of the altar, where the sacrifice is prepared, embodies the vertical structuring of the cosmos. In contrast, the horizontal axes of the world are visualized as the four directions of the compass by the woven blanket of the *miso* (altar). In form of the four spouts of the jar, they point to the four directions of the Tahantinsuyu, the Quechua name for the Inca Empire.

Many Andean highlanders believe that destructive powers or illnesses originate from powers whose existential needs have been left unattended. Their hunger and thirst especially must be appeased with burnt offerings and libations. Seen against this background, the Christian stress on monotheistic exclusiveness seems problematic. The restriction on the god of heaven and his saints led to the renunciation of the autochthonous and under-worldly powers and domains, which are essential for people who live from the land. Whilst Santiago the Great or the Right Santiago are worshipped in church, the *Lloque* Santiago or left one, associated with destructive lightning, is not. Instead he must be appeased in private rituals, which is problematic for various reasons. First, these rituals are not performed for prevention but only as a response to previous devastation. Secondly, their familial performances lack the magnitude and power of communal performances. They have to be conducted in private or clandestinely and hence are deficient in public support. And thirdly, the proper knowledge to perform theses ceremonies is on the wane since many ritual specialists do not command the immense ritual knowledge of their forefathers.

Looked at in this way, rituals that answer natural catastrophes point to concepts and practices that have been pushed aside. The destruction represents the vengeance of a forgotten power that must be identified and ritually appeased. The ritual involved can be seen as a criticism and corrective of a repressively imposed official world view and an excessive postulate of celestial rule, which cannot be aligned with the experiences of the indigenous, rural population of the Andes.

Woven festive blanket from the Cuzco region depicting black and white Santiagos. At the top a dark Santiago appears next to the drawing and quartering of Tupac Amaru II in Cuzco (1781), the leader of the last big indigenous uprising. Myth tells that his body parts were buried at the four corners of the empire and that once underground they will unite again. Then the Inca will resurrect and re-establish his rule. At the centre dual felines with big whiskers, probably related to hail and lightning, are depicted. Below a white Santiago appears with horses and Andean flamingos. These birds are symbols of the remote mountain lakes associated with water and rain. The saint is seen as patron of horses, which is why they are branded on his day. The latter is connected with Peru's national day, which explains the national banners.

Interesting is the selection of the places where the offerings are sent to. They have strong and contrasting symbolic connotations. Potosi is the legendary silver city of the Bolivian *altiplano*. Its silver mines represent the epic wealth of the region (Absi 2003), being at once the symbols of inhuman might and violence during colonial times. In the ritual Potosi is related to the Left Santiago, embodied as the death-bringing left lightning. The offering of the Right Santiago is sent to Cuzco, also called the Golden City. It was the political and religious capital of the Empire of the Inca, the lawful masters of the country and the models for just and ideal governance.

Polarized concepts like Quechua and Aymara, Spaniards and indigenous, legitimate and illegitimate violence, sun and moon, gold and silver, above and below, are symbolized in profound and multiple ways in relation to lightning and the conceptualization of Santiago. He brings destruction, death and traumatic experience, but he also creates fertile rain and leaves his powers in *altumisayuq*s or ritual specialists, in the curing 'bullets of Santiago' and the procreating *ylla*s or stone totems of cattle. Thus the dual Santiago also contains the chance to overcome mutilation and ruin. The postulate of two antagonistic principles is "one of the classic means of creating equilibrium" and "the guarantee of stability" (Maybury-Lewis 1989: 8/113). By worshipping the split figure of Santiago, ambivalent forces are balanced and experienced meaningfully. On the one hand the saint may embody the archetype of the Spanish invader, who charges the Incas in his search for precious metals. On the other hand he is one of the most important patron saints of rural life, and of the indigenous population in particular. Behind this idea seems to be the wise insight that the same energies can bring good and bad. The effect depends on how one succeeds in controlling their intensity and direction. Extreme forces in nature are experienced as expressions of the same antithetical powers that also lie at the heart of the structural violence of social reality. Hence, the analysis of the cultural interpretation and ritual contestation of natural disasters can help us understand fundamental cosmological, ethnic or social-political structures and deep-rooted historical processes.

5. Bibliography

Absi, Pascale (2003): Les Ministres du Diable: Le travail et ses représentations dans les mines de Potosí, Bolivie. Paris: L'Harmattan Connaissance des Hommes.

Aguiló, Federico (1985): Enfermedad y salud según la concepción Aymara-Quechua. Sucre, Bolivien.

Albornoz, Cristobal de ([1584]1989): Instrucciones para descubrir todas las guacas del Perú y sus camayos y haziendas. In: Molina, Cristóbal de (ed.): Fábulas y mitos de los Incas. Madrid: Historia 16, pp. 161-198.

Allen, Catherine J. (1997): When Pebbles Move Mountains: Iconicity and Symbolism in Quechua Ritual. In: Howard-Malverde, Rosaleen (ed.): Creating Context in Andean Culture. Oxford: Oxford University Press, pp. 73-84.

Anónimo Jesuita ([1590]1968): Relación de las costumbres antiguas de los naturales. Biblioteca de Autores Espanoles, tomo 209 (Cronicas Peruanas de Interes Indigena). Madrid: Ediciones Atlas, pp. 151-189.

Arriaga, Pablo José de ([1621]1992): Eure Götter werden getötet: Ausrottung des Götzendienstes in Peru. Darmstadt: Wissenschaftliche Buchgesellschaft.

Arroyo Aguilar, Sabino (1987): Algunos aspectos del culto al Tayta Wamani. Lima: Seminario de Historia Rural Andina, Universidad Nacional Mayor de San Marcos.

Bouysse Cassagne, Thérèse (1988): Lluvias y Cenizas: Dos Pachacuti en la Historia. La Paz: HISBOL.

Bolin, Inge (1998): Rituals of Respect: The Secret Survival in the High Peruvian Andes. Austin, Texas: Audio Forum.

Casaverde Rojas, Juvenal de (1970): El mundo sobrenatural en una comunidad. In: Allpanchis, Vol. II. Cuzco: Instituto Pastoral Andino, pp. 121-243.

Castro, Americo (2004): España en su historia: Ensayos sobre historia y literatura. Madrid: Trotta.

Choy, Emilio (1958): De Santiago Matamoros a Santiago Mata-indios. In: Revista del Museo Nacional. Lima 27, pp. 195-272.

Claros Arispe, Edwin (1991): Yllapa, Gott und Kult des Blitzes in den Anden. Regensburg: S. Roderer.

Cobo, Bernabé ([1653]1964): Historia del Nuevo Mundo y Fundación de Lima. Obras Vol. 2 (edited by Mateos, Francisco). Madrid: Ediciones Atlas.

Cortés Ortiz, Maria; Pantoja Revelo, Gonzalo (1995): Mito de San Francisco y la comida. In: Nueva Revista Colombiana de Folclor 3, 15, pp. 45-54.

Duviols, Pierre (1976): La Capacocha. In: Allpanchis, Vol. IX. Cuzco: Instituto Pastoral Andino, pp. 11-58.

Flannery, Kent V., Marcus, Joyce and Robert G. Reynolds (1989): The Flocks of the Wamani: A Study of Llama Herders on the Punas of Ayacucho. New York, Berkeley: Academic Press.

Flores Ochoa, Jorge A. (1976): Enqa, enqaychu, illa y khuya rumi: Aspectos mágico-religiosos entre pastores. In: Journal of Latin American Lore 2, 1, pp. 115-134.

Franco, Alberto (1944): Leyendas del Tucumán. Buenos Aires: Editorial Nova.

Gade, Daniel W. (1983): Lightning in the Folklive and Religion of the Central Andes. In: Anthropos 78, 5/6, pp. 770-788.

García Miranda, Juan José (1989): Los carnevales: ritual y relaciones de inter-
   cambio. In: Antropología 7, pp. 55-69.
Girault, Louis (1988): Rituales en las regiones andinas de Bolivia y Perú. La
   Paz: Escuela Profesional Don Cosco.
Gretenkord, Barbara (1997): Indianische Einflüsse im Architekturdekor der Ko-
   lonialbauten im Hochland der Anden. In: Bosse, Monika and André Stoll
   (eds.): Theatrum mundi. Figuren der Barockästhetik in Spanien. 2nd edition.
   Bielefeld: Aisthesis, pp. 215-222.
Guamán Poma de Ayala, Felipe ([ca.1614] 1980): El primer nueva Corónica y
   buen gobierno. (edited by Pease G.Y., F.) Caracas: Biblioteca Ayacucho.
Harris, Olivia (2000): The Mythological Figure of the Earth Mother. In: Harris,
   Olivia: To Make the Earth Bear Fruit: Ethnographic Essays on Fertility,
   Work and Gender in Highland Bolivia. London: Institute of Latin American
   Studies, pp. 201-219.
——— (1983): Los muertos y los diablos entre los Laymi de Bolivia. In: Chun-
   gara 11, pp. 135-152.
Holguín, Diego Gonzalez ([1608]1989): Vocabulario de la lengua general de
   todo el Peru llamada Lengua Qquichua o del Inca. Lima: UNMSM.
Huhle, Rainer (1994): Vom Matamoros zum Mataindios oder Vom Sohn des
   Donners zum Herrn der Blitze: die wundersamen Karrieren des Apostels
   Jakobus in Amerika. In: Schönberger, Axel and Klaus Zimmermann (eds.):
   De orbis Hispani linguis litteris historia moribus. Festschrift für Dietrich
   Briesemeister zum 60. Geburtstag, Vol. 2. Frankfurt am Main: Domus
   Editoria Europaea, pp. 1167-1196.
Isbell, Billy Jean (1985): To Defend Ourselves: Ecology and Ritual in an
   Andean Village. Illinois: Waveland Press.
Lefranc, Harold Hernández (2006): El trayecto de Santiago Apóstol de Europa
   al Perú. In: Investigaciones Sociales X, 16, pp. 51-92.
Lehmann-Nitsche, Robert ([1928]1995): Coricancha. El templo del sol en el
   Cuzco y la imágenes de su altar mayor. Buenos Aires, In: Ostermann de
   Petricevic, Denise (ed.): La enigmática etnoastronomía andina, Vol. 2. La
   Paz: Centro de Cultura, Arquitectura y Arte Taipinquiri, pp. 155-230.
Lévi-Strauss, Claude (1968): The Savage Mind. Chicago: University Press.
Lira, Jorge A. (2008): Diccionario Quechua-Castellano. Lima: Universidad
   Ricardo Palma.
Maybury-Lewis, David (1989): The Quest for Harmony. In: Maybury-Lewis,
   David and Uri Almagor (eds.): The Attracion of Opposites. Thought and So-
   ciety in the Dualistic Mode. Michigan: University Press, pp. 1-17.
Mendizábal Losack, Emilio (2003): Del Sanmarkos al retablo ayacuchano. Dos
   ensayos pioneros sobre arte tradicional peruano. Lima: Univ. Ricardo
   Palma.
Minelli, Laura L. (1999): The Inca World. Oklahoma: University of Oklahoma.

Mitchel, William (1991): Peasants on the Edge. Crop, Cult and Crisis in the Andes. Austin, Texas: University Press.

Mishkin, Bernard (1940): Cosmological Ideas among the Indians of the Southern Andes. In: Journal of American Folklore 53, 210, pp. 225-241.

Montesinos, Fernando ([1642]1882): Memorias antiguas historiales y políticas del Perú. Madrid: Impr. de M. Ginesta.

Murúa, Fray Martin de (1946): Historia del origen y genealogía de los reyes Incas del Perú. Madrid: CSIC, Ed. Constantino Bayle.

Nuñez del Prado Béjar, Juan V. (1970): El mundo sobrenatural de los quechuas del sur del Perú a travez de la comunidad de Qotobamba. In: Allpanchis, Special Issue 'Mundo sobrenatural andino', pp. 56-119.

Ossio, Juan M. (2002): Contemporary Indigenous Religious Life in Peru. In: Sulliva, Lawrence E. (ed.): Native Religions and Cultures of Central and South America. Antrhopology of the Sacred. New York: Continuum, pp. 200-220.

Polia Meconi, Mario (1994): Cuando Dios lo permite. Encantos y arte curanderil. Lima: Prometeo Ed.

Polo de Ondegardo, Juan ([1584]1916): Los errores y supersticiones de los indios sacados del tratado y averiguación que hizo el Licenciado Polo. Lima: Imprenta San Martí y Cía.

Rösing, Ina (2008): Defensa y perdición: la curación negra en los Andes bolivianos. Madrid: Vervuert.

————— (1990): Der Blitz. Drohung und Berufung. Glaube und Ritual in den Anden Boliviens. München: Trickster Verlag.

Rostworowski de Diez Canseco, Maria (1983) Estructuras andinas del poder. Ideología religiosa y politica. Lima: Inst. de Estudios Peruanos.

Schäfer, Axel (2007) Santiago Yllapa. Der Weg eines Apostels in die Anden. Unveröffentlichte Magisterarbeit, Berlin.

Soto Ruiz, Clodoaldo (1993): Quechua. Manual de enseñanza. Lima: Inst. de Estudios Peruanos.

Steele, Paul R. with Catherine J. Allen (2004): Handbok of Inca Mythology. Oxford: ABC-CLIO.

Tschopik, Harry (1968): Magia en Chuquito: Los Aymara del Peru. México: Inst. Indigenista Interamericano.

Tschudi, Johann J. von (1853): Die Kechua-Sprache. Vienna: Kaiserl.-Königl. Hof- und Staatsdr.

Urton, Gary (1981): The astronomical system of a community in the Peruvian Andes. Urbana/Illinois: University of Illinois.

Vega, Garcilaso de la (1983): Wahrhaftige Kommentare zum Reich der Inka. Berlin: Rütten & Loening.

Véricourt, Virginie de (2000) Rituels et croyances chamaniques dans les andes boliviennes. Paris/Montreal: L´Harmattan

Yaranga Valderama, Abdón (1979): La divinidad Illapa en la región andina. In: América Indígena 39, 4, pp. 697-720.

In the shadow of an unreconciled nature: Muslim practices of mourning and/as social reproduction in Uganda

Dorothea E. Schulz

1. Introduction: mourning and/as social reproduction

In this chapter, I want to reflect on the kind of research needed to understand how people deal with premature, violent death in a situation where they face a series of continued social and natural disasters.[1] This perspective on death and mourning is relatively new in the scholarly literature on disasters. Scholars working on disasters have dealt with the occurence of death primarily in material and quantitative terms, that is, by taking death numbers and the extent of damage as indicators to differentiate "mere crises" from "veritable" disasters or catastrophes (e.g. Tobin and Montz 1997; see Macamo and Neubert, this volume). A definition of the United Nations Disaster Relief Organisation (UNDRO) moved beyond this quantitative perspective by integrating quantitative and qualitative criteria to assess the degree of damage and loss inflicted on a society by a disaster, and also the society's capacity to recover from it (UNDRO 1987 quoted by Plate, Merz and Eikenberg 2001: 1). The recent literature on disasters comprises a number of perspectives. In the sciences, numerous studies investigate the causes and effects of disasters (e.g. Alexander 1997; Blaikie et al. 1994; Tobin and Montz 1997; Clausen, Geenen and Macamo 2003), whereas others inquire into prevention possibilities and measures (Dombrowsky 2001). Yet other authors discuss the conditions that lead to the breakdown of a society's capacity for resilience and coping (e.g. Geenen 2003; Wisner, Blaikie and Cannon 2004). Many of these analyses center on the strategies of "social engineering" on which people may draw to rebuild and use social structures and institutions. "Vulnerability" (and, though to a lesser extent, "resilience") have become hallmark concepts for this research agenda and its focus on questions of security and disaster prevention. Scholarly investigations that address disasters from a social science perspective (e.g. Casimir 2008; Göbel 2008; Hoffman and Oliver-Smith 2002; Oliver-Smith and Hoffman 1999) have drawn attention to

---

1 I thank Ute Luig, Jochen Seebode, Artur Bogner and Anne Wermbter for their insightful comments on an earlier version of this chapter.

cultural and religious signification practices that allow disaster-stricken populations to interpret their experiences and losses (Groh, Kempe and Mauelshagen
2003) in ways that allow them to explore possibilities for future prevention
(Frömming 2001; 2005a and b; 2006). Schlehe discusses how disasters generate
certain moral judgements that may, at least temporarily, foster gestures of solidarity and mutual support; yet these gestures are subsequently replaced or neu-
tralized by actions ruled by feelings of envy (Schlehe 2006; 2008). All these
studies offer important insights into the ways in which people deal with disasters. Nevertheless more research, if possible from a comparative perspective, is
needed to account for the various symbolic and expressive resources that allow
people to make sense of the losses and damages inflicted on them, and to construct visions for the future out of their past experiences. Also rare are studies
that illuminate the emotional dimension of people's responses to disaster and
loss (e.g. Harms, this volume). Given the great currency that research on emotions holds in current anthropological and neurobiological approaches, the lack
of attention to the emotional repercussions of disaster experience is all the more
surprising. The few studies that address issues are either anthropological accounts of people's fearful anticipation of a disastrous event (see Frömming
2005a and b); or they deal, very often from a literary perspective, with mythological or literary accounts that interpret the occurrence of disaster as an expression of the revengeful hatred of the Gods or Spirits. Entirely absent are studies
that ask how those affected by a disaster deal with the death of their relatives,
and in what practices of leave-taking they engage to re/define the relationship
between the dead and those who survived. The now-classical account by Renato
Rosaldo (1989) of the loss, rage and hatred he felt in response to the deadly accident of his wife Michelle Rosaldo has not generated a separate line of inquiry
in the anthropological scholarship on disasters. Inquiring into the subjective,
emotional dimensions of people's efforts to come to terms with disasters are not
only timely; they are also relevant to scholarship interested in people's coping
strategies. As recent research on emotions suggests, emotions inflect not only
people's interpretations of disasters, but also their capacities to deal and "cope"
with these events (see the chapters of Schild, Schäfer and Harms, this volume).
Inquiries into the emotional repercussions of disasters could also open up new
lines of convergence between on one side, research on disasters and, on the
other, recent sociological theorizing on the multidimensional nature of violence.
Sociological accounts of violent experiences inflicted through war and other
forms of aggression explicitly address the emotional and physical dimensions of
these experiences (see, e.g. Eng and Kazanjian 2003; Luig 2010). Although disasters such as earthquakes or tsunamis occasion the sudden occurrence of uncountable numbers of deaths, the emotional and physical aspects of these experiences of unanticipated violence have not been dealt with adequately. Nor have
social science studies on disasters taken into account the productive and future-

oriented potential of mourning (Butler 2004). To fill this lacunae, I propose that we take a closer look at practices of mourning, and of taking leave from one's deceased relatives. These practices, I suggest, can be understood as a way of coming to terms with the past by making it relevant to the present (see Jewsiewicki and White 2005: 1).

As a point of entry into my discussion of Muslim practices of mourning in Uganda, let us consider two recent events, both of them – though to different extents – reported and commented upon in newspapers and in reports posted on the internet. Both of them are of import to debates among Muslims over how to properly take leave from your beloved ones in case of their sudden death. The first event concerns a mudslide that happened in the Bududa region, near Mount Elgon in Eastern Uganda, in spring 2010. The second event was the hotly debated "sea burial" of Osama bin Laden following his murder by a special US task force in May 2011.

> Floods block Uganda mudslide rescue.
> Heavy rains in eastern Uganda have triggered major floods, displacing thousands of people and hampering the rescue of communities devastated by landslides that destroyed whole villages. (...) Last Monday, rains triggered massive landslides that buried three villages in the Bududa region, burying what officials estimate to be several hundred people. Rescue crews in the remote district, unreachable by road, are using hand tools to dig through thick rivers of mud left in the wake of the disaster. Villagers in the Bududa region have recovered 92 bodies so far but were unable to continue searching on Friday because of the rain, said Kevin Nabutuwa of the Uganda Red Cross.
> (from a news post on Aljazeera.net on March 6, 2010)[2]

The second event, the hasty "sea burial" of Osama bin Laden in the North Arabian Sea, arranged within hours after his death in Pakistan, does not seem to have an immediate connection to what in the the scholarly literature is commonly treated as a disaster. Still, the event similarly highlights the importance of the physical presence of a body for rituals that allow the bereaved to take leave from the person they lost to death. Below, I will return to the positions taken by different Muslim scholars and intellectuals with regard to the "properly Islamic" treatment of bin Laden after his death. For now, suffice it to note that the controversy around the deposition of his body illustrates that the presence or absence of a corpse may become a charged political issue, putting into relief divisions among the living, as well as their competing efforts to endow the deceased person, and death itself, with a peculiar significance.

As Jewsiewicki and White argue in their introduction to a collection of articles on the significance of practices of mourning for "divisions in political

---

2   http://www.aljazeera.com/news/africa/2010/03/20103562651560742.html, last accessed January 21, 2012.

times", the "phenomenon of mourning has taken on new importance in many parts of Central Africa, in part because the necessary conditions for its occurrence are not always met". One of these conditions, they explain, is the physical presence of the body of the deceased, a presence that, in "normal" times, allows relatives to turn mourning into a ritual that enables them to achieve a sense of closure with the past. As they put it, "defining the status of the deceased means making important decisions about how to move on, since "the moment of mourning is not only a moment for weighing the acts and deeds of the deceased, but also a way of testing more generally the criteria for becoming recognized as an ancestor." (Jewsiewicki and White 2005: 1-2). If, on the other hand, relatives cannot perform rituals that ensure the deceased person's transformation from a relative to an ancestor, "the ghost of the deceased remains obsessively in the present" (2005: 2).

Jewsiewicki's and White's observation sadly applies to Uganda, too. After all, in numerous regions of this country so many bodies have remained unidentified or missing over the last four decades, and this because of politically motivated assassination, civil war, or other, natural disasters such as floods and their devastating social "side-effects". One could therefore argue, following Jewsiewicki and White, that the absence of dead bodies, as well as other reasons that prevent people from taking leave from their deceased relatives in accustomed and socially sanctioned ways, only "reinforces the widespread perception that political crises from the past are still unresolved" (2005: 2). In this way, past experiences of violence in Central Africa and in the Great Lakes Region remain part of the present.

If we take Jewsiewicki's and White's point that mourning – and, I would add, other rituals of taking leave – constitute important sites for the social production of meaning, this raises a series of questions with regard to Uganda, a country whose populations have been haunted, though to different degrees, by decades-long histories of human and natural disasters, such as civil war, famine, and AIDS. Given these multiform manifestations of violence whose specters reach uncannily into the present, it is worth exploring in what practices of leave-taking from deceased relatives people engage; and how these practices differ from conventional ways of doing so. How do people reassess and revise practices of leave-taking, if death has occurred through what people consider unnatural or anti-natural causes – or if the body is entirely absent? And how do people make sense of these unnatural deaths that, as in so many other societies form the anti-thesis to a "good death" (cf. Bloch and Parry 1982).

Rather than addressing these questions more broadly, with regard to "the" population of the Bududa district, located about 40 miles southeast of the commercial town Mbale in Eastern Uganda, I am interested in how Muslims, as an internally stratified minority group of the population of this area make sense of these human and ecological disasters. Mudslides have a longer history in the region around Mount Elgon, but over recent decades, the risk of their occurence has increased considerably, as a consequence of local populations' increasing encroachment on areas formerly covered by trees and other soil-retaining flora (Radio Netherlands Worldwide 2011). Yet even if people living in the area around Mount Elgon consider mudslides a regular and unavoidable, though disastrous occurence, the question remains as to how they explain this form of "natural" disaster. Do they account for its occurence by reference to the forces of nature – and, if they do so, what exactly do they understand by "natural forces", and on what conventional understandings of a – protecting yet also potentially revengeful – nature do they draw when articulating these understandings? How do they understand the complex interplay between natural resources and human intervention? And finally, do their accounts reveal more general understandings of a nature in a state of social, natural and moral turmoil?

2. Muslims in Uganda: the existing literature

There are several reasons to address these questions with respect to African Muslims in the Mount Elgon region in eastern Uganda.[3] Firstly, empirically grounded accounts of Muslim religious practice and reasoning in Uganda is virtually non-existent. Much of our anthropological and historical knowledge on Islam in East Africa is based on research on the Swahili world of the East African coast and on its various transnational influences, especially from the Arab-speaking world and the Indian sub-continent. Religious practices by Muslims inhabiting the "hinterlands" of the East African coast have received comparatively little attention. Classical writings on Islam in East Africa, such as that by Trimingham (1964) tended to draw a distinction between an allegedly more orthodox "oriental" Islam of the Swahili civilization, and the less orthodox (or "unorthodox") "African" or "Bantu" Islam of the East African hinterlands that was allegedly characterized by its assimilation of "pre-Islamic" elements and practices.[4] With this classificatory divide, Muslim religious practice in the interior of East Africa (and in analogy to depictions of the low- status "Bantu" Islam in coastal society), was implicitly dismissed as not representative of "true" Islam. Although this classical approach has been replaced since the late 1970s by a perspective that dismisses the opposition between an "unorthodox" "Bantu"

---

3 My discussion will not consider views articulated among the Asian Muslims because being mostly engaged in commerce, they are a negligible minority in the rural district of Bududa.

4 A remarkably similar dichotomous view of Muslim religious practices was articulated by scholars working on Islam in West Africa. Here, the dichotomy between a traditional, largely apolitical, "hybrid" and "Black" Islam on one side, and a "radical", "Arab" Islam characterized by the political and reformist agenda of its representatives provided the basis for the divide et impera strategy of the French colonial administration (see Harrison 1988).

and an "orthodox" Swahili Islam, nuanced anthropological accounts of Muslim religious practice and institutions in East Africa's interior are still rare. It is this lacuna that a historically informed ethnography of Muslim religious practices in the area around Mbale could fill.

A second reason for directing one's focus toward the Muslim population in the Bududa district relates to the perspective from which the existing scholarship on African Muslims in Uganda has addressed their situation, past and present. There exists a considerable body of literature that focuses on the role played by educational policy in establishing systematic inequalities in positions of wealth and power by (African) Muslims as opposed to Christians in Uganda. This literature, produced mostly by scholars who work from a political science perspective, highlights that Muslims in Uganda have for a long time occupied a marginal position in Ugandan society and politics. Their marginalization, so the argument goes, was importantly rooted in the educational system established by Christian missionaries prior and during the British Protectorate. These systemic reasons for Muslims' marginal political and social positions were not altered significantly even when, under Idi Amin's regime in the 1970s, a particular group of (northern) Muslims reached political power and, at least momentarily, bent the power balance to the advantage of Muslim interest groups. The common denominator of this scholarship is thus the argument that in Uganda, religious identity, in combination with ethnic and regional identities and social status categories, fueled a history of struggle and confrontation since the colonial period from which Protestants and Catholics emerged as winners, while African Muslims remained the "loosers" of political and economic competition. Closely related to this privileging of a view of Muslim-Christian relations as an instance of identity politics and struggles over access to the resources of the state are studies that examine Muslims' persistent marginality with reference to their long-standing history of factionalism and internal leadership struggle. This interpretational scheme is certainly pertinent and sheds light on important dimensions of Muslims' historical engagements with the politically dominant position of Christians in the national community. But this interpretation does little to help us account for the conventions and understandings that have shaped Muslim religious practice in this region of East Africa. A focus on Muslims' positions in nation-state politics grants us little insight into their everyday life and concerns, and into how they structure, and possibly rework, their daily relationships and interactions with those who favor another religious tradition and creed. To date, several intricate anthropological accounts document how Christians and adherents to other religious traditions organize their relations with the transcendent world (e.g. Reynolds White 1997; von Weichs 2009). Because of the invisibility of Muslims in anthropological studies on Uganda, and also due to the preoccupation of the few existing studies on Muslims in Uganda with their political and economic position, we do not have a thorough knowledge of the funerary rituals of Muslims and of other ways in which they engage the world of the ancestors and invisible powers. Another unfortunate side-effect of the preoccupation of scholars with the competition between Muslims and Christians is that they implicitly portray Christians and Muslims as living in totally separate social and moral universes, rather than exploring their possible interactions, mutual influences, and practices of borrowing from each other (but see Ahmed 2008; Scharrer 2010). As a result, little is known about the extent to which Muslims and non-Muslim groups of the population draw on shared understandings with regard to the world of invisible powers; nor do we have an appropriate understanding of the actors that shape inner-Muslim debate and of controversies among Muslims about doctrinal and ritual matters with regard to burial practices.

3. Mediating the worlds of the living and the dead: Muslim rituals of mourning

Scholars of Islam have noted the great import of issues of death and mourning for observant Muslims (e.g. Halevi 2007), and have documented certain elements characteristic of a proper Muslim burial, such as the ritual washing of the body, the wrapping of the corpse in white cloth and its subsequent emplacement in a tomb, and finally the farewell ritual during which participants ask the deceased for foregiveness and pray on his or her behalf (e.g. Denny 2006). As Hirschkind (2008) has recently observed with respect to Egypt, the recurent mention of death in sermons and in other public interventions by Muslim clerics, scholars and other intelletuals, does not point to a macabre preoccupation with a "culture of death" and matyrdom inherent to Islam (or endemic to Muslim societies), as this has been alleged in popular press publications and certain critics of Islam in the aftermath of September 11 (e.g. Hirsi Ali 2007, quoted in Hirschkind 2008:39). Rather, Hirschkind demonstrates, the important place that many Muslims accord to death and to questions of how to prepare the corpse for burial and about proper conduct, dress and speech during these rituals, reveals deep eschatological concerns and a sustained effort to render death meaningful, and give it a place within moral and political life. Many of these concerns and questions, it seems, occupy non-Muslims as well. Still, the responses given to these questions are – at least in some respects – inflected by distinctly Islamic doctrines and understandings of the nature of the afterlife, and of the path that leads to it.

That practices of burial, mourning and of commemorating the deeds of the decesased person are essential to the ways in which Muslims make sense of the past and envision possible paths of future action, was vividly illustrated in the earlier-mentioned controversy surrounding bin Laden's "sea burial". See, for instance, the following excerpt from an IOL/ South Africa Time news post, dated May 10, 2011.[5]

> Clerics slam US for burial at sea.
> (...) Osama Bin Laden was given an Islamic burial at sea, US officials revealed, in a move which will rob the terror leader's followers of a shrine. The US claimed the sea burial was handled in accordance to Muslim laws which demand the body should be washed, wrapped and buried within 24 hours. A defence official said Bin Laden's corpse was taken aboard an American aircraft carrier and buried in the North Arabian Sea, although the exact location was not revealed. (...) The body was placed in a white sheet and then a weighted bag, and a military officer read "prepared religious remarks", which were translated into Arabic by a native speaker, before the body was eased into the sea. (...)
> Islamic teachings say burials such as Bin Laden's are permissible if the person has died while travelling at sea and they are too far from land to permit a burial. But Bin Laden died on land and should have been buried with his head pointed towards Mecca, clerics said.
> Radical preacher Omar Bakri Mohammed, who is banned from returning to Britain, said: "The Americans want to humiliate Muslims through this burial."
> Dubai's highest official of religious law, grand mufti Mohammed al-Qubaisi, said: "Sea burials are permissible for Muslims in extraordinary circumstances. This is not one of them. "They [the US] can say they buried him at sea, but they cannot say they did it according to Islam. " (...)
> Islamic scholar Abdul Sattar al-Janabi, who preaches in Baghdad, added: (...) "It is not acceptable and it is almost a crime to throw the body of a Muslim man into the sea."
> But Mohammed Qudah, a professor of Islamic law at the University of Jordan, suggested burial at sea was not forbidden.

These conflicting accounts about the Islamic or un-Islamic nature of the "sea burial" highlight a divide that some Muslims present as an opposition between "Muslims" and their "enemies". The accounts also point to divisions among Muslim groups, each of which claims (or refuses) the legacy of Osama bin Laden's call for violence. For the US administration, too, the stakes were high. By avoiding a burial of Osama bin Laden at a known place that could become a commemorative site, they sought to contravene attempts to portray Osama bin Laden's master-minded murder of more than 2000 people during the attack on the World Trade Towers as part of a legitimate *jihad* waged against unbelievers. Not mentioned in this news post is another detail that supports this reading[6]: According to one Islamic doctrine, a person who suffered matyrdom in the name of Islam must not be washed prior to his burial. Bin Laden thus did not receive the

---

5   http://www.iol.co.za/news/world/clerics-slam-us-for-burial-at-sea-1.1063632,   last   accessed on September 22, 2011.

6   I thank Rüdiger Seesemann for bringing this detail to my attention (personal communication, Uppsala, June 2011).

treatment deemed proper for a martyr. By ordering that bin Laden's body be washed prior to his deposition into the sea, the US political leadership *denied* bin Laden the status of a martyr.

Given the charged meanings that proper burial rites hold in the Islamic traditions, it is imperative to gain a better understanding of Muslim practices of dealing with death in Uganda, and especially regarding so-called "natural disasters".

A few scholars working on Muslim societies in Africa suggest that issues of proper burial rites may play a key role in Muslim scholarly debates (e.g. Seesemann 2005: 106f; Hanretta ms.). Still, comparatively little attention has been devoted to the question of how Muslim burial practices differ across societies, and how they interlock with the ways in which people make death relevant to those who live on.[7] The existing scholarship on Muslims in Uganda, with its scant references to mourning rituals (e.g. Kayunga 1994), suggests that questions of proper Muslim burial rites and subsequent rituals of commemoration have played an important role in controversies among Muslims since at least the early twentieth century. My own field research points to two important aspects of contemporary burial practices in the Mount Elgon area. Firstly (and not surprisingly), the physical presence of the dead body is of paramount importance to Muslim rituals of leave-taking in and around Mbale. If the body cannot be retrieved and properly buried, if the final prayer cannot be spoken over the deceased so that s/he may be relieved from her responsibility, then the spirit of the dead will continue to roam freely and haunt those who failed to perform the proper burial practices. Secondly, debates over proper Muslim mourning rituals have been invigorated over the past thirty years, especially with regard to what is locally referred to as "prayer over the dead" (i.e. the last "farewell" prayer) and to the commemoration ritual on the 40th day after the death. These controversies pit against each other Muslims who favor conventioanl ritual elements and those Muslims who, sometimes as a result of intellectual trends from the Arab Peninsula, seek to reform these conventional local practices of Islam that they consider as "un-Islamic". These reform-minded Muslims are located mostly in Mbale town, where they find a sound institutional basis in the Islamic University of Uganda and from where they seek to educate Muslims in the surrounding countryside about what they themselves consider proper Muslim burial and commemoration practices. Further research is needed to understand the complexities of this reformist endeavor as well as the controversies it generates. Also needed is an account of the actual burial and mourning practices in which Muslims in this area engage. Moreover, the seemingly unending sequence of human and "natural" disasters in Uganda's postcolonial history should prompt

---

7   But see Bowen (1992) for an insightful discussion of variations and commonalities in another core Muslim ritual in Morocco and Sumatra.

us to explore Muslims' constructions of the relationship between the social, moral and "natural-environmental" life worlds in this area. Yet, rather than focussing exclusively on Muslim ways of dealing with premature and "unnatural" or even "evil" death, we need to be open to the idea that their rituals emerge in – and reflect on – a field of religious practices constituted by Christians, Muslims, and those adhering to "traditional"[8] religious practices, all of whom live in the Mount Elgon region. Muslim ritual practices that address death and bereavement need to be understood with reference to more broadly shared cultural understandings about how relations between the world of the living and the world of invisible forces should be mediated and maintained. And it should give us a better understanding of what people around Mbale consider as being "nature" or "culture".

4. The interlocking of social and natural worlds

How do Muslims in the area around Mbale account for the series of natural and social disasters that have haunted their recent past? To what extent do they draw on religious explanations to do so?

The literature I discussed so far further expands the argument by Parry and Bloch (1982) that funerary and other death rituals aim at the transition of the deceased relative to the status of an ancestor, and hence as a being in a transcendent world whose workings are beyond human perception and cognition. Simultaneously, death rituals center on moral and social restoration after the period of disorder caused by death and disaster (see Oliver-Smith and Hoffmann 1999) , and thus on ensuring social and moral continuity (Bloch and Parry 1982). This perspective can be productively applied to an exploration of how Muslims in the Mount Elgon area come to terms with experiences of disorder and disaster. Here I want to suggest that Muslim rituals of mourning should be also understood as practices aiming at establishing and maintaining a balance between the social and natural worlds and at constructing these worlds as part of the same moral universe.

In the absence of ethnographic literature that would allow me to illustrate this perspective specifically with respect to Muslims, I want to draw on comparative studies conducted in different regions in Uganda. Susan Reynolds White, Heike Behrend, Mikael Karlstroem, Raphaela von Weichs, and Sverker Finnstroem, all offer important documentation about how people treat, and thereby construct, the social and the "natural" environment as part of the same moral universe. Their studies focus on different ethnic groups and attendant po-

---

8   By using the notion of "traditional religion" here, I refer to religions other than Christianity and Islam. I do not imply a view of these practices as static, unchanging and unhistorical, but understand them as – ever changing- manifestations of people's efforts to make sense of the world and rework the conditions of their daily existence.

litical traditions: the Banyole in Bunyole, Eastern Uganda; the Acholi of the area around Gulu, in northern Uganda, and the Baganda of the kingdoms of Buganda and of Bunyoro in southern Uganda, who (with the exception of the Baganda) do not count among themselves significant numbers of Muslims. Still, I suggest, their insights into the intertwining of social and "natural" life worlds, and into rituals of reconstituting and regenerating these close connections, may also hold true for other regional practices and understandings – and may thus hint at the direction into which the ethnographic study of Muslim religious conceptions and practices could be taken in the future.

Heike Behrend (1999) studied the Alice Lakwena Holy Spirit Movement that, as a kind of guerilla resistance movement, opposed the newly established government of Museveni and his National Resistance Movement. She maintains that the Holy Spirit Movement drew much of its appeal and mobilizing force from the claim that at the center of the social disaster accompanying the deadly civil war was a breach effected by humans of the moral pact that existed between them and nature, its animals, plants and its invisible forces. A mutilated nature mirrored a social world that had turned upside down. Excessive killings of animals by looting bands of soldiers, and also the offence caused to rivers and lakes by the great numbers of mutilated and drowned bodies floating in them, had brought nature in a state of revolt – against humans. Only a return to the kind of morality expected from humans, and their agreement to keep on to certain standards of human interaction, could invert the current state of moral disorder – and reinstore a balance between the social and the natural side of the moral universe. To reinstaure order, humans had to perform sacrifices for the offended nature, and to appease the spirits of the dead who had remained unavenged. If they did so, nature would renew the pact and join the struggle of those human beings willing to make up for the sins of humanity and struggling to restore moral and social order by punishing the offenders. As Behrend recounts, following the call of the Holy Spirit Lakwena animate and inanimate elements of the natural world, such as bees and other animals, stones, and rivers came to support humans in their battle against the forces of evil.

Behrend's account of the Lakwena Spirit Movement reveals how the Acholi among which she did fieldwork construct nature as a mirror of the state of social disorder in which humans find themselves. Simultaneously, nature is a source for visible and invisible forces that may come to assist humans in their efforts to restore social and moral order. Also significant in this peculiar construction of the relationship between social and natural worlds – and between moral and natural disorder – is the role of invisible, revengeful spirits, *cen* and *yok*.

Behrend (1999) and Finnström (2003) both show that Acholi constructions of their social and moral universe are based on the belief in the existence of *cen*, revengeful spirits (or, as Finnström (2003: 218) ventures, spiritual revenge. *Cen* may be the spirits of people whose deaths remained unavenged or for whose death the killer did not perform the necessary purifying (and redeeming) ritual. A *cen* comes back to haunt the killer and, possibly, also those who saw and touched the body of the victim (Finnström 2003:218). The revenge of these spirits who resent that they have not been treated in a proper manner may also manifest itself in the persistent curse with which particular clans or groups are inflicted and that bring them repeated death, misfortune and illness.

But *cen* and other spirits may also be set off by other human infrictions of norms, and roam freely in the zones of wilderness outside of human habitation and civilization. By positing that the *cen* and other potentially harmful spirits, err around freely in the natural environment of human habitation and need to be pacified to prevent them from attacking human beings, Acholi construct these realms of "wilderness" as areas that may pose a danger to human beings, but that can also be tamed and (temporarily) harnessed by performing certain rituals of exchange.

Along similar lines, Raphaela von Weichs (2009) shows in her dissertation on the kingdom of Bunyoro-Kitara that sacred kingship played a key role in constructing a particular relationship between the social and the natural universe. The institution of sacred kingship embodied, and was responsible for, the social, physical and moral regeneration of the entire population: the king's physical body reflected and mapped the body politic. The king was said to hold a very particular kind of spiritual power (*mahona*), a productive and generative, but also potentially dangerous force, that harmed those who were not allowed, or not able, to handle and control it. Even the king, the quintessential holder of this spiritual power, had to regularly perform specific rituals that, while constituting a breach of conventional norms of sociality and proper etiquette, allowed the king to contain this force and to harness it for productive ends. Anyone else, however, who came into contact with this force without being authorized or able to control it, would suffer from this contact: he or she would turn mad, go out into and roam in the wilderness or threaten to come back and haunt other human beings. To cure this afflicted person from the anti-social dangers associated with *mahona*, the person needed to be exorcised through special rituals that would ensure that the dangerous spiritual force could be safely relegated to the realm of the wilderness.

Underlying this concept of sacred kingship was a close alignment of notions of social and natural well-being: the king was responsible for the regeneration of the community, and also for the protection of natural environment. Placed under the king's protective control was a certain demarcated territory in which animals were not allowed to be hunted down. At the center of these protected areas was usually a spirit shrine.

Sacred kingship was the power that, through rituals of restoration and through shrines associated with certain spirit cults, helped maintain a balance between humankind and the forces of nature.[9] Different spirit cults and mediums existed, whose exact relationship oftentimes reflected different, historically layered, political alliances between different influential clans. Many of these spirit mediums and related cults formed part of the realm of political and ritual power associated with the king. People associated many of these spirits with the wilderness, where they believed them to roam beyond control of human beings. Some of them, such as the ancestor spirits, were feared to take revenge if humans infringed on certain norms, such as when they neglected ancestor spirits and ignored their demands. Also, most importantly, people believed that the revenge of these spirits could take the form of various physical afflictions, such as illnesses and mental disturbances. But their revenge would also manifest itself in natural disasters such as drought, floods and famines. Any instance of this form of disaster had to be countered by performing rituals through which the offended spirits could be placated and called upon to restore social and physical well-being.

Mikael Karlstrom's work on the Baganda kingdom similarly demonstrates the central importance of ritual practice aiming at (re)establishing the 'pact' between humans and invisible forces that was believed to ensure moral, social and physical harmony. As he demonstrates with regard to the *gwembe lumba* ceremony, a set of mourning rituals dedicated to a deceased family head, this ceremony was essential to ensuring social and moral continuity because it allowed family members and other clans to take leave from the former family head and to simultaneously establish and celebrate his successor (Karlström 2004: 598-599).

Susan Reynolds White (1997: chapters 4, 5) demonstrates how the Banyole seek to ensure social and moral continuity by continually constructing and reconstructing the relations between the living and the dead through daily practice and ritual. Reporting on the uncountable little gestures and remarks through which people acknowledge the presence and relevance of deceased family members to the well-being of those who live on, her analysis extends beyond rituals directly related to burial and mourning practices, and shows the on-going, life-long relevance of these constructions of ancestorship to people's everyday life.

All the authors whose work I just summarized interpret death and mourning rituals, whether performed by the Baganda in Central Uganda, by the Acholi in Northern Uganda, or the Banyole in Eastern Uganda, as a set of practices that conventionally enabled the transformation of deceased relatives into protective and benign ancestors. By engaging in the funerary rituals, the living secured for

---

9   Von Weichs differentiates between different types of spirits, yet also observes that it is not always possible to clearly distinguish between them, their functions, and also the rituals and other cult practices associated with each of them (von Weichs 2009: chapter 3).

themselves the protection of the dead. These rituals, whether performed on behalf of a family head or even a clan elder or king, involved the symbolic construction of social and moral continuity. From these studies emerge several related insights. First, in all these regional religious traditions, the social world and
the natural world are constructed as two faces of the same medal that mutually
constitute and mirror each other. Secondly, in all these religious traditions, invisible forces play a central role in mediating the relations between the living
and the dead. Whether personified in vengeful and in other spirits that interfere
with human lives, these invisible forces are believed to have the power to cause
havoc among the living people but may also protect them. Thirdly, indicative of
people's belief in the working of these invisible forces are ritual practices that
are intended to communicate with these spirits, to pacify them, and to render
them benign and protective of humans' actions and aspirations. Thus all these
rituals are, broadly speaking, about the establishment and reproduction of social,
moral and natural order. Each of them comprises acts that highlight death as a
turning point, that allow people to mourn and commemorate the deceased, and
thereby to actively *make* the deceased relative a member of a world of invisible
forces.

5. Mourning as mediation and as a call for remembrance

So far, I argued that Muslim rituals of mourning not only establish and mediate
a particular relationship between the living and the dead, but also serve to produce (and work upon) the social and natural worlds as part of the same moral
universe. Yet understanding Muslims' ways of coming to terms with death in
contemporary Uganda implies more than an inquiry into Muslim rituals of death
and mourning. Rather, given Uganda's postcolonial history, the questions that
spring to one's mind are: How do Muslims deal with the *premature* death of
their beloved ones? How do they deal with and make sense of death, in a situation where continued civil strife, exacerbated by continued instances of economic and ecological crisis, has changed the very conditions for social and
moral restoration, and threatens to implode the terms on which a balanced, har-
monious relationship between social and natural life worlds was conventionally
founded? And in what discursive, material, and symbolic practices do these
changing conditions for ensuring social and moral continuity manifest themselves? To illustrate the direction into which these questions might leave us, let
us take a look at another mourning practice drawn from Uganda's recent political past.

    In a provisional analysis of the National Resistance Movement regime es-
tablished by Museveni in 1986, Michael Twaddle, a prominent analyst of
Uganda's political history, reports on a rather gruesome, if telling, local practice
of remembering violent death. In September 1986, in the aftermath of the civil

war that brought the NRM and Museveni to power, smallholders in the Luwero triangle just north of Uganda's capital Kampala, where hundreds of thousands had reportedly been killed by soldiers of Obote's army, were "setting up memorials to dead relatives so that the sheer extent of the slaughter there should not be forgotten." These memorials often took the form of skulls mounted on poles or laid out on tables. Twaddle interprets this memorialization not only as a reflection of the human devastation brought to this country, but also as an indication of a widespread perception that the task of rehabilitation was seen as a "moral matter as well as a material concern" (Twaddle 1988: 314).

As I want to suggest, the practice of mounting the skulls of one's murdered relatives for public rememberance might be telling in respects other than those proposed by Twaddle. One evident significance of this practice – the one Twaddle alludes to – is that the skulls are made into a memory site: they speak of past atrocities inflicted on a victimized civil population and, more generally, on the body politic. Also evident is the specific, partial nature of this memory site. Its particular location and the victims it called to remember was a reflection of the new political leadership of the NRM regime and of its attempt to legitimize its imposition of law and order by reference to atrocities committed under the second regime of Obote.

But the practice also implies that one of the most important element of a person's bodily remains has been turned into an object of public visibility and scrutiny, and is thereby, in a sense, de-personalized and objectified. Although in all likelihood, people performed specific rituals intended to place the souls of the deceased prior to publicly displaying emblematic parts of their bodies,[10] I would argue that the act of public exposition did deprive the skulls of their untouchable (and in this sense, "sacral") nature. Not only did the head, and especially the chin, historically hold a special symbolic significance in certain religious and political traditions of Uganda.[11] The skulls, by visualizing violent, premature death that occurred at a massive scale, also seem to speak of the presence (and danger) of the souls of the deceased who never received proper burial in the ways it was habitually conducted. Even if rituals of placating unredeemed souls or revenging spirits were performed on the skulls prior to their public exposition, it is likely that, in moments of crisis or existential insecurity, survivors feel

---

10 Ute Luig, in her recent research on the commemoration of violent death in Cambodia collected numerous reports on practices of "placating" the spirits of those who had been murdered before turning their bodily remains into public memory sites (Ute Luig, personal communication, June 2011).

11 As Weichs observes, the chin of deceased kings of Bunyoro-Kitara believed to be the seat of his special spiritual powers; its was therefore severed from the head of the deceased king after his death and buried separately (2009: 11). Even if this applied to royal families, the practice in itself points to the special significance attributed to the skull as a site of spiritual power.

that violent death foreclosed the succesful transformation of their decesased relatives into peaceful ancestor spirits To these relatives, the ghosts of the deceased might therefore continue to linger in the realm of the living and interfere with their affairs.

If considered in this light, the act of mounting skulls on the roadside is not only a reminder of past atrocities and injustices and a material practice that articulates calls for retribution and/or reconciliation. At the risk of overstating its multilayered significance, I suggest that the public posting of skulls also hints at the perception of survivors of the impossibility – and of their own incapability – to lay their deceased relatives forever at rest. What I find particularly striking in this context are the ways in which skulls, as an emblematic part of personal individuality and intimacy, are positioned next to the roadside, in the realm associated with travel and the space, where different, potentially harmful, spirits that move outside of human civilization and control, may exert their powers. For survivors to decide to expose a skull, this intimate and symbolically charged part of the human remains of their dead relative, to the realm of uncontrollable spiritual intervention and revenge, could be taken as an indication that those who survived live with a sense of loss, and this in several respects: a loss of relatives; of control over maintaining a balance between the social and natural universe; and of the resulting fear of surrendering to the spirits that control the world from their locations in the invisible realm of forces.

Thus, in a paradoxical twist, the public posting of skulls, although intended by certain actors and political interest groups as a reminder of past atrocities and injustices, may simultaneously feed into the sense of non-closure experienced by those who survived the murderous interventions of Obote's army.

6. Conclusion

Drawing on my discussion of possible significances of the skull memory sites created along the roadsides in the infamous Luwero triangle, I want to return to the argument by Jewsiewicki and White (2005) and by de Boeck (2005) that the political significance of mourning has been changed radically in recent decades. As these authors argue, in social contexts in Africa where people are confronted with an exponential explosion of death caused by ecological disaster, civil war, Aids, violence and everyday thuggery, mourning practices, and the struggles on which they reflect, reveal a new relationship to death. At stake is not just the explanation of sudden, unexpected death, such as in case of Aids where witchcraft accusations and alleged breaches of norms of moral conduct play an eminent role in accounting for these deaths. Rather, as de Boeck argues for the case of Congo, the relationship between the living and the dead, and between the real and the imaginary has been changed.

Exploring the (possibly changing) practices by which Muslims in Uganda deal with premature, violent death is a powerful means to probe to what extent the portrayal of these authors captures the politics of death and mourning in Uganda. Specifically, this analysis allows us to assess whether, as de Boeck argues, the very relationship between living and dead has changed as a consequence of the new pervasive presence of death.

In view of the postcolonial history of Uganda, we need indeed to make room for the possibility that traditional understandings of "regular" modes of dying and of dealing with death loose their validity in the face of the sheer amount of premature, violent death caused by ecological disasters, civil war, and ethnic strife. Yet, to understand people's changing abilities to deal with this situation, and to explore how their abilities reflect on the specific position they occupy in a national body politic, I propose to proceed from an analytical angle that inverts de Boeck's focus on the apparent encroachment of death on the world of the living. As I argued in this chapter, it is time to detect in Muslim practices of mourning signs of their on-going efforts – and abilities – to ensure social and moral continuity, and to claim a place within national and local memory politics. For this, closer attention is needed to the emotional force and dimensions of mourning, as a way of dealing with the past and of envisioning a path into the future. Interpreting Muslim practices of dealing with massive, premature death in terms of continuity (with the past and the future), rather than of radical rupture, also allows us to reach beyond explorations of the agency and "coping" strategies of those affected by natural and social disasters. In this sense, a focus on Muslim rituals of mourning and leave-taking in Uganda offers a window onto the directions into which future research on the emotional ramifications of disasters could take us.

7. Bibliography

Ahmed, Chanfi (2008): The Wahubiri wa Kislamu (Preachers of Islam) in East Africa. In: Africa Today 54, 4, pp. 3-18.

Alexander, David (1997): The Study of Natural Disasters, 1977-1997. Some Reflections on a Changing Field of Knowledge. In: Disasters 21, pp. 284-304.

Behrend, Heike (1999): Alice Lakwena and the Holy Spirits. Oxford: James Currey.

Blaikie, Piers, Cannon, Terry, Davis, Ian and Ben Wisner (1994): At Risk. Natural Hazards, People's Vulnerability, and Disasters. London: Routledge.

Bloch, Maurice and Jonathan Parry (1982): Introduction: Death and the Regeneration of Life. In: Bloch, Maurice and Jonathan Parry (eds.): Death and the Regeneration of Life. Cambridge: Cambridge University Press, pp. 1-44.

Bowen, John R. (1992): On Scriptural Essentialism and Ritual Variation. Muslim Sacrifice in Sumatra and Morocco. In: American Ethnologist,19, 4, pp. 656-671.

Butler, Judith (2004): Precarious Life. Precarious Life: The Power of Mourning and Violence. London, New York: Verso.

Casimir, Michael J. (ed.) (2008): Culture and the Changing Environment. Uncertainty, Cognition and Risk Management in Cross-Cultural Perspective. New York, Oxford: Berghahn Books.

Clausen, Lars, Geenen, Elke M. and Elisio Macamo (eds.) (2003): Entsetzliche soziale Prozesse. Theorie und Empirie der Katastrophe. Münster: Lit Verlag.

Denny, Frederick (2006): An Introduction to Islam. Upper Saddle River, N.J.: Pearson Prentice Hall.

De Boeck, Filip (2005): The Apocalyptic Interlude. Revealing Death in Kinshasa. In: African Studies Review 48, 2, pp. 11-32.

Dombrowsky, Wolf R. (2001): Katastrophenvorsorge als gesellschaftliche Aufgabe. Die globale Dimension von Katastrophen. In: Plate, Erich J. and Bruno Merz (eds.): Naturkatastrophen. Ursachen, Auswirkungen, Vorsorge. Stuttgart: Schweizerbart, pp. 229-246.

Eng, David L. and David Kazanjian (eds.) (2003): Loss. The Politics of Mourning. Berkeley: University of California Press.

Finnström, Sverker (2003): Living with Bad Surroundings. War and Existential Uncertainty in Acholiland, Northern Uganda. Uppsala.

Frömming, Urte Undine (2006): Naturkatastrophen. Kulturelle Deutung und Verarbeitung. Frankfurt/New York: Campus Verlag.

———— (2005a): Der Zwang zum Geständnis. Friedensrituale und Mythologie im Kontext von Naturkatastrophen auf Flores (Ostindonesien). In: Anthropos 100, pp. 379-388.

———— (2005b): Von Traumfischern und Sibirientestern. Gemischte Gedanken einer Ethnologin zu TV-Erlebnisdokumentation. In: Journal Ethnologie; Themenschwerpunkt: Visuelle Anthropologie, edited by Ulrike Krasberg. www.journal-ethnologie.de.

———— (2001): Volcanoes: Symbolic Places of Resistance. Political Appropriation of Nature in Flores, Indonesia. In: Wessel, Ingrid and Georgia Wimhöfer (eds.): Violence in Indonesia. Hamburg: Abera.

Geenen, Elke M. (2003): Kollektive Krisen. Katastrophe, Terror, Revolution – Gemeinsamkeiten und Unterschiede. In: Clausen, Lars, Geenen, Elke M. and Elisio Macamo (eds.): Entsetzliche soziale Prozesse. Theorie und Empirie der Katastrophe. Münster: Lit Verlag, pp. 5-23.

Göbel, Barbara (2008): Dangers, Experience, and Luck: Living with Uncertainty in the Andes. In: Casimir, Michael J. (ed.): Culture and the Changing Environment. Uncertainty, Cognition, and Risk Management in Cross-cultural Perspective. New York, Oxford: Berghahn Books, pp. 221-250.

Groh, Dieter, Kempe, Michael and Franz Mauelshagen (eds.) (2003): Naturkatastrophen. Beiträge zu ihrer Deutung, Wahrnehmung und Darstellung in Text und Bild von der Antike bis ins 20.Jhdt. Tübingen: Gunter Narr Verlag.

Halevi, Leor (2007) Muhammad's Grave: Death Rites and the Making of Islamic Society. New York: Columbia University Press.

Hanretta, Sean ms. Developing Love and Death: Ghanaian Muslim Weddings and Funerals in an Era of Reform (book manuscript, in preparation)

Harrison, Christopher (1988): France and Islam in West Africa, 1860-1960. Cambridge: Cambridge University Press.

Hirschkind, Charles (2008): Cultures of Death. Media, Religion, Bioethics. In: Social Text 96, 26/3, pp. 39-58.

Hirsi Ali, Ayaan (2007): "Violence Is Inherent in Islam — It Is a Cult of Death," Interview, *Evening Standard* (London), 7 February 2007.

Hoffman, Susanna M. and Anthony Oliver-Smith (eds.) (2002): Catastrophe and Culture. The Anthropology of Disaster. London: James Currey.

Jewsiewicki, Bogumil and Bob W. White (2005): Introduction. In: African Studies Review, Focus: Mourning and the Imagination of Political Time in Contemporary Central Africa, 48, 2, pp. 1-9.

Karlström, Mikael (2004): Modernity and its Aspirants. Moral Community and Developmental Eutopianism in Buganda. Current Anthropology 45,5, pp. 595-619.

Kayunga, Sallie Simba (1994): Islamic fundamentalism in Uganda. The Tabligh Youth Movement. In: Mamdani, Mahmood and Joseph Oloka-Onyango (eds.): Uganda: Studies in Living Conditions, Popular Movements, and Constitutionalism. Frankfurt: Brandes & Apsel, pp. 319-363.

Luig, Ute, (2010): Über das Erinnern von Gewalt und die Verarbeitung des Schmerzes am Beispiel von Flüchtlingen und Ex-Kämpferinnen. In: Curare Zeitschrift für Medizinethnologie und transkulturelle Psychiatrie, 33, 1 and 2, pp. 60-71.

Oliver-Smith, Anthony and Susanne M. Hoffman (eds.) (1999): The Angry Earth. Disaster in Anthropological Perspective. New York: Routledge.

Plate, Erich J., Merz, Bruno and Christian Eikenberg (2001): Naturkatastrophen. Herausforderung an Wissenschaft und Gesellschaft. In: Plate, Erich J. and Bruno Merz (eds.): Naturkatastrophen. Ursachen, Auswirkungen, Vorsorge. Stuttgart: Schweizerbart, pp. 1-45.

Radio Netherlands Worldwide (2011): Disaster strikes again at Mount Elgon despite Warnings. By Arne Dornebaal. http://www.rnw.nl/africa/article/disaster-strikes-again-mount-elgon-despite-warnings (accessed 21.01.2012).

Reynolds White, Susan (1997): Questioning Misfortune. The Pragmatics of Uncertainty in Eastern Uganda. Cambridge: Cambridge University Press.

Rosaldo, Renato (1989): Culture & Truth: The Remaking of Social Analysis. Boston: Beacon Press.

Scharrer, Tabea (2010): Narrating Islamic Conversion – Erzählungen religiösen Wandels in Ostafrika. Doctoral Dissertation, Dept. of Anthropology, Free University, Berlin.

Schlehe, Judith (2008): Cultural Politics of Natural Disasters. Discourses on Volcanic Eruptions in Indonesia. In: Casimir, Michael J. (ed.): Culture and the Changing Environment. Uncertainty, Cognition and Risk Management in Cross-Cultural Perspective. New York, Oxford: Berghahn Books, pp. 275-300.

————— (2006): Nach dem Erdbeben auf Java. Kulturelle Polarisierungen, soziale Solidarität und Abgrenzung. In: Internationales Asienforum 37, 3-4, pp. 213-237.

Seesemann, Rüdiger (2005): Islam and the Paradox of Secularization: The Case of Islamist Ideas on Women in the Sudan. Sociologus 55, 1, pp. 89-118.

Tobin, Graham A. and Burell E. Montz (1997): Natural Hazards. Explanation and Integration. New York: Guilford Publishing.

Trimingham, Spencer (1964): Islam in East Africa. Oxford: Clarendon Press.

Twaddle, Michael (1988): Museveni's Uganda. Notes Toward a Provisional Analysis. In: Hansen, Bernt and Michael Twaddle (eds.): Uganda Now: Between Decay and Development. London: James Currey. pp. 313-335.

Von Weichs, Raphaela (2009): Die Rückkehr des Königs von Bunyoro-Kitara. Praktiken und Diskurse der Politischen Kultur in West-Uganda. Diss. Cologne: University of Cologne University/Institute of African Studies.

Wisner, Ben, Blaikie, Piers M. and Terry Cannon (2004): At Risk. Second Edition. Natural Hazards, Peoples' Vulnerability, and Disasters. London: Routledge.

Negotiating culture: indigenous communities on the Nicobar and Andaman Islands in the post-tsunami media coverage 2004/05

Brigitte Vettori

1. Introduction

The news on the 26th December 2004 was shocking. At 7:58 a.m. local time a gigantic marine earthquake occurred in the Indian Ocean which reached 9.1 on the Richter scale. It caused a tsunami which killed between 230,000 and 400,000[1] people in South East Asia, taking different statistics into account; several hundred thousands were injured, and around 1.7 million coastal inhabitants became homeless at one go. Not only were the Nicobar and Andaman Islands, situated near the epicentre, badly affected, but many coasts around the Indian peninsula, in Indonesia, Sri Lanka, Thailand and on the Maldives, were heavily devastated[2] as well. While the local populations and many tourists were directly exposed to the tsunami, their friends and relatives, as well as the general public, watched these events unfold in front of their television sets. They listened to the news about the tsunami and its thirty-metre high waves, which destroyed shore lines, even whole islands, their material culture, like the houses, hotel complexes, cars, boats or artefacts being drawn into the sea. In addition to the many people killed, immaterial culture, like local knowledge of peoples from different and heterogeneous communities, was lost.

This chapter analyses interpretations of the disaster by international media reports in 2004/05 in relation to the indigenous communities of the Andaman and Nicobar Islands. It discusses the hypothesis of the irretrievable loss of culture that the tsunami has brought about in these indigenous communities, which often appeared in press reports, and asks whether it is justified. Examples of post-tsunami media coverage are compared with a long interview with a village

---

1 The statistics vary a great deal with regard to those injured by the Tsunami. Some statistics include persons who died later due to injuries caused by the Tsunami.

2 The following countries (on a descending scale) were affected by the tsunami in 2004: Indonesia, Sri Lanka, India, Thailand, Myanmar, Somalia, Maldives, Malaysia, Tanzania, Seychelles, Bangladesh, South Africa, Kenya, Yemen and Madagascar (see Weber, Georg, Andaman Association). This article also includes statistics about victims from abroad.

headman from the Nicobar Islands. This interview not only provides an insight into the dramatic events during the tsunami on the island of Trinket, but also includes information which allows inferences of the period before the tsunami and rejects essentializing press reports. On a broader level the interview supplies arguments for a more theoretical discussion of how far culture in general can be extinguished by disasters.[3]

2. Outline of the research topic and the research situation.

My enquiry took place while I was working and doing research as an anthropologist in the post-tsunami context. Since June 2005 I had been working as a coordinator for the Vienna-based Sustainable Indigenous Futures Fund (SIF-Fund). The SIF-Fund, which arose out of the cooperation of such diverse institutions as a university institute, a record label and an NGO, were engaged in post-tsunami projects on the Andaman and Nicobar Islands. Because my work for the SIF-Fund was limited to twenty hours a week, I used the opportunity to undertake additional research in my free time, exploring new research questions in line with the project. I soon decided not to concentrate on the analysis of specific post-tsunami projects or on those from international organizations. Instead, I focused on the vividly discussed topic of culture and the loss of culture among the indigenous communities of the Andaman and Nicobar Islands in post-tsunami media reports. I opted for an enquiry in the manner Ulf Hannerz had once suggested in the *European Journal of Cultural Studies*:

> Insofar as academic scholarship on culture carries any intellectual authority outside our own institutions, we would do better to keep a critical eye on the varieties of culture speak both among ourselves and in society at large (...). (Hannerz 1999: 396)

I therefore collected more than a hundred articles from predominantly Indian and European media reports over a time span of one year after the disaster, as well as radio reports about indigenous groups in the Gulf of Bengal hit by the tsunami.[4] At the same time, I also observed how people in my work surroundings spoke about the topic. The following headline in the *Süddeutsche Zeitung* underlines the style of most of the press reports:

---

3 This article is based on the research results of my unpublished thesis for the Austrian diploma (see Vettori 2006).
4 In the bibliography, I list only those media reports which I used for my analysis in this article.

Doomed. The population of the Nicobar Island lost by the tsunami, 15 islands, several thousand people and her whole culture. [5] (Steinberger in *Süddeutsche Zeitung*, 2005-12-09)

Is it possible to lose one's whole culture, I asked myself, while putting this article down on the pile of all the other reports about the Nicobar and Andaman Islands? The message prevalent in all those reports was a clear and easily understandable one: the threat of the loss of culture, or its destruction already in an island paradise with happy but isolated natives. During my working hours I had to tackle the ambivalent situation that my colleagues in the SIF-Fund mostly agreed with such media reports, even while describing those indigenous groups with whom we were working as passive recipients of aid. But at the same time we expected from our Nicobarese project partners that they should carry out those projects which were funded by the SIF-Fund with a high degree of self-determination.[6] As an anthropologist I had to question such essentializing reports, despite the fact that they might have stimulated people's willingness to give donations to the victims of the tsunami and increased knowledge of SIF-projects at the same time. I decided to discuss the topic of 'culture talks' – or, in the words of Hannerz, of 'culture speak' – with my colleagues on the one hand, while critically analysing the post-tsunami media reports of 2004/05 about indigenous groups on the Andaman and Nicobar Islands for my diploma thesis on the other.

First of all, I compared the contents of the media reports with the relevant scientific literature and with information I was given by the human ecologist and specialist in research on the Nicobar Islands, Simron Jit Singh, who was also active as a consultant for the SIF-Fund. In addition I had the opportunity to discuss the situation on the islands directly with some of the people concerned. Because it was not possible for outsiders to stay on the islands in the Bay of Bengal under a law of 1956, the Andaman and Nicobar Protection of Aboriginal Tribes Regulation (ANPATR), which the Indian Government had passed to protect the indigenous communities, the SIF-Fund decided to invite a group of six Nicobarese from the partner organisation, the Nicobar Youth Association (NYA), to Vienna. This personal exchange, which took place in September 2005, was necessary for their cooperation in the project, but it also provided valuable insights into the life of the Nicobarese both before and after the tsunami. The discussions became the key to answering my research question as to how far the tsunami might have irretrievably destroyed the culture of the indigenous communities or, more generally, how far culture can be extinguished by disasters.

---

5   This text as well as the passages in the journal was translated by Ute Luig.
6   Concerned were mainly projects on the Central Nicobar Island regarding either immediate emergency relief or long-term capacity-building or livelihood projects.

Before I discuss these questions in more detail, I will give a description of the islands in the Bay of Bengal and the different groups which live there, information that will be enriched throughout the article. The following information, measurements and estimates all refer to the time before the tsunami.

3. Portrait of the Andaman and Nicobar Islands

The group of the Andaman and Nicobar Islands, 700 km long, represents with its capital of Port Blair the most remote territory in the Indian Union. 1,200 km of sea separate the east coast of the Indian mainland from the total of 306 islands and 206 rocks in the Bay of Bengal[7] that occupy an area of 8,249 km^2; the area of the Nicobar Islands makes up only a quarter of this (1,841 km^2). The journalist Oliver Lehman once compared the space of the Nicobar Islands with the area of Vienna and its forest hinterland (Wienerwald) (see Lehmann in *Universum*, 2005-08/07). The highest mountain is the Saddle Peak on the northern Andaman Islands at 730m, followed by Mount Thullier at 642m on the southern island of Great Nicobar.

The Andaman and Nicobar Islands have belonged to India since her independence in 1947. Since 1956 they have been classified as a union territory with two districts: the Andaman Islands, with Port Blair as their capital, and the Nicobar Islands, with their centre on Car Nicobar. Prior to the Indian government, the 78 years of British rule (1869-1947) installed a colonial system of administration on the islands, which was taken over and adapted by the Indian government. Denmark was another important colonial power (1756-1848), but Austria (1778-1783) and Japan (1942-1945) were only intermittent players, soon to be superseded by more continuous colonial governments. In the Indian Parliament in Delhi, the Andaman and Nicobar Islands are represented by just one member. Alongside exist different systems of self-administration on the islands (see Singh 2003). On the Nicobar Islands the Tribal Councils have taken over important functions regarding consultation and decision-making in the economic, political and judicial domains. They consist of several elected (male and female) village chiefs and are subordinate to the Chief Captains of the northern, central and southern Nicobar Islands, with whom the speaker of the Tribal Councils exchanges views. The Lieutenant Governor mediates in the union territory between the Indian government and the respective local structures (see Vettori 2006: 22-25).

The Indian population census for 2001 gives a total population in the archipelago of 356,152, of which 314,084 live on the Andaman Islands and only

---

7   Only 36 islands are inhabited.

42,068 on the Nicobar Islands.[8] On 12 of the 24 inhabited islands, 200 persons of the local group of the Shompen were counted before the tsunami on the southerly situated island of Great Nicobar. Also before the tsunami 30,000 Nicobarese[9] were distributed on the other islands, as well as 10,000 settlers mainly from India. While the indigenous groups on the Nicobar Islands constitute the majority of the population, on the Andaman Islands they make up only 0.1 to 0.2 per cent of the local population. According to estimates there are about 200 persons from the group of the Jarawas in West Andaman, about 28 persons from the group of Great Andaman on Strait Island, 100 Sentinels from the North Sentinel Islands and about 98 Ongi from Little Andaman.[10]

According to the Indian constitution, all indigenous groups with the exception of the Nicobarese belong to the so called PTGs, the Primitive Tribal Groups or Primitive Tribes. In comparison to other indigenous groups the Nicobarese are regarded as more progressive, which is why they are classified as a Scheduled Tribe.[11] According to this definition the Primitive Tribal Groups as well as the Scheduled Tribes are characterised by a pre-agrarian economy, a poor degree of literacy and a stagnant or even declining population (see Delius 2005: 28; Pandya 2006: 10-13).[12] However, everyday practice proves that most of the indigenous groups have adopted the modern practices of hunters and gatherers, as well as agriculture or cattle-breeding to different degrees. The production of and trade in copra was one of the most prominent sources of income for the Nicobarese before the tsunami. Scheduled Tribes and Primitives Tribes do not belong to the category of Indian castes. Despite developments in the direction of more equality initiated by the Indian state, they are still regarded as 'Untouchables', 'Outcasts' or 'Dalits' in everyday life (see Pandya 2006: 10-13; Tharoor 2005: 117-159). Besides 'traditional' religions, Hinduism as well as Christianity and Islam are practiced on the islands.

The translation of the Indian term 'Primitive Tribes' or 'Backward Classes' into the German word 'primitive' or 'backward' calls for misunderstandings. Although the classification of people into more or less developed groups is gen-

---

8 The last census took place in 2011. However, in the article I only use figures for the period before the tsunami (see Government of India, Ministry of Home Affairs and Office of the Registrar General & Census Commissioner 2011).

9 Until John Richardson, who came from Car Nicobar and in 1950 became the first Bishop of the Andaman and Nicobar Islands, subsumed the different communities under the name 'Nicobarese', these communities were known under different names. "Richardson wished to inculcate a collective identity among the inhabitants of the Nicobars, however dispersed and distinct the islands and the islanders were" (Singh 2003: 84).

10 Regarding the estimates of the different indigenous groups, I refer to a SIF letter from Simron Jit Singh of 12th July 2005.

11 The notion of Scheduled Tribes depends on the fact that the different groups are recorded in a schedule attached to the Indian constitution (see Tharoor 2005: 153).

12 There are further criteria which I will not discuss here in detail.

erally to be rejected, these terms are widespread in the Indian idiom and are therefore found in neutral and differentiated texts concerned with this topic. In German parlance, however, such a classification does not exist; therefore the meaning of a phrase like 'primitive society' is much more pejorative in German everyday speech and understanding than in Indian expression and devalues these societies in comparison to supposedly more developed groups.

4. The relativity of victim statistics

The largest of the six indigenous groups on the archipelago also had the highest number of victims as a result of the tsunami. According to information provided by the local Nicobar Youth Association (NYA), 10,000 out of 30,000 Nicobarese died during the tsunami. Indian statistics listed far fewer victims and missing persons and were not in agreement with these local statistics. The estimates in both statistics neglected the victims among the group of more or less legally immigrated Indians. When the six Nicobarese were guests in Vienna, Rasheed Yusuf, a member of the NYA and speaker of the Tribal Council of Nancowry, reported that 4,975 persons had been killed on the island of Katchal (see Yusuf during a group discussion, September 2005, in Vettori 2006: 39). Simron Jit Singh intervened immediately, saying "But this is not the official number," and Rasheed Yusuf answered: "Maybe, but this is the official number from us" (Yusuf and Singh during a group discussion, September 2005, in Vettori 2006: 39). It is, however, certain that fewer were killed in the Andaman Islands than in the Nicobar Islands. The Indian Daily, *The Hindu*, also reported on December 30, 2004 that five out of six indigenous communities in the archipelago had survived nearly undamaged. The Nicobarese, however, were badly hit (Nambath in *The Hindu*, 2004-12-30/A).

5. Indian post-tsunami media coverage in 2004/05

"How are the Jarawas, the Ongis and the Sentinelese tribes doing? Have the six aboriginal tribes survived the killer tsunami that slammed the Andaman and Nicobar Island?" (*The Hindu*, 2004-12-29/B). This and similar questions were being asked by journalists in the first days after the disaster. The media had no easy task at the beginning to collect information about the situation on the islands because NGOs and journalists were not easily allowed to travel there due to the Andaman and Nicobar Protection of the Aboriginal Tribes Regulation (ANPATR). This law protects those territories on the Andaman and Nicobar Islands which are inhabited by indigenous communities. Nearly the whole territory of the Nicobars and a large area in the Andaman Islands are affected by this regulation (see Andrews and Sankaran 2002: 34; Singh 2003: 104-105). Because the ANPATR was also maintained in times of disaster, some sixty NGOs and several journalists were stuck in Port Blair. Only a few were allowed to travel into the disaster area with a so-called Tribal Pass. In this situation the journalists had to rely on information at second or third hand. In the beginning they obtained the majority of their information from the military base on Car Nicobar:

> The tsunami, generated by the first quake, left at least 10 soldiers dead at the IAF [Indian Air Force] base at Car Nicobar. Flights from Port Blair airport could not be operated since the runway developed huge cracks under the onslaught of the tsunami. The seaport was also damaged, upsetting maritime operations as well (…). (Pandit in *The Times of India*, 2004-12-27/B)

However, there was no information in the beginning about the fate of the indigenous communities on the archipelago: "No word on rare tribes", ran one headline in *The Hindu* three days after the disaster (*The Hindu*, 2004-12-29/B). And the daily *The Times of India* wrote:

> An enormous anthropological disaster is in the making. The killer tsunami is feared to have wiped out entire tribes – already threatened by their precariously small numbers – perhaps rendering them extinct and snapping the slender tie with a lost generation. (Dutta, Mago in *The Times of India*, 2004-12-29/B)

It was not only feared that the "most ancient indigenous communities in the world" (Khosa in *Indian Express*, 2004-12-31/A) had been extinguished, but also that exotic research objects had been lost. Before the first information became available about the number of dead, the daily *Asian Age* speculated about the possibility of the indigenous people being revived in a laboratory:

> The ancient Andamanese Tribes can be theoretically recreated even if their last few surviving members had been swept away in the Sunday's tsunami waves because their genes have been immortalised. Their loss if confirmed – would be an immense anthropological disaster. But it will not be a total loss to the scientific world – thanks to a thoughtful act by scientists at the Centre for Cellular and Molecular Biology [CCMB] here. Due to their efforts, the genes of the Andamanese Tribes will continue to be available for research even if the Tribes become extinct. The CCMB scientists have in the past collected the blood samples, prepared the cell lines and immortalised them, which means that they can be multiplied and perpetuated in test tubes. The cell lines carefully preserved in the laboratory freezer can be used for scientific studies and, theoretically, to revive the ancient tribes if ethics permit. (*Asian Age*, 2004-12-30/A)

In the Indian press, descriptions of the different Primitive Tribes could be found in which residents on the mainland were reminded of 'their natives' on the remote islands. These descriptions were accompanied by drawings of scantily dressed primitives (see Dutta and Mago in *The Times of India*, 2004-12-29B). Individuals like the ecologist Panka Sekhsaria and the documentary filmmaker Usha Deshpande, who had been in contact with the indigenous groups before the tsunami and who could talk about their experiences with these groups, were also interviewed by the press. In the article "Andamans in happier times" positive descriptions could be read about the indigenous groups in the archipelago, a paradise which had been devastated by the tsunami (Pendse in *The Times of India*, 2005-01-01/C):

> (...) Usha Deshpande, who made her film, Song of Silence, for the Films Division in 1984, remembers the jovial Nicobarese faces, their enthusiasm and straightforward nature. Deshpande distinctly remembers leaving her camera equipment unattended in an open jeep on Car Nicobar Island. There was never a feeling that anything would be taken from the jeep. 'They are such open, honest people,' she says.

However, in other media one could find reports about indigenous groups which were less friendly. In a headline of the daily journal *The Hindu* one could read "All primitive tribes safe". The report went on to explain how this statement originated: "A Coast Guard pilot today spotted the Sentinalese in the North Sentinalese island during one of several low-flying sorties. The Sentinalese had actually thrown stones at the aircraft" (Nambath in *The Hindu*, 2004-12-31/C). Another article in the *Guardian Weekly* with the headline "Top official seized by starving islanders" described a critical situation on Great Nicobar: an administrator was held by indigenous people because he had something to eat, since Indian settlers as well as Nicobarese had waited for help for several days to no avail (see Harding in *Guardian Weekly,* 2005-01-13/07). Occasionally, journalists tried to describe the difficult situation of the islands' inhabitants factually without using images of passive and/or exotic or wild natives: "The villagers allege that all relief operations were being concentrated in Port Blair, while Car Nicobar and other islands in the region have been ignored completely" (Chandramouli in *The Times of India*, 2005-01-02/B). The changes in the geology, flora and fauna were also described with greater neutrality: "Coral reefs may take years to recover" (*Maharashtra Herald,* 2004-12-30/A), read one headline in the daily *Maharashtra Herald*, while Suresh Nambath wrote in *The Hindu*:

> The Nicobar group of Islands will never be the same again. Their shapes have changed. (...) some islands have become smaller, yielding ground to the sea (...). (...) the sand bands in some of the islands have given way to the rising sea, there have been cases of two islands lying in the place of one. (Nambath in *The Hindu*, 2004-12-31/B)

What was of great interest to scientists and journalists alike was the fact that some indigenous groups on the archipelago had fared better during the tsunami than others:

> (...) primitive tribes of negrito origin such as the Great Andamanese, Onges, Jarawas and Sentinalese are not known to have been affected. The Shompens,

though they inhabit the Nicobar Island, were also relatively safe. But the Nicobarese, who like the Shompens are of Mongoloid stock, faced the full impact of the killer waves. Indeed, officials believe that no Nicobarese was untouched. (Nambath in *The Hindu*, 2004-12-30/A)

To explain this situation, post-tsunami media reports alluded to the unspoilt way of life of these 'primitive tribes'. The anthropologist and Andaman specialist Vishvajit Pandya published a scientific article in *Anthropology News* in which he reported on his post-tsunami research among the Ongees,[13] whom he had known since the 1980s. He explored the reason "(...) why the Andamanese were being depicted as stereotypically dependent on 'collective memory' and as having saved themselves resorting to their 'folklore' or 'primitive tribal wisdom'" (Pandya 2005: 12). In his article he described how the Ongees used their knowledge in order to protect themselves from the forces of nature by citing one member of the society:[14]

We saw the water and knew that more land would soon become covered with sea and angry spirits would descend down to hunt us away. But our ancestral spirits would come down to help us if we continued to be together and carried our ancestral bones (...) with us to ensure assistance from good spirits! See, if the water decreases then it has to come back and claim land, and much in the same way if humans outnumber spirits, then spirits too want to add to their community by bringing death upon humans. (Pandya 2005: 12)

When the water was receding on the morning of 26 December, the Ongees met at the beach in order to throw stones into the sea. While the spirits were looking for victims in the rising water, the people used the occasion to bring themselves to safety (see Pandya 2005:13). In his account of his research results, Pandya confirmed their use of traditional knowledge in the disaster. He mentioned at the same time that Ongees also took more recent experiences into account when they needed help in the first days after the disaster:

Soon after the tremors, about 83 Ongees packed up and walked through the forest (...) [to a place] where they stayed for eight days in a shelter set up by the A&N Administration which also provided rations, water and medical care. But by January 9 the Ongees decided to move back into the jungle and set up their own self-selected campsite at a location they call Tandalu (...). (Pandya 2005: 12)

Since the settlement of Dugong Creek, which had been set up by the Indian administration and in which the Ongees had lived before the tsunami destroyed

---

13 I quote here Pandya´s spelling for the indigenous group Ongees as having the same meaning as Ongis which I used earlier.
14 I give only a short quote from a much more elaborate description about the interpretation of this Ongee member (see Pandya 2005:12-13).

it, no longer existed, they negotiated a new way of life with the Indian government:

> (...) in its context of destruction, the tsunami has given a chance to the Ongee to
> resist an imposed idea of settlement, and to express and design the place in ac-
> cordance to their culture. (Pandya 2005: 13)

In the post-tsunami media reports, little could be read about these negotiations over new living and settlement conditions, which transcended societies.
While the argument of lost tribal wisdom being responsible for the higher degree
of victims among the Nicobarese was brought into the discussion by European
journalists especially, a few others were looking for more reasonable causes for
this situation:

> (...) the Nicobarese (...) had been badly affected because they live in the plains
> towards the south, as opposed to other threatened tribes – the Sentinalese, Jarawas
> and others located on higher grounds. (Pandit in *The Times of India*, 2004-12-
> 30/E)

The fact that on such a flat island – Car Nicobar – a woman was missing her
son was not worth reporting during the disaster. But the very fact that on the
following day this woman gave birth to a daughter who was given the unhappy
name 'Tsunami' was celebrated by the media: "Born on the waves of disaster,
loved by all", read the headline in *The Times of India* (see *The Times of India*,
2005-01-02/A, see also *Indian Express*: 2004-12-31).

6. Cuttings from European post-tsunami media reports (2004/05)

In the European post-tsunami media coverage, one could often read the same
arguments which had already been reported in the Indian press. Has the tsunami
caused an anthropological disaster? asked the daily *Der Standard*: "Ethnic
groups who have lived totally isolated since thousands of years in the now totally devastated islands are threatened with extinction (...)" (*Der Standard,
2005-01-04*). The topic of the unfair distribution of aid was also discussed: reports talked of several hundred people being transported in ships and helicopters
to refugee camps. But only 3 per cent of those saved belonged to the native
population. A quote by the female Chief Captain of Central Nicobar Island underlined the situation: "The administration as well as the rescue crews will not
listen to us", deplored Ayesha Majid, a tribal leader on Nancowry Island. "They
shy away from us because we are natives. They just let us die" (*Der Standard,
2005-01-07*).[15]

---

15 The granddaughter of the former so-called Queen Rani Ishlon, Ayesha Majid played an
   important role as the speaker of the Tribal Council of Nancowry and as Chief Captain of
   the Nancowry Group of Islands in the post-tsunami events (see Singh 2003: 91-94).

According to an article in the weekly magazine *Der Spiegel* an official of the Indian Ministry of Justice alleged that nobody was killed on Car Nicobar during the tsunami: "He said that the people of Car Nicobar are like animals, they listen to their instinct and when they felt that the water was coming, they fled" (Brink-bäumer in *Der Spiegel* 2005, No. 4) A colleague from the same magazine took up this rumour, which was persistently repeated in the western media, and later called it "folkloristic nonsense" (see Traufetter in *Der Spiegel* 2005, No. 41). The journalists, however, more or less agreed that the rich cultures of the indigenous population and of the Nicobarese were acutely threatened, and even the cultural heritage of the Nicobar Island might have been largely extinguished at one go (see *orf.at* 2005-10-20/A). To what extent the transmission of cultural knowledge might be limited in future on the Nicobar Islands was discussed in another article in *Der Spiegel* in which one of the six Nicobar visitors who had been in Vienna in 2005 were quoted as follows:

> 'The ones who died in the waves were mainly the elder who transmitted our customs to the youth' (narrated Rasheed Yusuf). They had been at home together with their children. Only those men and women survived who either worked on higher situated plantations or who had been with their boats at sea when the great wave arrived. 'Those who had been at home had no chance' (said the Nicobarese) (see Traufetter in *Der Spiegel* 2005, No. 41)

According to some journalists, the culture of the indigenous groups was already threatened before the tsunami. The daily *Die Presse* wrote that, as the indigenous cultures were already endangered by an decade of contact with the "Indian motherland", the tsunami may have contributed to the doom of one of the last of these cultures (see Simon in *Die Presse* 2005-12-12). Similar arguments were made by Karin Steinberger, who wrote in the *Süddeutsche Zeitung* that the Jarawas – who are called pejoratively Djunglees by Indian settlers – have undergone fundamental changes. She predicted a pessimistic future for this group since the title of her report read "On the Andaman islands hunters and gatherers meet with settlers, modern age meets stone age – a disastrous encounter for the aborigines." And again "They are a tourist attraction, the last one of their species, people from the stone age", being threatened by imported diseases and by the road which crosses their territory and takes their living space away. The negative impact of these changes was predicted by Steinberger in comparison with the example of Great Andaman society, the first to have contacts with the British and which at present consist of only fifty individuals. She concluded that following the imposed dependency on alcohol and consumer goods there would be a loss of identity and finally the end. "First the Great Andaman, then the Onge, Jarawas, and soon it will be the turn of the Sentinelese. All in all, a question of time."

She describes the Nicobarese as "much more open-minded than the Andamanese." She also quotes Singh, who according to her wrote that the Jarawas had arrived at a point where the Nicobarese had been 1200 years ago. At present the Nicobarese all buy TV sets – a "trend that will cost them dear" (see Steinberger in *Süddeutsche Zeitung*, 2005-08-21/20).

Yet the Nicobarese had very different problems in the post-tsunami area:

> Down in the South, on the Nicobar Island, the tsunami caused havoc, nothing could stop it. The wave was able to build itself up. 500m broad is the space of death which it left behind. It drags itself from bay to bay and looks always the same: torn houses, foundations, ships on land, cars in the water. And corpses. (Brinkbäumer in *Der Spiegel* 2005 No.2)

And *Der Standard* reported on the crocodiles which got stuck into the corpses and about the danger of epidemics due to the cholera which had been introduced by tourists in 2002 (see *Der Standard* 2005-01-04).

The dominant topic in the European press, however, was the loss of culture on the islands, which were also described as "a nearly forgotten world threatened by its near end" (see *Der Standard* 2005-01-04). Regarding the past life of the Nicobarese, the "treasure of the Nicobar Islands" was evoked as being doomed, as well as what were possibly the last documents of this culture, which were recorded in a book by the human ecologist Simron Jit Singh and the journalist Oliver Lehmann, *The Nicobar Island* (see Lehmann in *Universum*, 2005-08/07). Focussed on Nicobarese rituals and customs and published in two languages, this book was interpreted in diverse ways. Already the translations of the texts hinted at different viewpoints concerning the same topic. The English title *The Nicobar Island: cultural choices in the aftermath of the tsunami* emphasized the challenge for the Nicobarese to negotiate their cultural futures in a vastly devastated island. As Singh pointed out, with this book he intended to open a "window into the past" (Singh 2006: 43) for the Nicobarese. Among German-speaking readers, however, the book with the title *The Nicobars: cultural heritage after the tsunami* evoked admiration for the destroyed paradise. The editor of the book and journalist Oliver Lehmann reported in an Ö1-Radio feature and in the magazine *Universum* that in November 2005, 500 books had been sent to the Nicobar Islands:

> ...in order to equip ca. 500 extended families to give them a kind of manual of there own culture. The book is (...) written in German and English and contains 300 pictures. It is thought of as a kind of operating manual for a devastated and lost culture. (Lehmann in *Ö1: Von Tag zu Tag*, 2005-12-27)

In this context, Lehmann also referred to some kind of support for consciousness-raising (ibid.). "The pictures will serve as some kind of cultural memory and as a practical stimulus for their use..." (Lehmann in *Universum*

2005-08/07) and thus contribute to the cultural reconstruction of Austria's unique former colony (see Lehmann in *Universum* 2005-11). The times around 1778 when the ship *Joseph and Maria Theresia* arrived in the Nicobar Islands and "the natives signed a document that declared four islands (from 1778 till 1783) an Austrian crown colony", were also recalled with pleasure (see *orf.at*, 2005-10-20/B).

In an article in the Bavarian broadcasting corporation online referring to Singh and Lehmann's published book in 2006 Spaeth concluded: 'There is not much left of Nicobarese culture: 228 pages, in which their festivals, their past, their lost paradise are protected from being forgotten'. (...) Spaeth in *BR-online*, 2006-01-07).

According to other accounts, artefacts and everyday objects survived which had been collected during the *Novara Expedition* by Karl Ritter von Scherzer when in 1858 the Austrian research vessel *Novara* anchored for a short time in the Nicobar Islands (see Steger in *Bridges*, 2005-12-05, Kaspar 2002a and b).[16] A comment by *Der Spiegel* noted that these artefacts were stored in the Noah's Ark of Nicobarese culture (see Traufetter in *Der Spiegel* 2005, No. 41), a reference to the Museum of Ethnography in Vienna, which was also visited by our project partners. *Orf.at* reported at the time: "Stone age society looks for traces in Vienna" (*orf.at*, 2005-10-20/A).

7. The village chief of Trinket describes his experience of the tsunami

I will now confront the above examples of post-tsunami media reports with an interview with Joseph Portifer.[17] I conducted this interview during the visit of this 32-year-old village chief from the village and island of Trinket in Vienna in September 2005. During our discussion he suddenly started to describe the events on the day of the tsunami:

You know what? Actually I was a bit annoyed on 25th, Christmas Day: All the people [in Trinket village] had been invited by my mum and my papa to our house to a cocktail party. (...) And you know, (...) my father did not ever do that! And I was (...) thinking: Why is my father doing all this? (...) every Christmas it was just a simple party but (...) this Christmas he spent a lot of money. (...) I just did not know what was happening. (...) Maybe they felt something; they invited everybody and were just partying the whole night. (...).

Yes, and (...) actually it did not work out for me this Christmas: (...) My girlfriend did not turn up [because she had a job in Nancowry Island]. That's the main thing. (...) I was a little annoyed. So (...) I went to the hill house: On top of the hill there is a house and I was staying there. I went to a friend of mine (...). I was

---

16 See Novara Expedition for brief information about this expedition.
17 Portifer is also written Fortifer (see Singh 2003: 22).

just not feeling good. (...) my friend was listening to the music and I stood up to
[go to] sleep.

Next morning [he laughs] (...) I got up and the earth was shaking (...). And I said:
What is this? Last night I was not drinking like this! (...) Then I opened the door. I
ran outside. [Opposite the house] there is a pond, a small lake. And it was just
exploding from below: (...) 'tuch, tuch' (...) and I said [to myself]: What is hap-
pening? Again I went inside. I took the needle from the machine. I started working
[it] into my hand like this: [he showed how he stung himself several times in his
hand]. I couldn't believe [it] myself. But when all the blood was coming, I put
down the needle and said, No, I am not dreaming!

My friend was sleeping and I gave him a kick. I said: Get up, get up...! (...) I took
the boy [the friend] with me [and] than I ran to the village. (...) I told everybody
that this is not an ordinary earthquake. Let's run to the hill! (...) I said: Come,
come, come! [But] they said: No, no, no. (...) At least for five minutes (...) the
house was shaking. (...) They were still in the house, because an earthquake
sometimes means it comes and it goes. They [people] don't run away. (...) they
are used to it. (...).

Then after five minutes, the sea started drying up. (...). Just drying! (...) [There
was] no water, the fish were jumping around, dying! Just five minutes [later] (...)
the water was coming back. Unbelievable! I said: Come on, come on, come on!
Let's go to the top of the hill! My mother said: No, I am going by boat. My father,
my mother, my sister, my brother-in-law ... my small sister, my sister's four sons,
my other sisters, [her] two daughters, my mummy, my daddy, all were there in one
boat [because there they felt safer]. (...) They all went in one boat but they did not
start the engine. If they had started the engine, they [would] have got away [from
danger].

We ran to the hill – 175, 179 people who listened to me. They all ran to the hill.
(...) everybody was listening to me, but these people who were in the church. (...)
I (...) [couldn't] get in the church and start shouting. (...) So I went into the vil-
lage and started shouting. And all the people who heard me came back to the
[hill], all came with me and then again I ran [down]. (...)

I started (...) untying all the cows and all [the pets]. (...) there were ladies [who
before were] milking [the cows], because it was in the morning. (...) So I untied
all the cows, (...), and the water was coming to this level [points at his knees] very
fast. (...) [To this level] and even further! (...) I ran and I saw one child (...) and I
said: Come, come with me! (...) The water was not so strong at that time. I could
have easily caught him. [But] he just waved to me and then he disappeared (...) in
the flood! (...)

And then these people, they were shouting on the hill, they were shouting for me:
Chief come up, chief come up, chief come up! We don't know what to do... (...) [At
the same time] I was seeing my friends almost floating in the thing [sea] and I
said: Come, come, come! But they said: No, we'll have a bath in this thing! (...)

Actually they did not know [what was happening...]. It was Christmas – everybody was drunk!

Then again, when I went to the hill, I was thinking of the father – the priest. And again, I ran back. Still the sea was not so rough. I mean, I could handle it. (...) I went into the water; I was searching for the priest. The priest was there in the (...) back yard. I said: Father come, come! The water is rising. He said: No. You go and see to the people [so] that they don't feel hungry. You don't come with me. You stay with the people. I said: No father you come! (...) He [answered]: No, no, no. And then the (...) [place] was uprooted by the waves. (...) I couldn't see the father [anymore]. (...) The wave took him. But [before] he was not [standing] far away from me. That's why I was wondering: (...). Why did the wave not take me, when twice I was there? Everybody is going. And the wave where I am standing, the wave is not so strong? (...). I was thinking, I had to be saved and I was saved. Otherwise I would have also gone.

And then I ran to the top, to the top-most hill, and I was looking at the boats and [they were] just dusting in the water, in the sea. (...) The waves were coming (...) on both sides. I could see my father and my mother, all. I could not do anything. I was really... and at that day also, my mobile was not working; everything, all the connection gone. (...) I could not do [anything]..., I could hear them shouting. But I was really helpless. I could not do anything in the end. (...). – I went up the hill. I was trying to contact the administration in Kamorta [but there was no connection]. From Nancowry [Island] everybody was thinking that in Trinket [Island] (...) we [were] all finished. Some of our brothers (...) landed up in Kamorta Island by the sea. Others..., everybody has gone. (...)

We were [then] standing on the top of the hill. Just the peak of the hill! (...) and the water came. (...) everybody was up there: 170 people were up the hill. So the water, you know, it was coming like this [shows that the water nearly touched his ankles] and then again it went back. And then it came back, three times! (...) and then it didn't [come anymore]... (...) You just could stay there [on the peak] (...) in a congested way. (...). [But] it was not only us who were on that peak: There were the cows, the pigs, the hens, the goats, the dogs... [he laughs]. All [of] them! It was [the place of] the surviving. They also were there. Can you imagine? And after about three o'clock, (...) when the sun was setting, everything was coming into normal. I was just thinking: things are going to be normal. So the water (...) went back, (...) but not totally. (...) I told (...) all the girls stay here, all the boys come with me!

So I went down the hill and I went again to the village, with the boys. There I saw quite a lot of wise [people] just managing to come up to the hill: (...) The boys who were there, they were just struggling to come out. And I just said: Come – take them. All the boys took them. And then we took them up the hill and then again we came back. We were searching [throughout] the village. (...) What to search and how to search, we did not understand. Because we could not see anything: No houses, no trees. Only some big, big trees were there. And on that big tree, six or seven boys were there on that tree. (...) [and also some] girls. Some of the girls hair was hanging on (...) the branches (...) and [they were] shouting. (...)

We had to climb up with the help of a rope and cut their hair. And some of the other boys were looking out for coconuts. We gave them those coconuts and then evening time came.

All the babies, small, small babies – they [the mothers] were caring with them. About four or five babies. I was just asking [myself]: Why didn't they take us all away? If they had taken some of us, why not full [all] Trinket? Because, I could not feed the babies; the babies were just crying very hungry. I just did not know what to do. (...) The mothers were there, only feeding them with their breast feeding and (...) the sun was too much for them. (...) there was no cover anymore, where they could sleep. Nothing! I was totally lost. I was wondering what to do. I could not think of anything. And they were asking me: Chief what are we to do now? We want to eat something. We are very hungry! I was thinking to myself: I am not god. How can I give them something to eat? If I had been god, I would have brought something for you all! I was just thinking to myself; but what to do? They were also confused, they were hungry. (...) I saw some animals (...) and I said: Kill the animals. Go and kill the animals. (...) All the boys, all the youth was..., you know, they were really very cooperative at that time! They just went out, killed the animals and gave it to everybody. (...) we made a fire and smoked the animals. For about two days we were lying there, on that grassy land without any... [But] one thing, one thing I really liked: It did not rain! If it would have rained, I think that the babies would have died. (...)

Then we had some good police officers. Because everybody was saying [that] there is no[body] [any]more in Trinket. Nobody is there in Trinket. (...) But there was one police officer, who was my friend. He said: I don't believe it. I have to go there! Until I go there and I see my friend's body, I won't believe you. He went there [and] he got two or three hundred people back, (...) back to Kamorta [Island].

Now we were saved back in Kamorta, but not in the camp. People were starving. (...) In Trinket I did not have a thing [anymore]. My coconut trees and everything was destroyed. So I couldn't feed them. It was a very tough time for me (...). Decision making, making decisions was really very tough. So I started to take all my people to the medical [centre] from the government, just to have them checked. (...) Some of them were admitted [got the admission to stay]. (...) some of the babies did not have any clothes. I myself I did not know what to do. I just went to a shop and I took whatever I want[ed] and then I just came out, (...) without paying... So this type [the shop owner] went and reported to the police! (...) [Before] I went with the boys I said: Take what[ever] you want and especially things for the babies. The boys took everything from the shop [he laughs] and they came. They came back and then the police came to me and said: What are you doing? I said: I just don't know what I am doing; you just shut up and get out of here! The police said: No, I don't go out of here. I am going to help you! And the police also helped me to get more articles from this shop. The police were helpful. (...) I got things for most of the people. I got them blankets to cover themselves, and then I came back.

And then I was totally... My friends who were [living] in Kamorta, they were saying: Portifer, why don't you take rest! Your people are safe now. Come to my house and stay. I said: No. If I have to stay, I stay with my people. Whatever they are eating, I will also eat. I don't need your... You are my friend, that's okay. But (...) I am concerned about my people. Let me be with my people. So for the next two or three weeks I was like that. (...)

When my people got strong[er], I had a lot of talks with the administration. [I needed] a truck to clean up the land, where we are staying now. We had shifted from Trinket to Kamorta. (...) My father had [some] land in Kamorta. So after two weeks my people got stronger. I took my people there (...). I took all the gents (...) to clear up the land so that we can stay there, in Kamorta Island. (...) And after three, four weeks everything was ready. I took all my people and shifted there. [We are living there now,] three hundred people. [They all came] not only from Trinket village but [also from] Trinket Island (...).

Actually I was not strong for the past six to seven months. If anybody would have asked me for this story, I would just have burst out crying. [But] it has changed, because it has to change: if I cry, my people also will cry. So I have to change [and] my people also. I have to tell them that everything is real.

(Interview with Joseph Portifer, September 2005)[18]

8. Culture change in Nicobarese societies before the tsunami

The interview with the chief of Trinket had a deep emotional impact on me. His authentic narrative impressed me more than any media report had ever done. With more emotional distance, the interview also allows a different kind of reading. While transcribing the text, I realised the many pieces of information inherent in the text which describe life on the island of Trinket and on Kamorta before the tsunami – details which in sum bear testimony to the fact that the Nicobarese islands did not have such a traditional and isolated way of life, as many news reports gave the impression, either through actual descriptions or by referring indirectly to images of an unspoilt paradise. The interview, which was conducted in *English*, also repudiates the theory of the total extinction of culture at one go. As becomes apparent, the interview contains many remarks about en-dogenous and exogenous forms of social change on the Nicobarese Islands which have taken place over a period of decades or even centuries, and not only in the events[19] after the tsunami.

---

18 For a longer version of this interview see Brigitte Vettori 2006: 32-35.

19 The historical information which I use in this chapter comes from diverse resources, e.g. Andrews and Sankaran 2002, Icke-Schwalbe and Günther 1991, Kaspar 1987, 2002a, 2002b, Singh 2003, Singh, Pandit and Sarkar 1994. For more detailed explanations, see Vettori 2006, here 18-19, 123-130.

Portifer said that his father had spent a lot of *money* on his *cocktail party,* which indicates that the Nicobarese were already accustomed to money before the tsunami, although at present they do not always use money when they go shopping. They still practice the exchange of goods, though the values are expressed and written down in rupees. This system of exchange was displaced by a money economy after its introduction in 1945 by the Indian trading company R. Akooje Jadwet & Company. Since 1956, when the Indian government protected part of the Nicobar and Andaman islands with the ANPATR, this trading company, like others from neighbouring Myanmar or Malaysia, was no longer allowed to trade with the Nicobarese. Instead the Indian government allowed the foundation of two Nicobarese-Indian trading companies, which still possess branch shops in which food and luxury items can be bought. Some of these shops have had the label *Fair Price Shops* since the 1960s. With a 'family identity card', products like rice and sugar can be bought at reduced prices. Portifer hinted at a shop from which he had taken cloths for babies and adults, which confirms the existence of shops, as well as the fact that Nicobarese normally wear clothes.

When Portifer visited his friend, he listened to music. The radio or the cassette recorder was presumably bought by his friend in one of the shops, just described, or from illegal traders who, despite the ANPATR regulations, live on the Nicobar Islands. With a radio you can listen not only to music but also to regional or international news.[20] Communications between the islands and with the outside world take place, except for personal contacts, via telephone. The fact that Portifer possessed a *mobile* may reflect his position as a village chief. But without doubt mobiles and the necessary networks already existed before the tsunami. This underlines the fact that the Nicobarese in no way lived such a traditional life as is often assumed. This assumption is confirmed by numerous examples from many other social spheres as well.

For instance, the fact that Portifer had to open doors before he went outside indicated that he did not sleep in a very traditional house, since these round buildings on stilts have no doors. Also the forms of mobility hint at their familiarity with modern consumer goods. Before the tsunami, there were not only canoes, but also motorboats and boats with outboard motors. Portifer's girlfriend presumably travelled by such a boat to another island, where she had begun her new job. And also the *alcohol,* whose importation is forbidden, was most likely smuggled to the islands in such a boat.

Among the consumer goods, alcohol, which is not a new phenomenon, plays an important part since its consumption has been on the increase since the tsu-

---

20 The Indian government has radio stations on the Andaman Islands whose programs can be received on the Nicobar Islands. At times one can also listen to broadcasts from neighbouring countries.

nami. Although Portifer's father's *cocktail party* was without doubt a special event, the fact that Portifer made a connection between the shaking ground and alcohol proves that it was in no way unusual for the islanders to drink alcohol. Likewise his removal of a needle from a machine in order to reassure himself that he was not dreaming demonstrates the use of *technical equipment and machines* before the tsunami by the Nicobarese in order to facilitate their workload. On Kamorta, where a road from the port existed, it was even possible to drive a lorry.

As Portifer's report shows, village chiefs have far-reaching responsibilities towards their communities, but the office of village chief did not always exist. Earlier on, there were no official structures that attributed specific responsibilities to members of the community that transcended the horizons of their own families. But since the seventeenth century, the Nicobarese have adapted the customs of their trading partners by introducing the office of the *captainship*. They chose a captain who negotiated the rate of exchange for a particular quantity of goods for a larger community with the captain of the incoming ship. The name of Chief Captain (or in brief Chief) for a village chief still indicates the origin of this spelling. The selection of such village chiefs through the polls was only introduced by the daughter of the famous Rani Ishlon, Rani Lachmi, in 1982. The title of 'Queen' was given to 'Rani' Ishlon by the British as recognition of her authority over the people of Nancowry. Respect towards the 'royal family' has continued to be shown under the Indian administration, which replaced the British in 1947.

In Portifer's report, one learns about the Indian administration predominantly through his mentioning of the *police*.[21] The Indian government improved the health system on the islands in the context of its social aid programmes and five-year development plans with the aim of so-called economic uplift. In the interview, Portifer talked about a *state-run health centre* to which he brought his people for treatment after their arrival from Kamorta. Parallel to treatment by local healers, who can still be consulted, biomedical treatment thus already existed before the tsunami. Portifer added further changes which had already occurred before the integration of the island into the Indian administration. He mentioned the cows whose ties he loosened. These animals were the offspring of those cows which the British kept on a farm in Kamorta in the nineteenth century. When the British administrators left the islands, they handed these animals over to the Chiefs of Trinket. The cows then became feral, but were later domesticated again. Some of these cows were slaughtered by Portifer's village

---

21 According to the homepage of the Andaman and Nicobar Police, the first police station was founded on Car Nicobar in 1974. On the Andaman islands the history of the police goes back to the year 1858, when a team of about a hundred was recruited for the control of local prisoners (see Andaman and Nicobar Police).

community when they were without water and food for two days on the island, receiving no help.

A further example of important cultural changes in the Nicobar Islands is the membership of the majority of the Nicobarese in the Anglican Church and their celebration of *Christmas and Easter.* Christianity was brought to the islands in the early twentieth century. The first Anglican Church was built on Car Nicobar in 1936. Portifer told me in an earlier part of the interview that he joined the Church because it gave him the chance of a good education on the Indian mainland:

> *I converted to christianism because I was being sponsored by the government to study on the mainland. [...For this] you needed a birth certificate and (...) a baptism certificate (...). It was like that those days, so I had to (...) switch from one religion to the other (Interview with Portifer, September 2005, in Vettori 2006:128).*

But at the age of fifteen Portifer had to break off his education, which he had started when he was nine years old:

> *Then I had to return home. My father said: Don't study, you have to stay now because one son died and I cannot lose another son. (...) For one year I was staying in the house and then I sneaked out again for studies. I just ran away from home (...). I got my admission through a friend from Nancowry. (...) Then my father came to know that I am in Port Blair. So he sent somebody to get me back. (...) I just wanted to have a little more... just – you know... – but he did not allow me to do all these things. (Interview with Portifer, September2005, in Vettori 2006:128)*

He wished very badly to continue his studies and to participate in the Indian way of life. But in the interview he also recounted how this way of life estranged him from his own culture:

> *One thing I can tell you – I do not want to get into it deeply but let's blame something for it: You also may be knowing it, everybody is knowing it – the main disturbances of cultures: one is by religion, the other is education. These two things now they change everything (Interview with Portifer, September2005, in Vettori 2006:128)*

Somebody like Portifer, who has lived off the islands for some time, had problems in acquiring a comprehensive understanding of those rituals which are periodically performed at so-called traditional celebrations. Quite often, these rituals are only understood by the elderly or initiated persons from the community and performed under their guidance:

> *We (...) have come into contact with the world, we know everything, we know what is what. [Now] after getting into these things, to go back into our culture, I find it very difficult. (...) I have tried it, but see: one thing is – in our culture, there*

is so much of art you have to do, to be in our culture. Without knowing these arts, you cannot be very much in the culture. (...) There is some kind of art that you have to know during these big festivals. So you have to sit down with the older people, you have to learn from them in that festival. So what our generation is doing nowadays: if this festival comes – because we are going back to our culture now: (...) no matter if we miss schooling for one year or we miss college for one year, we are going for that festival, just to bring back the culture. (...) So people are finding that this culture should be kept. Without knowing your culture, without knowing yourself, you cannot do anything. That's the main thing. (Interview with Portifer, September 2005, in Vettori 2006:129)

In contrast to other Nicobarese, Portifer was introduced to me as a person who valued tradition and who had the desire to revive such old cultural practices again after the tsunami. At the same time, he was open to innovation. As the above text reveals, he tries to make a balance between old and new, between the activation of traditional festivals (which without doubt contain new elements) and the advocacy of education outside the islands. However, there are also schools in the Nicobar Islands. Many of them were destroyed by the tsunami but later rebuilt. As an example of the adaptation of strange cultural traditions to local needs, football must be mentioned. It was already popular[22] many years before the tsunami. Portifer told me how a team captain introduced new regulations:

I was a good football player in school. (...) And I was playing football in Trinket. (...) I am still the captain of the team. (...) Actually we (...) don't actually keep a person – even if he's a good player – if he is married. We then just say: Okay my friend, it's time, you have to leave the team. (...) That's my rule, as a captain I keep the rule like that. (...) [The] A and B [Team] play against each other only for practice. But for a big game we play against (...) another island. (Interview with Portifer, September 2005, in Vettori 2006: 129)

Thomas Francis from the village of Munak on Kamorta also talked about places in which traditional competitions like stick fights, but also volleyball were performed. But the sea swept away most of these sports and festival places:

There was a football ground. [But] there is no more ground on Trinket (...) and there is no more ground on Nancowry. Anywhere there is no more ground. (...) It was a flat land and everything submerged into the sea. (Interview with Portifer, September 2005, in Vettori 2006: 130)

The high waves also destroyed large numbers of coconut trees, which forced the Nicobarese communities to look for new possibilities to replace the trade

---

22 The Indian government first promoted sports like football and volleyball, as well as activities like music and dance, in the third decade of its development plans (1970 to 1980s) (see Singh 2003: 258).

with copra through new trade products. Regarding the island of Trinket, there will be no recovery for at least ten years, and to make it worse the inundation of the whole island during the tsunami made the water resources saline.

9. Cultural changes among the Jarawas on the Andaman Islands: a brief comparison

As my interpretation in the last section showed, it needs only a single interview to prove continuous cultural change in the Nicobarese society. Selected parts of a lecture given by the anthropologist Vishvajit Pandya in Vienna in 2006 allow a brief comparison between Nicobarese society and the Jarawas on the Andaman Islands. In contrast to Karin Steinberger (see Steinberger in *Süddeutsche Zeitung 2005-08-21/20)*, Pandya does not see contact between the Jarawas and other societies as fatal, but stresses: "Jarawas interpret historical contacts with outsiders with ease, and not with stress of change" (Pandya 2006: 37). In his lecture he focussed on Jarawa perceptions of the fact that other societies associate clothes with civility, while nudity by contrast is seen as primitive. In this context Pandya noticed how Jarawa perceptions are adapted to those of other societies without passivity on their part (see Pandya 2006: 31-44).

"Jarawa hunters are practicing a new form of gathering, both at the roadside as well as at Port Blair" (Pandya 2006: 36). Along the tarmac road which crosses their territory, Jarawas accept as 'modern gatherers' clothes and other presents by passers-by. Since 1999 they have also attended a clinic in Port Blair where they ask for treatment and where they collect presents from the hospital staff and other attendees. At the same time they continue their life in the same way as they did before, that is, as hunters and gatherers (see Pandya 2006: 33-36). However, they have more chances to collect presents along the tarmac road in their territory when they present themselves as primitives (with few clothes), whereas in the hospital in Port Blair they are expected to wear long dresses, presenting themselves as *former* primitives. The Indian staff are then proud that the Jarawas have left behind their phase of nakedness as a first step away from 'primitiveness' and towards a civilised way of life (see Pandya 2006: 31-37).

> Jarawas, by switching back and forth as body without clothes and body in clothes, structure a relation where the non-Jarawa are visually induced and seduced by the Jarawas re-presenting themselves as naked primitives. A process of material mapping that is perpetuated by the Jarawas in which *ways of seeing a lack of nakedness* is effected by *ways of showing nakedness*, creating Primitiveness as performance. This is a performance where Jarawa subjects are objectified for the outsiders by Jarawas themselves. (Pandya 2006: 37, emphasis in the original)

The Jarawas have understood the societies with which they are in contact or, as Pandya puts it: "The 'primitive mentality' has grasped the governmentality to operate in a world, (...) where the islanders have a cultural strategy to co-exist with the policy and welfare authority" (Pandya 2006: 44). The criticism which Pandya formulates at the end of his lecture with regard of non-indigenous societies can also be applied to many media reports about the Nicobar and Andaman Islands:

> It is we who constantly cry out that the primitives' indigenous knowledge system, their environment and above all their identity have to modify in accordance with *our* vision in which the tribal always remains an unchanging, backward, stigmatised, non-political noble savage. (Pandya 2006: 44, emphasis in the original)

10. Conclusion

Did the tsunami destroy the culture of the indigenous societies on the Andaman and Nicobar Islands beyond retrieval? I say no. Culture cannot be defined as a stable, closed and homogenous totality that can be lost as a whole. Without doubt the tsunami put the indigenous societies under pressure. The inhabitants of the islands have had to cope with the direct consequences of the disaster and with the increased cultural flows that Appadurai (1996) has identified as ethno-, techno-, finance-, media- and ideoscapes. However, life in Nicobarese society does not come to a standstill. Some individuals want to preserve old traditions, while others use this situation to enforce new ideas. These contradictory forces existed already before the tsunami. However, the intensity of contacts within and outside society and the urgency in handling of socio-economic and ecological changes have definitely increased. But in other places too, like where I worked in Vienna, changes have been noticed since the tsunami: suddenly a record label and a research institute have become involved in social projects. Together with a local and an Austrian NGO, they tried to establish a culture of cooperation within the SIF-Fund. The usual roles of scientists and practitioners became shaky and needed to be renegotiated. Even scientific lectures and publications were adapted to the events. Disasters do not extinguish cultures at one go. But usual ways of life and forms of existence are questioned and are negotiated again in different places and spaces – e.g. in different regions and in social spaces, as in families, in the Tribal Council or in project meetings. The example of the tsunami in South East Asia, which I discuss here, is only one example. A comparison could be made with the disaster in Japan in 2011. We read in the media about the 'stoic Japanese' who carry their fate with dignity. In many places fund-raising and discussions about a possible phasing out of nuclear power have taken place.

To have an anthropological perspective on culture and on *culture talk* in a situation of disasters will be important in the future as well. My own research into the post-tsunami media coverage in 2004/05 show that many reports on disasters are characterized by ethnocentric and evolutionary values. Journalists are classifying these societies as being on the most primitive level of unilinear development when they draw images of indigenous people suggesting that they react *like animals* because of their instinctive behaviour (see Brinkbäumer in *Der Spiegel,* 2005, no. 4). These descriptions recall the theories of the early evolutionists in the nineteenth century, which is why I have called this form of news coverage evolutionary journalism. Although these theories were superseded in anthropology long ago, they are still being perpetuated by journalists in the 21st century. In 2003 the daily *Die Presse* had already compared the changing life-style of the Nicobarese with a 'dangerous journey' leading from a subsistence economy via a market economy to a modern way of life (see Langenbach in *Die Presse,* 2003-04-26). Due to the tsunami, one of these evolutionary stages has been overcome, as indicated by an article in *Der Spiegel,* which suggested that the deadly wave had uprooted this aboriginal society and thrown it directly into modernity (see Trauffetter in *Der Spiegel,* 2005, no. 41). Such views and descriptions, combined with images of unspoilt island paradises, not only influence readers' perceptions. An evolutionary journalism can indirectly do harm to the people they depict as proud, nude primitives, since such descriptions of passive backwardness motivate paternalistic forms of aid which do not comply with the needs and customs of the disaster-stricken societies.

It is necessary to pinpoint the difference between the evolutionism of the nineteenth century and today's evolutionary journalism. The unilinear development of primitive societies to a technological society was discussed in the nineteenth century as a form of cultural progress. In the post-tsunami media, unilinear models of social development are still being maintained. However, when indigenous societies distance themselves from earlier and in this sense more primordial ways of life, this is referred to as the 'break-up of culture'. The development into whatever form of modernity – that is, into an industrial society embedded in global processes – is not considered desirable for all societies from a journalistic point of view.

Is it permissible for some societies to develop their ways of life and courses of action as they like, while others have to be protected in their alleged primitiveness? Knowing that this conclusion cannot be persuasive, some authors write: "We cannot treat the Nicobarese like museum objects" (Lehmann 2006: 16). And Lehmann suggests further that, so that they do not make the same mistakes as Western societies, in the post-tsunami period they should be informed about the risks and dangers which are part of a modern way of life in a global market economy. In their discussions with the experts, the Nicobarese should gain an understanding of the possible consequences of their decisions – actually before planned activities are started (see Lehmann in *Universum,* 2005-08/07).

The answer to the question to what extent these conversations were and are in agreement with their partners and to what extent they will have a sustainable effect on cultural decisions among the members of heterogeneous groups in post-tsunami events is beyond the topic of my research. With regard to Joseph Portifer's account of indigenous disaster management on the islands of Trinket and Kamorta and Vishyavit Pandya's research on the reactions of the Ongees to the tsunami and its direct consequences, it can be concluded that in future indigenous communities too will actively react to direct challenges. In addition we have to assume that the people of the Andaman and Nicobar Islands will creatively integrate the different techno-, finance- and ideoscapes of their counsellors into long-term cultural negotiations on the islands. Regarding post-tsunami media coverage, it should be pointed out that creative ways of embedding complex events and ways of life into a more differentiated and thus more progressive form of reporting are rather rare. My research indicates that journalists are still practicing a tradition of essentializing, ethnocentric and paternalistic writing and that only a few accept the challenge of communicating the complexity of the disaster, as well as the complexity of the negotiations over cultural futures. It is therefore time to go beyond evolutionary journalism and develop new forms of textual strategies and cultural communication.

11. Bibliography

Andrews, Harry V., Vasumathi Sankaran (2002): Sustainable Management of Protected Areas in the Andaman and Nicobar Island. New Delhi: ANET, IIPA, FFI.

Appadurai, Arjun (1996): Modernity at Large. Cultural Dimensions of Globalisation. Minneapolis, London: University of Minnesota Press.

Delius, Ulrich (2005): Indiens ferne Inseln. Die Andamanen und Nikobaren nach der Flutkatastrophe. In: Waibel, Michael, Tanja Thimm and Werner Kreisel (eds.): Fragile Inselwelten. Tourismus, Umwelt und indigene Kulturen. Bad Honnef: Horlemann, pp. 27-32.

Hannerz, Ulf (1999): Reflections on varieties of culture speak. In: European Journal of Cultural Studies 2, 3, pp. 393-407.

Icke-Schwalbe, Lydia and Michael Günther (1991): Andamanen und Nikobaren – ein Kulturbild der Inseln im Indischen Meer. Dresden, Münster: Lit.

Kaspar, Franziska (2002a): Die österreichische Kolonie auf den Nikobaren 1778 – 1783. Eine ethnohistorische Untersuchung des kolonisatorischen Unternehmens Österreichs im Indischen Ozean Ende des 18. Jahrhunderts mit einer Bewertung der Ethnographica aus dem 19. Jahrhundert. Diss. A., Wien: Universität Wien, Institut für Kultur- und Sozialanthropologie.vol. I, 1-205.

——— (2002b): Die österreichische Kolonie auf den Nikobaren 1778 – 1783. Eine ethnohistorische Untersuchung des kolonisatorischen Unternehmens Österreichs im Indischen Ozean Ende des 18. Jahrhunderts mit einer Bewertung der Ethnographica aus dem 19. Jahrhundert. Diss. A., Wien: Universität Wien, Institut für Kultur- und Sozialanthropologie. vol. II, 206-389.

——— (1987): Nikobaren. Inselgruppe im Indischen Ozean. Österreichische Expeditionen im 18. und 19. Jahrhundert. Wien: Museum für Völkerkunde.

Lehmann, Oliver (2006): 'Kein Mensch ist eine Insel...'. Vorwort. In: Singh, Simron Jit (ed.): The Nicobar Island. Cultural Choices in the aftermath of the tsunami. / Die Nikobaren. Das kulturelle Erbe nach dem Tsunami. Published by Oliver Lehmann. Wien: Czernin Verlag, pp. 12-18.

Pandya, Vishvajit (2006): Housing and the Welfare of the Primitive: Perpetuated and Sustained Views from an Area. Unpubl. lecture at the Institute for Social Ecology, IFF Vienna, University of Klagenfurt, 4th July 2006, pp. 1-44.

——— (2005): 'When Land became Water'. Tsunami and the Ongees of Little Andaman Island. In: Anthropological News 46, 3, pp. 12-13.

Singh K. S., Pandit, T.N. and B.N. Sarkar (1994): People of India. Andaman and Nicobar Island. Vol. 12. Madras, New Delhi, Bangalore, Hyderabad: East-West Press.

Singh, Simron Jit (2006): The Nicobar Island. Cultural Choices in the aftermath of the tsunami. / Die Nikobaren. Das kulturelle Erbe nach dem Tsunami. Published by Oliver Lehmann. Wien: Czernin Verlag.

——— (2003): In the Sea of Influence. A World System Perspective of the Nicobar Island. Lund Studies in Human Ecology 6. Lund: Lund University.

Tharoor, Shashi (2005): Eine kleine Geschichte Indiens. Frankfurt a.M.: Surkamp.

Vettori, Brigitte (2005-2007): Workjournals 1-5, Vienna.

——— (2006): Kultur verhandeln. Vom Wandel nikobaresischer Gesellschaften vor und nach dem Tsunami 2004. Diplomarbeit, Wien: Universität Wien, Institut für Kultur- und Sozialanthropologie.

Post-tsunami-Media Reports 2004/05

From India
Asian Age, Mumbai
2004-12-30/A   An island tribe can be created again.

Indian Express, Pune
2004-12-31   Birth amidst destruction.
2004-12-31/A   Khosa, Aasha: Endangered tribes survived tsunami.

Maharashtra Herald, Pune
2004-12-30/A   coral reef may take years to recover.

The Hindu, Hyderabad

2004-12-29/B   No word on rare tribes.
2004-12-30/A   Nambath, Suresh: Death toll in Andamans put at 10,000.
2004-12-31/B   Nambath, Suresh: Where forests saved the people.
2004-12-31/C   Nambath, Suresh: All primitive tribes safe.

The Times of India, Pune
2004-12-27/B   Pandit, Rajat: 45,000 Nicobar inhabitants cut off from mainland.
2004-12-29/B   Dutta, Sanja, Chandrika Mago: tsunami may have rendered threatened tribes extinct. A World Lost Forever?
2005-01-02/A   Born on the waves of disaster, loved by all.
2005-01-02/B   Chandramouli, Rajesh: Indifference in aid, allege residents of Car Nicobar.
2005-01-01/C   Pendse, Sanjay: Andamans in happier times.

European Reports
Der Spiegel, Germany
2005 - Nr. 2   Brinkbäumer, Klaus: Ende der Zeit am Ende der Welt.
2005 - Nr. 4   Brinkbäumer, Klaus: Die Suchenden von Car Nic.
2005 - Nr. 41   Traufetter, Gerald: Fenster in die Vergangenheit. Der Tsunami löschte große Teile ihrer Kultur aus. Stammesführer von der Inselgruppe der Nikobaren suchen deshalb im Wiener Völkerkundemuseum nach ihren Wurzeln.

Der Standard, Austria
2005-01-04   Fast vergessener Welt droht das Ende.
2005-01-07   'Sie lassen uns hier einfach sterben.' Indische Behörden verweigern indigener Bevölkerung auf Nikobaren weiterhin Hilfe.

Die Presse, Austria
2003-04-26   Langenbach, Jürgen: "Navigator ohne Kompass"
2005-12-12   Simon, Anne-Catherine: Auch Boote können Böses tun. Ethnologie. Der Untergang nach dem Tsunami: Ein Buch zeigt die Kultur der Nikobaren.

Guardian Weekly, Great Britain
2005-01-13/7   Harding, Luke: Top official seized by starving islanders. India. Nicobar aid effort criticised.

Süddeutsche Zeitung, Germany
2005-08-21/20 Steinberger, Karin: Mit den Schiffen kam der Untergang. Auf den Inseln der Andamanen im indischen Ozean treffen Siedler auf Jäger und Sammler, trifft Neuzeit auf Steinzeit – eine unheilvolle Begegnung für die Ureinwohner.
2005-12-09   Steinberger, Karin: Der Untergang. Das Volk der Nikobaresen verlor durch den Tsunami 15 Inseln, viele tausend Menschen und seine gesamte Kultur.

Universum, Austria
2005-08/07   Lehmann, Oliver: Der Schatz der Nikobaren.
2005-11   Lehmann, Oliver: Besuch von Österreichs Kolonie in den
                  Tropen.

Radio Broadcast

Österreichischer Rundfunk – Ö1, Vienna
2005-12-27   Simron Jit Singh und Oliver Lehmann zu Gast in der
                  Anrufersendung „Von Tag zu Tag'. Moderator: Andreas
                  Obrecht.

Internet Sources

BR-online (Bayerischer Rundfunk-online)
2006-01-07   Spaeth, David
                  http://www.bronline.de/kultur/literatur/lesezeichen/20060108
                  /20060108_1.html.

Bridges
2005-12-05   Steger, Philipp: The Nicobar Island: Linking Past and Future.
                  In: Bridges, OST (Office of Science & Technology at the
                  Embassy of Austria in Washington, DC.), Vol. 8.
                  http://www.ostina.org/content/view/196/222/.

orf.at
2005-10-20/A   Steinzeitvolk sucht Spuren in Wien. Zwischen Identität und
                  Modernität. http://www.orf.at/051019-
                  92456/92457txt_story.html.
2005-10-20/B   Österreichs erste Kolonie. Maria Theresias Asien-Abenteuer.
                  http://www.orf.at/051019-92456/92457txt_story.html.

Further Internet Sources

Andaman & Nicobar Police.
http://police.and.nic.in (accessed 21.11.2006).

Government of India, Ministry of Home Affairs, Office of the Registrar General
& Census Commissioner, India
http://censusindia.gov.in/2011-common/censusdataonline.html (accessed
09.10.2011).

Novara-Expedition
http://www.novara-expedition.org (accessed 21.11.2006).

Weber, Georg, Andaman Association: Tsunami. The 2004 Indian Ocean Earthquake and tsunami.
http://www.andaman.org/mapstsunami/tsunami.htm#statistics, last updated 1.
January 2009 (accessed 16.10.2011).